Sustainable Cacao Cultivation in Latin America

This book examines cacao cultivation with a view to improving the sustainable management and production of this crop.

Theobroma cacao is an important species originating in the Ecuadorian Amazon and its product chocolate is consumed worldwide. Cacao cultivation is an industry supporting over ten million people, and so it is vitally important, in this changing climate, that the species is properly and sustainably managed and cultivated. This book brings together a wide range of experts from across the globe to examine cacao cultivation, from the basic aspects of reproduction, genetic improvement, nutrition and pest management to agroforestry, industrialization, and marketing in a global food system. Case studies are drawn from across Latin America, but the research reflects the nature of a crop that is cultivated in over 60 countries and processed, manufactured, and consumed worldwide.

This book will be of great interest to students and scholars of agronomy, sustainable agriculture, and crop science.

Luz Cecilia García is a Professor at the Faculty of Agronomic Engineering of the Technical University of Manabí, Ecuador.

Naga Raju Maddela is a Professor at the Faculty of Health Sciences of the Technical University of Manabí, Ecuador.

Freddy Zambrano Gavilanes is a Professor at the Faculty of Agronomic Engineering of the Technical University of Manabí, Ecuador.

Carolina Aguilar Duarte is the Cocoa Director of the MOCCA (Maximizing Opportunities for Coffee and Cocoa in the Americas) program funded by Lutheran World Relief and the USDA.

Earthscan Food and Agriculture

Genome Editing and Global Food Security
Molecular Engineering Technologies for Sustainable Agriculture
Zeba Khan, Durre Shahwar, and Yasmin Heikal

University Engagement with Farming Communities in Africa
Community Action Research Platforms
Edited by Anthony Egeru, Megan Lindow, and Kay Muir Leresche

Globalisation and Livelihood Transformations in the Indonesian Seaweed Industry
Edited by Zannie Langford

Principles of Sustainable Aquaculture
Promoting Social, Economic and Environmental Resilience, 2nd edition
Stuart W. Bunting

The Spatial Organization of Urban Agriculture in the Global South
Food Security and Sustainable Cities
Ada Górna

Digital Communication for Agricultural and Rural Development
Participatory Practices in a Post-COVID Age
Edited by Ataharul Chowdhury and Gordon Gow

Regenerative Farming and Sustainable Diets
Human, Animal and Planetary Health
Edited by Joyce D'Silva and Carol McKenna

Sustainable Cacao Cultivation in Latin America
Edited by Luz Cecilia García, Naga Raju Maddela, Freddy Zambrano Gavilanes, and Carolina Aguilar Duarte

For more information about this series, please visit: www.routledge.com/books/series/ECEFA/

Sustainable Cacao Cultivation in Latin America

Edited by Luz Cecilia García, Naga Raju Maddela, Freddy Zambrano Gavilanes, and Carolina Aguilar Duarte

Routledge
Taylor & Francis Group
LONDON AND NEW YORK

earthscan
from Routledge

First published 2025
by Routledge
4 Park Square, Milton Park, Abingdon, Oxon OX14 4RN

and by Routledge
605 Third Avenue, New York, NY 10158

Routledge is an imprint of the Taylor & Francis Group, an informa business

British Library Cataloguing-in-Publication Data
A catalogue record for this book is available from the British Library

Library of Congress Cataloging-in-Publication Data
Names: García, Luz Cecilia editor | Duarte, Carolina Aguilar editor | Gavilanes, Freddy Zambrano editor | Maddela, Naga Raju editor
Title: Sustainable cacao cultivation in Latin America / edited by Luz Cecilia García, Naga Raju Maddela, Freddy Zambrano Gavilanes, and Carolina Aguilar Duarte.
Description: Abingdon, Oxon ; New York, NY : Routledge, 2025. | Series: Earthscan food and agriculture series | Includes bibliographical references and index.
Identifiers: LCCN 2024028241 (print) | LCCN 2024028242 (ebook) | ISBN 9781032464596 paperback | ISBN 9781032464572 hardback | ISBN 9781003381761 ebook
Subjects: LCSH: Cocoa trade--Latin America | Sustainable agriculture--Latin America | Cacao--Latin America
Classification: LCC HD9200.L292 S87 2025 (print) | LCC HD9200.L292 (ebook) | DDC 338.1/7374098--dc23/eng
LC record available at https://lccn.loc.gov/2024028241
LC ebook record available at https://lccn.loc.gov/2024028242

ISBN: 978-1-032-46457-2 (hbk)
ISBN: 978-1-032-46459-6 (pbk)
ISBN: 978-1-003-38176-1 (ebk)

DOI: 10.4324/9781003381761

Typeset in Times New Roman
by SPi Technologies India Pvt Ltd (Straive)

Contents

Figures

Tables

About the editors

Luz C. García Cruzatty received her PhD in Forest Sciences from the Universidad Austral de Chile (UACH), Valdivia, Chile in 2014. Since 2014, she has been a full-time professor at the Faculty of Agronomic Engineering, Universidad Técnica de Manabí (UTM), Portoviejo, Ecuador. She also worked as a Director of the Institute of Investigation at UTM from 2017 to 2022. Her line of research is 'Conservation and Improvement of Plant Genetic Resources'. She directs a program of 'Conservation and Genetic Improvement of *Theobroma cacao*' in which higher genotypes are being developed with better production, disease tolerance, drought tolerance, and excellent organoleptic characteristics. To her credit, there are more than 30 scientific publications, and she has participated as a speaker at several national and international conferences. She has received some distinctions and 13 awards, such as best graduate of the doctoral program (UACH), scholarships from MECESUP (Programa de Mejoramiento de la Calidad y la Equidad de la Educación Superior, Govt. of Chile, Chile), and SENESCYT (Secretaría de Educación Superior, Ciencia, Tecnología e Innovación, Govt. of Ecuador, Ecuador) for her doctoral study.

Naga Raju Maddela received his MSc. (1996–1998) and PhD (2012) in Microbiology from Sri Krishnadevaraya University, Anantapuramu, India. During his doctoral program in the area of Environmental Microbiology, he investigated the effects of industrial effluents/insecticides on soil microorganisms and their biological activities. He has been working as a Faculty in Microbiology since 1998, teaching undergraduate and postgraduate students. He worked as a Prometeo Investigator (fellowship received from SENESCYT) at Universidad Estatal Amazónica, Ecuador during 2013–2015, and he received a 'Postdoctoral Fellowship' (2016–2018) from Sun Yat-sen University, China. He also received external funding from 'The China Postdoctoral Science Foundation' in 2017 and internal funding from 'Universidad Técnica de Manabí' in 2020. He has participated in national/international conferences and presented research data in China, Cuba, Ecuador, India, and Singapore. Currently, he is working as a Full Professor at the Facultad de Ciencias de la Salud, Universidad Técnica de Manabí,

Portoviejo, Ecuador. He has been actively publishing scientific articles, books (authored and edited), and chapters since 2007. As of now, he has published 85 articles, 15 books, and 40 book chapters.

Freddy Zambrano Gavilanes received a Master's in Business Administration in Public Management from Pitágoras Unopar University in Brazil (2018), a Master's in Bioenergy (2015), and a PhD in Agronomy (2017) from the State University of Londrina, Brazil. He has been a principal professor at the Faculty of Agronomic Engineering at the Technical University of Manabí (UTM) since 2018, teaching courses in Soil Conservation and Experimental Design. He teaches in the Academic Master's Program with a research trajectory in Agronomy, specializing in Plant Health, in the Biotechnology Master's Program, in the Academic Master's Program with a professional trajectory in Agricultural Engineering at UTM, and in the Renewable Energies Master's Program at the University of Los Hemisferios in Spain. His research focuses on studying growth-promoting bacteria, the use and production of biofertilizers, microalgae and cyanobacteria cultivation, oil extraction from oleaginous plants, biofuel production, and byproducts. Additionally, he conducts studies on various crops such as cotton, castor bean, physic nut, common bean, corn, and cacao. He is the coordinator of the research group 'Sustainable Agriculture and Bioenergy' at UTM and is part of the research groups 'Soil and Water Conservation Management' and 'Plant Species for Biodiesel and Bio-kerosene' at the Agronomic Institute of Paraná in Brazil. He has around 50 publications (scientific articles and book chapters).

Carolina Aguilar holds an MBA from de GADEX program of the University of Cadiz, Spain. She has more than 20 years of experience in economic development programs, health, education, and humanitarian assistance in Latin America. With Lutheran World Relief, she has been the key to developing the coffee and cocoa portfolio of the region, with more than 14 years of designing proposals to innovate the cacao sector in Central America, Peru, Ecuador, and Colombia building private–public partnerships that have helped build more sustainable programs. She is part of the team in charge of the Lutheran World Relief's Cocoa Toolkit and Cacao Movil platform as the editor and content designer. Under the MOCCA program, Carolina is the Cocoa Director supervising the six prioritized countries' cocoa strategy for the region.

Contributors

P. Abirami
Department of Environmental Sciences
JSA College of Agriculture and Technology
Tamil Nadu, India

Carolina Aguilar
Directora de Cacao del Programa Maximizando Oportunidades en Café y Cacao en las Américas (MOCCA)
USDA, Technoserve & Lutheran World Relief
Baltimore, United States

Rolando León Aguilar
Facultad de Ingeniería Agronómica
Universidad Técnica de Manabí
Portoviejo, Ecuador

Marlon Arturo Cedeño Álava
Universidad Técnica de Manabí
Portoviejo, Ecuador

Freddy Amores
National Institute of Agricultural Research (INIAP)
Ecuador

M. Ashwini
Department of Environmental Sciences
JSA College of Agriculture and Technology
Tamil Nadu, India

Jose Ricardo Macias Barberan
Faculty of Zootechnical Sciences
Research Institute, Faculty of Basic Sciences
Department of Physics Technical University of Manabí
Portoviejo, Ecuador

Shruti Bhargav
Chandigarh University
Mohali, India

Anne-Gaël Bilhaut
IRD, Institut de recherche pour le développement
Alemania N32-188 y Guayanas, 170519, Quito, Ecuador

T. Bowya
Department of Agricultural Microbiology
Tamil Nadu Agricultural University
Tamil Nadu, India

Soraya Peñarrieta Bravo
Facultad de Ingeniería Agronómica
Universidad Técnica de Manabí
Portoviejo, Ecuador

A. Brundha
Department of Agricultural Microbiology
Tamil Nadu Agricultural University
Tamil Nadu, India

Hayron F. Canchignia-Martínez
Laboratorio de Microbiología del Departamento de Biotecnología, Ciencias Agrarias
Universidad Técnica Estatal de Quevedo
Quevedo, Ecuador

Mirian Capa-Morocho
Department of Agronomy
Faculty of Agricultural Sciences and Renewable Natural Resources
Universidad Nacional de Loja
Ciudad Universitaria Guillermo Falconí
Loja, Ecuador

Jessenia Castro
Facultad de Ingeniería Agronómica
Universidad Técnica de Manabí
Portoviejo, Ecuador

Rossana Castro
Facultad de Posgrado
Maestría en Sanidad Vegetal
Universidad Técnica de Manabí
Portoviejo, Ecuador

Ángel V. Cedeño-Moreira
Laboratorio de Microbiología del Departamento de Biotecnología, Ciencias Agrarias, Universidad Técnica Estatal de Quevedo
Quevedo, Los Ríos, Ecuador

Jagadeesh Chandra Bose K
Chandigarh University
Mohali, India

Jaime Vera Chang
Faculty of Industry and Production
Universidad Técnica Estatal de Quevedo
Empalme, Los Ríos Ecuador

Dorys T. Chirinos
Facultad de Ingeniería Agronómica
Universidad Técnica de Manabí
Portoviejo, Ecuador

Luis Vásquez Cortez
Facultad de Ciencias Agropecuarias
Carrera de Ingeniería en Agroindustria
Universidad Técnica de Babahoyo
Babahoyo, Los Ríos, Ecuador

Karime Montes Escobar
Departamento de Matemáticas y Estadística
Facultad de Ciencias Básicas
Universidad Técnica de Manabí
Portoviejo, Ecuador

Juan Fernández-Manjarrés
INRAE, – Campus Agro Paris-Saclay
UMR SADAPT, 22 Place de l'Agronomie
CS 80022, 91120 Palaiseau Cedex, France

Frank Intriago Flor
Department of Agroindustrial Process
Faculty of Zootechnical Sciences
University of Técnica de Manabí

Romina Fossati
Laboratoire ESE CNRS/AgroParisTech/ Université Paris-Saclay
Gif sur Yvette, France

Luz Cecilia García
Facultad de Ingeniería Agronómica
Universidad Técnica de Manabí
Campus Experimental "La Teodomira"
Santa Ana-Manabí, Ecuador

Marina García
Facultad de Ingeniería Agronómica
Universidad Técnica de Manabí
Portoviejo, Ecuador

Michelle Sánchez García
Yachay Tech University
Biology School
Imbabura, Ecuador

Felipe R. Garcés-Fiallos
Laboratory of Phytopathology, Experimental Campus La Teodomira
Faculty of Agronomic Engineering
Universidad Técnica de Manabí
Santa Ana EC130105, Ecuador

Patricia Zambrano Gavilanes
Universidad Técnica de Manabí
Facultad de Medicina Veterinaria
Portoviejo, Ecuador

Freddy Zambrano Gavilánez
Facultad de Ingeniería Agronómica
Universidad Técnica de Manabí
Campus Experimental "La Teodomira"
Santa Ana-Manabí, Ecuador

David Giband
Université de Perpignan Via Domitia
Perpignan, France

Andrés González-González
Laboratoire ESE CNRS/AgroParisTech/Université Paris-Saclay
Gif sur Yvette, France

Fernando Granja
Department of Agronomy
Faculty of Agricultural Sciences and Renewable Natural Resources
Universidad Nacional de Loja
Ciudad Universitaria Guillermo Falconí
Loja, Ecuador

Anne Gschaedler
Centro de Investigación y Asistencia en Tecnología y Diseño del Estado de Jalisco (CIATEJ)
A.C., 45019
Zapopan, Mexico

Vijai Kumar Gupta
Biorefining and Advanced Materials Research Center
Scotland's Rural College (SRUC)
and
Center for Safe and Improved Food, Scotland's Rural College (SRUC)
King's Buildings, West Mains Road
Edinburgh, United Kingdom

Pankaj Jaglan
Chandigarh University
Mohali, India

Javier de la Hoz-M
Statistics Department
University of Salamanca
Salamanca, Spain

Manuel Kirchmayr
Centro de Investigación y Asistencia en Tecnología y Diseño del Estado de Jalisco (CIATEJ)
Zapopan, Mexico

Takumasa Kondo
Corporación Colombiana de Investigación Agropecuaria – AGROSAVIA
Centro de Investigación Palmira
Carrera 36A con Calle 23
Palmira, Colombia

Shivam Kumar
Chandigarh University
Mohali, India

A. Lakshmi
Department of Environmental Sciences
JSA College of Agriculture and Technology
Tamil Nadu, India

Ricardo Limongi-Andrade
INIAP: Instituto Nacional Autónomo de Investigaciones Agropecuarias
Estación Experimental Tropical Pichilingue
Quevedo, Ecuador

Cristhian Verduga López
Postgraduate Institute
Master's Degree in Agroindustry
University Técnica de Manabí

Eugenia Lugo-Cervantes
Centro de Investigación y Asistencia en Tecnología y Diseño del Estado de Jalisco (CIATEJ)
Zapopan, México

Naga Raju Maddela
Universidad Técnica de Manabí
Facultad de Ciencias de la Salud
Portoviejo, Ecuador

Armelle Mazé
INRAE, – Campus Agro-Paris-Saclay
UMR SADAPT, 22 Place de l'Agronomie
CS 80022, 91120 Palaiseau Cedex, France

Komal Mittal
Chandigarh University
Mohali, India

Raul V. Mora-Yela
INIAP: Instituto Nacional Autónomo de Investigaciones Agropecuarias
Estación Experimental Tropical Pichilingue
Quevedo, Ecuador

Patricia Morán
Facultad de Posgrado, Maestría en Sanidad Vegetal
Universidad Técnica de Manabí
Portoviejo, Ecuador

Marlene Molina-Müller
Department of Agronomy
Faculty of Agricultural Sciences and Renewable Natural Resources
Universidad Nacional de Loja
Ciudad Universitaria Guillermo Falconí
Loja, Ecuador

Anthony A. Moreira-Morrillo
Laboratory of Phytopathology, Faculty of Agronomic Engineering
Technical University of Manabí, Experimental Campus La Teodomira
Santa Ana, Ecuador

Mithila V Nair
Chandigarh University
Mohali, India

Viviana Suárez Navarrete
Facultad de Ingeniería Agronómica
Universidad Técnica de Manabí
Portoviejo, Ecuador

Zoi Papalexandratou
ZOTO
Haagstraat 17, 9280, Belgium

Fernando Sánchez-Mora
Departamento de Ciencias Agronómicas
Facultad de Ingeniería Agronómica
Universidad Técnica de Manabí
Portoviejo, Ecuador

Shreya Sahani
University Institute of Biotechnology
Chandigarh University
Mohali, India

Carlos A. Salas-Macías
Departamento de Ciencias Agronómicas
Facultad de Ingeniería Agronómica
Universidad Técnica de Manabí
Portoviejo, Ecuador

R. Saravanan
Department of Nanotechnology
Pondicherry University
Puducherry, India

Jyoti Sarwan
Chandigarh University
Mohali, India

Robin Singh
Chandigarh University
Mohali, India

Zoila K. Solís-Hidalgo
Departamento de Protección Vegetal
Instituto Nacional de Investigaciones Agropecuarias – INIAP
Ecuador

R. Suganthi
Department of Environmental Sciences
Tamil Nadu Agricultural University
Tamil Nadu, India

Nazim Uddin
Chandigarh University
Mohali, India

Kerly Alvarado Vásquez
Facultad de Ciencias Agropecuarias
Carrera de Ingeniería en Agroindustria
Universidad Técnica de Babahoyo
Babahoyo, Los Ríos, Ecuador

Santiago C. Vásquez
Department of Agronomy
Faculty of Agricultural Sciences and Renewable Natural Resources
Universidad Nacional de Loja
Ciudad Universitaria Guillermo Falconí
Loja, Ecuador

Danilo I. Vera-Coello
Departamento de Protección Vegetal
Instituto Nacional de Investigaciones Agropecuarias – INIAP
Ecuador

Dulce Velásquez-Reyes
Centro de Investigación y Asistencia en Tecnología y Diseño del Estado de Jalisco (CIATEJ)
Zapopan, México

Edgar Zambrano
Vocero de la Comisión de los Productores de la Coalición Cacao
Calceta, Ecuador

Roberto Bravo Zamora
Facultad de Ingeniería Agronómica
Universidad Técnica de Manabí
Portoviejo, Ecuador

Solanyi Tigselema Zambrano
Faculty of Industry and Production
Universidad Técnica Estatal de Quevedo
Empalme, Los Ríos Ecuador
Doctoral Student at the American University of Europe
México-Cancún

Ezequiel Zamora-Ledezma
Ecosystems Functioning & Climate Change Research Group FAGROCLIM
Faculty of Agriculture Engineering
Universidad Técnica de Manabí (UTM)
13132 Lodana, Ecuador

Preface

Cacao (*Theobroma cacao* L.) is a diploid species (2n=2x=20) and one of the 22 species assigned to the genus Theobroma (Malvaceae family). *T. cacao* is the only species of economic importance (Nair, 2021).

Throughout history, cacao has been an item of economic and social importance in different cultures of America, especially within the Inca culture, where it was a symbolic product and fulfilled a function both for consumption and trade (Andrade Almeida et al., 2019). Currently, cacao remains an economically important agricultural product for millions of people around the world, being vital for the industry in the production of chocolate, cosmetics, and pharmaceuticals. It is grown by around 6 million farmers around the world, and it is the livelihood of more than 40 million people (Wickramasuriya and Dunwell, 2018; Muller et al., 2021).

Cacao is grown worldwide in the humid tropics within 20° north and south of the equator; production is concentrated in West Africa, Southeast Asia, and Latin America. Within West Africa, which accounts for more than 70% of the world cacao supply, the main producing countries are Ivory Coast, Ghana, Nigeria, and Cameroon. In Southeast Asia, Indonesia is by far the largest producing country followed by Papua New Guinea, and within Latin America, production is highest in Brazil, Ecuador, and Colombia (Lahive et al., 2019).

Despite the importance of this crop in various social, economic, and ecological aspects, its average yield is still low. According to projections made by Kozicka et al. (2018), until 2050 the average yield per hectare is expected to remain below 0.6 t ha^{-1}. Cacao productivity is related to various factors, such as the genotypes used, the edaphoclimatic conditions, agronomic management, water availability, and pest and disease control, among others.

Working in coordination with different actors of the cacao activity is important to establish policies that benefit the sector, considering the integration of different practices focused on increasing the sustainability of the crop. In addition, it is necessary to establish strategies to apply circular economy techniques to improve the quality of life of cacao farmers around the world.

Based on this argument, this work was developed to consolidate different investigations about cacao cultivation in tropical climates around the world. This book consists of four parts: *Part I: Introduction, Part II: Agroforestry systems, Part III: Plagues and disease management*, and *Part IV: Industrialization and by-product valorization*. To justify this, 14 chapters have been designed: Section I consists of two chapters: Chapter 1 about general perspectives on the cacao agribusiness, and Chapter 2 focuses on the history and research of cacao in Ecuador. Section II consists of five chapters, with one chapter about the nutrition of cacao and four chapters focused on agroforestry and the social benefits of cacao cultivation. Section III consists of two chapters, about the main diseases and pests that affect this crop. Section IV consists of five chapters about the post-harvest process and industrialization of cacao to produce chocolate. The chapters were contributed by 72 Academicians/Scientists/Researchers from nine different countries (Ecuador, Colombia, Spain, France, India, the United States, México, Belgium, and the United Kingdom) across the globe.

Foreword

Cacao originating from the American tropics in the Upper Amazon were used as a ritual beverage thousands of years ago. In the pre-Hispanic period, it reached Central America through exchanges between natives on the Pacific Ocean coast. In the 17th century, during the colonial period, it was exported to Europe, and consumption grew as chocolate was mixed with sugar. The increase in demand led to the cultivation of cacao in European colonies in Africa, bringing production closer to the consumer market. The two World Wars, interspersed with the economic depression in the United States, compressed demand, which began to recover in the second half of the 20th century. Currently, the value of the traded beans amounts to around 15 billion dollars annually, accumulating and multiplying by 10 in the retail supply chain of confectionery, chocolate, beverages, and cosmetics. Consumption continues to grow at a rate of 2.5% annually, and forecasts for the next decade reach up to 4%, definitely an optimistic scenario for cacao.

Global production now exceeds five million tons of beans, with origins from Africa (76%), Asia and Oceania (6%), and America (18%). New regulations aimed at mitigating negative externalities in environmental and social issues govern the international trade of agricultural products, particularly with Europe, the main importing market for cacao (absorbing 40% of the global supply). Companies involved in the cacao and chocolate agroindustry already include in their strategic plans programs to mitigate this issue with varying intensity depending on the producing region. They also include projects to encourage sustainable production. This is a response to more demanding and better-informed consumers regarding how the cacao used in the products they consume is produced. Consequently, a wave of change is approaching, whose force will influence the dynamics of cacao supply in favor of the origins of producing countries in America. Future trade forecasts reveal both an opportunity and a challenge.

The opportunity translates into these countries capturing a larger share of the global market through increased production and cacao exports. Growth forecasts in demand anticipate benefits in the trade balance, as well as the creation and distribution of additional wealth among supply chain operators, primarily small producers. As a reference, the value of cacao sales represents 10%

of the GDP of the Ivory Coast and 1% of Ecuador's GDP, the first and third global producers. The gap highlights the advantages of increasing production and exports from Latin American origins. On the other hand, the challenge poses that the increase in production must primarily stem from efficiency and productivity gains in established cacao orchards or those replacing other crops. These gains require the adoption of technological innovations aligned with the principles of sustainable cacao farming.

The described context clearly underscores the importance of the book 'Sustainable Cacao Cultivation in Latin America', making it a valuable reference to guide producers in the process of technological intensification that allows for increased cacao production per square meter of cultivated land. Productivity in Ecuador has made a significant leap from 0.28 to 0.60 tons/ha in the last 20 years. The substantial expansion of this indicator in the short term, narrowing the technological gap, should be a shared goal of cacao-producing countries in this part of the world.

In line with the aforementioned framework, the book's initial chapter is particularly timely as it envisions cacao production as a business. Like any business, it needs constant innovation to build a sustainable economy; otherwise, businesses become extinct. It is followed by an analysis of the genetic and agronomic quality of planting material, the foundation of the crop's productive potential, which represents 50% of the business's economic performance from the start. Then, the approach from the environmental dimension is apt for identifying, choosing, and substantiating the best cultivation practices to develop the productive potential of the chosen planting material, managing growth, nutrition, and health in the best way possible. Finally, addressing post-harvest, quality, and processes to add more value to production is a fitting conclusion that reinforces the thematic content of the publication.

Freddy Amores, M.Sc.
Senior Researcher
National Institute of Agricultural Research (INIAP), Ecuador
18 January 2024

Acknowledgments

We appreciate the support of the chapter authors for their valuable contributions and timely responses for the success of this project. We acknowledge their enthusiasm and unwavering support to ensure this volume is ready within the scheduled time. We also extend our gratitude to the anonymous reviewers for their constructive criticisms, which helped us improve the quality of this book, and for their valuable comments during the proposal development and final book preparation. Additionally, we thank the editorial and production teams of Taylor & Francis, as without their guidance, this project would not have been successfully completed. It has been an honor and privilege to work with all of them.

Editors

Part I

Introduction

1 Cacao Agribusiness in a Global Context

An Overview

Luz Cecilia García, Edgar Zambrano, Naga Raju Maddela, Michelle Sánchez García, Carolina Aguilar, and Freddy Zambrano Gavilánez

1.1 Introduction

Cacao cultivation provides employment to millions of people in countries such as Ivory Coast, Ghana, Indonesia, Nigeria, Ecuador, and others. Not only do farmers benefit but also the associated industries in processing and marketing, generating income for local communities. Producing countries heavily rely on cacao exports for revenue, making significant contributions to their national income and trade balance.

In many cases, cacao cultivation serves as a catalyst for rural development, improving infrastructure, education, and healthcare services in the cultivated areas. Initiatives exist to enhance the conditions of cacao farmers by promoting sustainable practices and fair trade programs. These efforts aim to ensure fair labor conditions, preserve the environment, and offer equitable prices to producers. However, there are also socioeconomic challenges associated with cacao production, such as poverty, child labor, price volatility, and precarious working conditions in certain regions. Addressing these challenges is crucial to improving the living conditions of cacao farmers and their communities.

Cacao cultivation and trade involve a complex network of stakeholders, ranging from small-scale farmers to large multinational corporations. Its future sustainability will depend on the comprehensive addressing of environmental, social, economic, and political challenges. Innovative solutions and coordinated actions across all links in the supply chain, along with the commitment of national governments and the international community, are required. The development of a thriving, sustainable, and inclusive cacao agribusiness with equitable benefits for all participants is an ambitious yet indispensable goal to ensure the future of millions of farmers and their livelihoods in cacao-producing countries. The road ahead is still long, but with political will, corporate social responsibility, citizen engagement, and a shared vision of sustainability, laying the foundations for a positively transformed sector is possible.

DOI: 10.4324/9781003381761-2

1.2 Global Cacao Production and Economy

Over the last decade, global cacao production has fluctuated between 4 and 4.9 million tons annually, with slight variations influenced by factors such as climatic conditions, crop-related diseases, and market supply and demand. Between 2022 and 2023, cacao production exceeded 4.9 million tons, involving millions of producers across different countries worldwide. Of this total, 73.4% is produced in African countries (such as Cameroon, Ivory Coast, Ghana, Nigeria, among others), 21.2% in American countries (primarily in Ecuador and Brazil), and 5.4% in countries from Africa and Oceania (including Indonesia, Papua New Guinea, among others) (ICCO, 2023).

The leading cacao-producing countries are (Figure 1.1):

1. Ivory Coast: It is the world's largest cacao producer, contributing approximately 50% to global production (2.23 million tons).
2. Ghana: Holding the second position in global cacao production, contributing around 17% of the total (0.75 million tons).
3. Ecuador: Despite variable production, Ecuador is another significant cacao producer, contributing about 8% to the global supply (0.37 million tons).

Other countries making smaller contributions to the global total but remaining significant players in cacao production and export include Cameroon (±6%), Nigeria (±6%), Indonesia (±5%), Brazil (±5%), and Papua New Guinea (±1%) (Figure 1.1).

Among the mentioned countries with significant cacao production, six out of eight have increased their production in 2023 compared to the data from ten

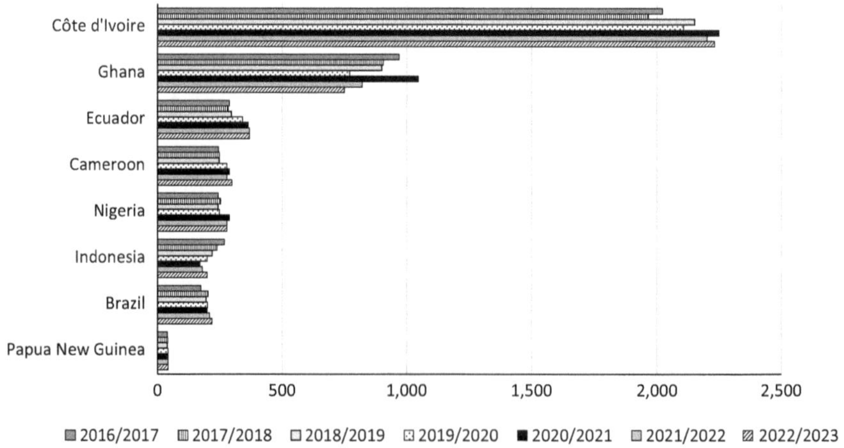

Figure 1.1 World cacao bean production by volume from 2016/2017/ to 2022/2023, by country (thousands of metric tons).

Source: ICCO; 2016–2023.

years ago, with notable increments in Ecuador, Brazil, and Cameroon by 22%, 21%, and 18%, respectively. Only Ghana and Indonesia have experienced a decrease in production, producing 29% and 30% less cacao in 2023 compared to 2014 (Figure 1.2).

With a global production of over four million tons annually, the cacao industry is a multibillion-dollar sector that significantly contributes to the global economy. The trade of cacao products, especially chocolate, represents a massive market with constant demand worldwide. The price of a ton of cacao beans has varied over the last ten years, reaching its lowest in 2017 at $2014 and its highest in 2023 at $3135 (Figure 1.2). According to the data in Table 1.1, Ivory Coast, the country with the highest global production (2,230,000 tons), would have traded approximately $7 billion in 2023. Ghana, the second-largest producer globally, has experienced a 29% decrease in production, implying a reduction in revenue by over $600 million (Figure 1.2, Table 1.1).

There is a growing demand for premium and high-quality chocolates, leading to an increased preference for chocolates with higher cacao content and more complex flavors. Few countries stand out in producing cacao varieties with highly valued taste and aroma profiles in the chocolate industry due to their distinctive sensory qualities and potential for crafting high-quality chocolates. Some of these countries include:

Ecuador: Recognized for its Arriba Nacional cacao, highly praised for its floral and fruity aroma, Ecuador is one of the leading producers of fine-flavor cacao. There are 626,962 hectares under cultivation in Ecuador,

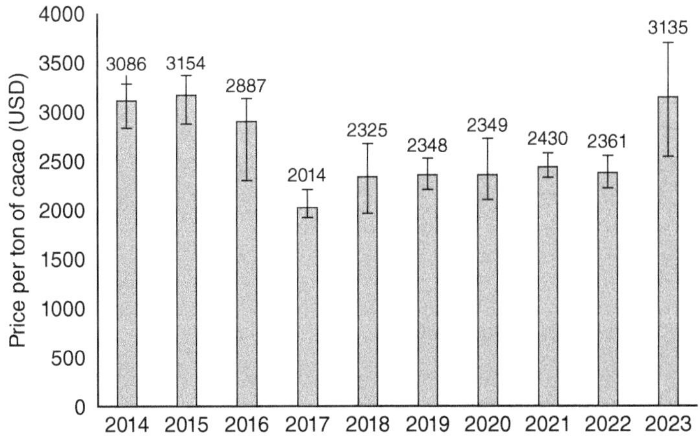

Figure 1.2 World cacao price average by metric tons from 2014 to 2023. The numbers on the bars are the yearly averages. The bars indicate the maximum and minimum values.

Source: ICCO; 2014–2023.

Table 1.1 Income generated by the eight countries with the highest cacao production in 2023

	Tons 2023	*Total (billion USD) 2023**
Côte d'Ivoire	2.230×10^3	6.99
Ghana	750×10^3	2.35
Ecuador	370×10^3	1.16
Cameroon	300×10^3	0.94
Nigeria	280×10^3	0.88
Indonesia	200×10^3	0.63
Brazil	220×10^3	0.70
Papua New Guinea	42×10^3	0.13

* Calculated using 2023 price ($3135.3).
Source: ICCO 2023.

with a production of 330,872 tons and an average yield of 0.61 tons of seed per hectare. The main producing provinces are Los Ríos (21.85%), Manabí (20.74%), Guayas (18.41%), and Esmeraldas (17.17%) (García et al., 2023, MAGAP, 2021).

Venezuela: Historically famous for its Criollo cacao, considered one of the best in terms of taste and aroma. The term Criollo defines cacao types that develop intense and distinct flavors, growing in Venezuelan plantations. These characteristics result from genetic contributions, ecological conditions where the trees grow, and post-harvest and processing management. Venezuela produces approximately 17,000 tons per year, with 70% being exported (Pérez et al., 2021).

Peru: Gaining recognition for its high-quality cacao, especially from the Piura region, it is considered to have fine flavor with complex flavor profiles. Although Peru has a total production of 120,000 tons per year, only a small percentage corresponds to fine-flavor cacao (Guimac & Verastegui, 2022).

Colombia: While cacao production is lower compared to other countries, around 56,000 tons per year, 95% of it is classified as fine-flavor due to its high organoleptic quality. However, Colombia only exports about 7,000 tons, approximately 12% of its production.

1.3 Cacao Cultivation Sustainability

According to the report 'Towards the sustainability of cacao supply chains: regulatory options for the EU', the global chocolate market in 2020 was estimated at around $103 billion, with an annual growth of 7%, and is projected to reach $162 billion by 2024. However, the study points out that despite 5 to 6 million people cultivating cacao globally and approximately 50 million relying on cacao production as their primary livelihood, the majority live in poverty. The income of cacao farmers contrasts with that of chocolate manufacturers, who receive around 40% of the final price of a bar, and retailers, about 35%.

Moreover, only three major companies concentrate on 60% of global cacao processing, and six companies constitute 40% of the global chocolate consumption market (Brack, 2019).

On the other hand, the study indicates that cacao-dedicated agricultural lands produce approximately half of what they should. This is largely due to the meager incomes of farming families, which don't allow them to invest in improvements related to sustainable practices, fertilization, irrigation, or replacing unproductive cacao trees with enhanced materials. Farmer poverty exacerbates other issues such as child labor or deforestation. Finally, the report proposes several measures directed toward the European Union (EU), the main consumer of chocolate (consuming 40% of global cacao, while the United States consumes 20%). Any regulation on cacao import would generate significant impacts on the countries of origin. Among the proposed measures are: 1) obligating companies to examine their supply chains and avoid illegally produced or unsustainably sourced cacao, 2) designing a system to minimize the risk of working with illegally or unsustainably produced cacao, 3) negotiating bilateral agreements with major producing countries by the EU or implementing a card system through which the EU would issue yellow or red cards to those not combating illegal practices in the supply chain; and 4) implementing public procurement policies to require institutions to adhere to sustainability criteria in cacao purchases (Brack, 2019).

Cacao (*Theobroma cacao* L.) is cultivated in systems of diverse complexities, ranging from monocultures with temporary shade to highly diverse agroforestry systems (SAF). The demand for cacao has increased in recent years due to the growth of cacao-based products, posing a challenge for the cacao industry to produce more cacao while ensuring its sustainability. Monocultures of cacao and SAFs are two contrasting ways of producing cacao; however, their impact on yields, farmers' livelihood contributions, and cacao quality remain understudied. In this regard, Mattalia et al. (2022) conducted a literature review (39 publications) comparing cacao monocultures with simple SAFs or complex SAFs, concluding that in simple SAFs, cacao trees produce more or equal yields than monocultures in only 52% of cases.

Niether et al. (2020) performed a meta-analysis of 52 articles directly comparing cacao SAFs and monocultures. Using an inductive and multidimensional approach, they analyzed differences in cacao and total system yields, economic performance, soil chemical and physical properties, incidence of pests and diseases, potential for climate change mitigation and adaptation, and biodiversity conservation. According to this analysis, cacao SAFs outperformed monocultures in most indicators. Cacao yields in SAFs were 25% lower than in monocultures, but total system yields were approximately ten times higher, contributing to food security and income diversification. Profitability was similar in both production systems. Cacao SAFs contributed to climate change mitigation by storing 2.5 times more carbon and adaptation by reducing average temperatures and buffering temperature extremes. Regarding soil parameters, no significant differences were found. The effect of the production system type on disease incidence depended on the fungal species. The few available studies comparing biodiversity showed higher biodiversity in cacao SAFs.

Schneider et al. (2016) published preliminary results of a field trial in Bolivia comparing cacao production systems in monoculture and SAFs under conventional and organic management. They concluded that the early productivity of cacao in monoculture systems was clearly superior to values found in SAFs (+86%), possibly due to greater light availability for cacao trees in monocultures. However, total yields of SAFs were significantly higher compared to monocultures (+135%), emphasizing the potential of SAFs to contribute to local food security and risk distribution in small farmer contexts. Yield gaps were observed between production systems under organic and conventional management (−47% in monoculture, −16% in SAFs), highlighting the need to develop knowledge on appropriate agricultural practices in organic cacao production in the agroecological and socioeconomic context.

1.4 Cacao Value Chain

The cacao value chain involves the total process of cacao production, from cultivation to consumption. It includes various stages from farming to harvesting, fermentation, drying, processing, and marketing. The cacao value chain emphasizes the interrelation of activities and participants in delivering cacao products to consumers. Among the stakeholders in the cacao value chain are farmers, traders, processors, manufacturers, and retailers, each contributing to the final cacao product. Understanding the cacao value chain helps identify challenges and opportunities within the industry and promotes sustainable practices.

The cacao value chain is crucial for various reasons. It supports the livelihoods of millions of farmers, contributes significantly to global economies, fosters employment, and provides raw materials for the chocolate and cosmetic industry. At the same time, it is helping to promote sustainable practices and environmental and social responsibility in cacao production.

The cacao value chain could be structured in several stages that encompass different functions from cultivation to consumption. These could include:

Cacao Farming:

- Cultivation: Farmers grow cacao trees, primarily in tropical regions.
- Harvesting: Ripe cacao pods are harvested, and the beans are extracted.
- Fermentation and Drying:
- Fermentation: Cacao beans are placed in piles or containers to ferment, this is a crucial step to developing flavor.
- Drying: Fermented beans are dried to reduce moisture content.

Processing and Grinding:

- Processing: Dried beans are cleaned, roasted, and winnowed to remove the shell.
- Grinding: The nibs obtained from processing are ground into cacao mass.

Cacao Butter and Powder Production:
Pressing: Cacao mass is pressed to separate cacao butter and cacao powder.

Manufacturing:
Chocolate Production: Cacao mass, cacao butter, sugar, and other ingredients are combined to make chocolate.
Confectionery Production: Chocolates and other cacao-based products are manufactured.

Distribution and Marketing:
Distribution: Finished products are distributed to retailers and consumers.
Marketing: Brands promote their products to attract consumers.
Retail and Consumption:
Retail: Products are sold in stores, online, or through various channels.
Consumption: Consumers enjoy chocolate products.

Waste Management:
By-products: Waste from processing may be repurposed for other uses.

Each stage has its challenges, from sustainable farming practices to fair trade concerns and ethical sourcing. The value chain interconnections have actions at one stage affecting others.

Latin America's cacao value chain has seen its production grow significantly over the last ten years and now accounts for approximately 21% of the global crop. The region has several advantages in light of market trends: first, it is the center of origin for cacao and is home to a rich genetic diversity that produces unique flavors that are well-suited for specialty products; farmers are relatively well organized, which better facilitates certification and traceability to meet consumer demand; a strong history and culture of consumption creates opportunities for domestic industries; and while primarily smallholder-driven, the region does have a sizeable number of larger plantations that drive technological innovation.

In Latin America, a great percentage of cacao is produced under SAFs and cultivated in buffer zones of national natural reserves. In essence, cacao produced under SAFs is grown under complementary shade trees in conjunction with other shrubs and crops in a system that has both environmental and income-producing benefits. For each farm, it is a question of how well-managed the SAF is and the level of benefits it delivers to the farmer (increased yield, long- and short-term supplementary income, and long-term conservation of soil and resources) compared to the possibility of what could be achieved using improved agroforestry practices.

First, improved management of cacao SAFs contributes to conservation. Well-designed cacao shade canopies support nutrient cycling and biomass production for the conservation of soil fertility; regulation of microclimates, and protection of cacao from pests, diseases, and extreme weather events; and carbon sequestration. Furthermore, cacao SAFs provide habitat and landscape connectivity that help conserve plant and animal biodiversity (Vaast & Somarriba, 2014).

1.5 Government Policies and Regulations on Cacao Cultivation in Ecuador

Ecuador, globally recognized for the quality of its cacao, has established various government policies and regulations over the years to boost and protect the cacao industry, considered one of the most important for the country's economy. These initiatives aim not only to strengthen the production and marketing of Ecuadorian cacao but also to promote sustainable practices and ensure equity throughout the production chain. The main regulations and government policies are detailed below, in accordance with Decree No. 811 (Presidencia de la República del Ecuador, 2019):

Promotion of Sustainable Production: One of the main government policies has been the promotion of sustainable cacao production. Through training programs and technical assistance, the Ecuadorian government has sought to improve agricultural practices by promoting the use of organic and sustainable methods. The adoption of these practices benefits not only the farmers but also contributes to environmental preservation.

Financial Incentives: The Ecuadorian government has implemented financial incentives to support cacao producers. These include preferential credit lines, subsidies, and financing programs aimed at facilitating farmers' access to capital needed to improve their plantations, acquire technology, and increase productivity.

Quality and Certifications: To maintain and elevate the quality standards of Ecuadorian cacao, strict regulations have been established regarding cultivation, harvesting, and post-harvest practices. Additionally, obtaining international certifications endorsing product quality and sustainability is encouraged, allowing producers access to more demanding and premium markets.

Marketing and Export: The Ecuadorian government has implemented policies to facilitate the marketing and export of cacao. These measures include streamlining customs procedures, participation in international fairs, and the promotion of the 'Cacao Nacional del Ecuador' brand, aiming to position the country as a global leader in high-quality cacao production.

Protection of Labor Rights: Recognizing the importance of ensuring fair labor conditions throughout the production chain, the Ecuadorian government has established specific regulations to protect the rights of workers on cacao plantations. These measures seek to prevent labor exploitation and ensure decent conditions for those involved in cacao production.

In summary, government policies and regulations on cacao in Ecuador reflect a strong commitment to sustainable development, product quality, and equity in the production chain. These initiatives benefit not only producers and exporters but also contribute to positioning Ecuadorian cacao as a global benchmark in terms of quality, sustainability, and social responsibility.

Despite significant advances in government policies and regulations in the cacao sector in Ecuador, there are still challenges and areas for improvement that could contribute to achieving a more efficient and competitive cacao production chain. The following is an analysis of some key aspects that could benefit from further attention:

Infrastructure and Technology: Investing in infrastructure and technology throughout the cacao production chain could significantly improve efficiency. The lack of modern infrastructure, such as roads and storage and processing systems, can impact the speed at which cacao reaches the market. Incorporating modern technologies in agricultural practices and post-harvest could also increase cacao productivity and quality.

Diversification and Value Addition: Although Ecuador is known for producing high-quality cacao, there is untapped potential in terms of diversification and creating value-added products. Promoting the diversification of cacao-derived products, such as premium chocolates, liqueurs, or beauty products, could open new markets and generate higher income for producers.

Research and Development: Investing in specific research and development for the cacao sector would address specific challenges, such as plant diseases, optimization of varieties, and more efficient agricultural practices. Innovation in these areas could lead to sustainable increases in cacao productivity and resilience, ensuring long-term competitiveness.

Access to Financing: Despite existing financial incentives, some producers may face difficulties in accessing credit. Improving access to loans, especially for small farmers, could be crucial to enable investments in technology, training, and infrastructure improvements.

Environmental and Social Sustainability: Despite ongoing efforts, environmental and social sustainability remains a critical area. Ensuring sustainable agricultural practices, promoting biodiversity, and ensuring fair working conditions will contribute to building an ethical and responsible production chain, aspects increasingly valued by global consumers.

Promotion and Marketing: While progress has been made in promoting the 'Cacao Nacional del Ecuador' brand, the marketing strategy could be improved to increase visibility and demand in international markets. Increased promotion could contribute to consolidating the reputation of Ecuadorian cacao as a high-quality and sustainable product.

In conclusion, achieving an efficient and competitive cacao production chain in Ecuador requires a comprehensive approach that addresses both technical and social and commercial aspects. Collaboration between the government, industry, and producers, along with continued investment in research and development, may be the key to overcoming challenges and positioning Ecuadorian cacao as a leader in the global market.

Ecuador, renowned for its superior quality cacao, faces a promising future in the cacao industry. However, to maximize its potential, it is essential to consider both future prospects and challenges that could impact the country's cacao production chain.

1.5.1 Future Perspectives and Challenges

1.5.1.1 Future Perspectives

Sustainable Growth: The global demand for cacao continues to rise, driven by increasing consumer preference for high-quality chocolate products. Ecuador is well-positioned to capitalize on this growth, especially with its focus on sustainable and organic production.

Diversification of Products: The diversification toward higher value-added cacao products, such as premium chocolates and derivatives, offers an exciting perspective. This approach would not only increase income for producers but also strengthen Ecuador's position in international markets.

Technology and Innovation: The adoption of advanced technologies, from agricultural techniques to post-harvest processes, promises to improve the efficiency and quality of cacao. Investment in research and development can lead to more resistant varieties, more efficient agricultural practices, and innovative products.

Sustainability and Social Responsibility: The growing environmental and social awareness worldwide creates an opportunity to further highlight the 'Cacao Nacional del Ecuador' brand. Continuing sustainable agricultural practices and ensuring fair working conditions will strengthen the position of Ecuadorian cacao as a leader in social responsibility.

1.5.1.2 Challenges to Overcome

Climate Change: Climate change poses significant challenges to cacao production. Variations in weather conditions can negatively impact yields and increase the spread of diseases. Adaptation and mitigation strategies are essential to ensure the sector's resilience.

Access to Financing: Despite the implementation of financial support programs, some farmers still face difficulties accessing credit. Improving access to financing, especially for small producers, will be crucial to stimulate investments and improvements in the production chain.

Global Competition: Other cacao-producing countries also seek to consolidate their position in the global market. Ecuador must remain competitive through product differentiation, consistent quality, and effective marketing strategies.

Education and Training: Continuous training and education in new agricultural practices and technologies are essential. Ensuring that farmers are aware of best practices will contribute to increased productivity and quality of cacao.

Infrastructure and Logistics: Improving transportation and logistics infrastructure is crucial to ensure that cacao reaches national and international markets efficiently. Investment in this aspect will facilitate competitiveness and reduce transaction costs.

In conclusion, the future prospects for the cacao chain in Ecuador are promising but not without challenges. Continued attention to sustainability, investment in technology, product diversification, and overcoming challenges such as climate change and global competition will be crucial to consolidating Ecuador's position as a leader in the production of high-quality and sustainable cacao.

1.5.1.3 Regulations in the European Union

The main regulations of the European Union are detailed below (Metsola & Kullgren, 2023):

EU Action Plan against Deforestation: The EU has shown a growing interest in addressing deforestation associated with supply chains of products like cacao. In 2019, the European Commission presented the 'EU Action Plan against Deforestation' to tackle imported deforestation and its connection to products consumed in the EU.

Due Diligence Regulation in the EU Supply Chain: The EU has discussed the possibility of introducing a due diligence regulation in the supply chain to address deforestation risks associated with imported products, including cacao. This regulation could require companies to conduct risk assessments and take measures to prevent the import of products linked to deforestation, effective since December 2023 with the European Union Deforestation Regulation (EUDR).

Certification and Sustainable Labeling: The demand for sustainable and certified products is increasing in the EU. Consumers are more aware of the relationship between their purchasing choices and deforestation. Certification of cacao as organic, fair trade, or meeting strong environmental standards can offer Ecuadorian producers an advantage in the European market.

Voluntary Partnerships: Some companies and associations in the EU have taken voluntary measures to address deforestation in their supply chains. Active participation in voluntary initiatives can be crucial for Ecuadorian cacao producers aiming to comply with European standards.

1.5.1.4 Challenges for Ecuadorian Cacao Regarding EU Regulations

Compliance with Regulations: Introducing stricter regulations in the EU with the EUDR, Ecuadorian cacao producers will face challenges in complying with new standards, especially those related to deforestation in production areas.

Need for Sustainable Practices: The pressure to adopt sustainable agricultural practices and avoid deforestation will become increasingly relevant. Ecuadorian producers will need to adapt and adopt sustainable approaches to ensure competitiveness in the European market.

Incentives for Sustainability: It will be crucial to establish incentives and support programs for Ecuadorian producers adopting sustainable practices and contributing to deforestation prevention.

The Ecuadorian cacao sector faces significant challenges in aligning with the growing environmental and social requirements of the EU. Complying with new regulations, adopting sustainable production methods, and accessing incentives will be fundamental to ensuring access to this key market in the future.

1.6 Conclusions

The global cacao agroindustry holds a prominent place in the world economy, serving as the livelihood foundation for countless farmers. However, this sector suffers from serious deficiencies in environmental sustainability and social equity that urgently require comprehensive solutions.

Countries in Africa, Asia, and Latin America that dominate cacao production generate substantial revenue through exports. However, the paradox is that many farmers live in poverty despite producing a highly valued and lucrative crop in global markets.

There are indeed promising paths to transform this reality, from sustainable intensification of crops and diverse SAFs to generating greater added value and certifying good agricultural practices. However, these changes demand political decisiveness, public–private coordination, investment in infrastructure and technological innovation, as well as strengthening local capacities along this intricate global supply chain.

The complex challenges ahead, from global climate change to stricter legislation in importing nations, will test the adaptability and resilience of this agroindustry. Technification, product differentiation, proactive adaptability, and the construction of solid comparative advantages will be crucial factors in capitalizing on emerging business opportunities and ensuring a sustainable future shared by all links in this value chain.

References

Brack, D. (2019). *Towards sustainable cacao supply chains: Regulatory options for the EU.* Coordinadora Estatal de Comercio Justo. https://www.fern.org/fileadmin/uploads/fern/Documents/2019/Fern-sustainable-cacao-supply-chains-report.pdf

García Cruzatty, L., García, B., & Sánchez, F. (2023). Variability and combinatorial ability in fruit characters of new *Theobroma cacao* hybrids in Ecuador. *Revista Brasileira de Ciências Agrárias, 18*(3), e3265. https://doi.org/10.5039/agraria

Guimac Cedillo, L. Y., & Verastegui Huanca, L. L. (2022). Características y producción de cacao: Caso de la zona productora de Amazonas – Perú. *Global Business Administration Journal, 6*(1), 3–13. https://doi.org/10.31381/gbaj.v6i1.4268

Mattalia, G., Wezel, A., Costet, P., Jagoret, P., Deheuvels, O., Migliorini, P., & David, C. (2022). Contribution of cacao agroforestry versus mono-cropping systems for enhanced sustainability. A review with a focus on yield. *Agroforestry Systems, 96*, 1077–1089. https://doi.org/10.1007/s10457-022-00765-4

Ministerio de Agricultura y Ganadería – MAGAP. (2021). Informe de Rendimientos Objetivos de cacao. Obtenido de. https://sipa.agricultura.gob.ec/index.php/informe-de-rendimientosobjetivos/rendimiento-de-cacao

Niether, W., Jacobi, J., Blaser, W. J., Andres, C., & Armengot, L. (2020). Cacao agroforestry systems versus monocultures: A multi-dimensional meta-analysis. *Environmental Research Letters, 15*(10), 104085. https://doi.org/10.1088/1748-9326/abb053

Pérez, E., Guzmán, R., Álvarez, C., Lares, M., Martínez, K., Suniaga, G., & Pavani, A. (2021). Cacao, cultura y patrimonio: un hábitat de aroma fino en Venezuela. *RIVAR (Santiago), 8*(22), 146–162. https://doi.org/10.35588/rivar.v8i22.4781

Presidencia de la República del Ecuador. (2019). Decreto Ejecutivo No. 811 – Política Pública del Plan de Mejora Competitiva del Cacao. https://www.agricultura.gob.ec/wp-content/uploads/2020/07/decreto-811.pdf

Metsola R., & Kullgren P. (2023). Reglamento (UE) 2023/1115 del Parlamento Europeo y del Consejo, del 31 de Mayo de 2023. Publicado el 9 de junio del 2023 en Diario Oficial de la Unión Europea Número 150. https://d2eb79appvasri.cloudfront.net/smartlex_boletines/74647.pdf

Schneider, M., Andres, C., Trujillo, G., Alcon, F., Amurrios, P., Perez, E., Weibel, F., Milz, J. (2016). Cocoa and total system yields of organic and conventional agroforestry vs monoculture systems in a long-term field trial in Bolivia. *Experimental Agriculture, 53*(3), 351–374. https://doi.org/10.1017/S0014479716000417

Vaast, P., & Somarriba, E. (2014). Trade-offs between crop intensification and ecosystem services: The role of agroforestry in cacao cultivation. *Agroforestry Systems, 88*, 947–956.

2 History, Origin, Cultivars, and Cacao Research in Ecuador

Freddy Amores

2.1 Introduction

The last two decades have witnessed a significant increase in cacao production in Ecuador, driven by the cultivation of high-yielding clonal varieties, particularly CCN 51. Its participation in the international market has also grown, making Ecuador the third-largest global producer with 411,000 tons exported in 2023, generating revenues of 1384 million dollars. Some 2% of the production is allocated for domestic consumption.

While the global demand for chocolate and cacao products is increasing at an annual rate of 2.5%, the additional supply of beans to meet this consumption pace amounts to more than 100,000 tons annually, necessitating increased productivity rather than expanded cultivation areas. Growing production and sales translate into additional wealth and well-being for participants in the cacao chain, mainly supplied by small-scale producers. In this context, the technological advancement of cultivation must be a continuous objective to achieve production efficiency. Investments in research and the development of new high-yielding varieties, as well as the modernization of cultivation practices, play a significant role in achieving this objective.

In August 1942, the governments of Ecuador and the United States signed an agreement to establish the Agricultural Experimental Station of Ecuador, an institution that in 1943 set up the country's first research center at Hacienda Pichilingue. Later, it took the name of Tropical Pichilingue Experimental Station (EETP). The initial team responsible for its operation was composed of Ecuadorian and American technicians, with a focus on cacao cultivation. In 1948, a new agreement was signed, strengthening the interest of both governments to collaborate on agricultural research. In 1952, the EETP was transferred to the Inter-American Cooperative Agricultural Service (SCIA), and in addition to cacao, research expanded to other crops: coffee, balsa, timber, oil palm, abaca, and soybean. In 1963, the EETP was transferred to the National Institute of Agricultural and Livestock Research (INIAP), established in 1959. Subsequently, the Boliche Experimental Station (1968) near Guayaquil, the Napo Experimental Station (1978) in the northern Amazon region, and the

DOI: 10.4324/9781003381761-3

Portoviejo Experimental Station were created, all of which currently work with cacao and other crops.

Cacao has the longest tradition of export and has been a significant source of foreign exchange for 400 years, first for the colonial territory and later for the Republic of Ecuador. Despite fluctuations, the role of this crop has been fundamental in sustaining the country's economy and a source of capital for the development of trade, banking, and industry, at least until the early 20th century when its share of export revenues approached 80%. In 2023, cacao exports represented 3% of total non-oil exports and a little over 1% of the GDP. Going forward, the growth of the cacao value chain will depend on further technological innovation in cultivation, in harmony with the principles of sustainable production and economics.

2.2 Theories and Evidence to Explain the Presence of Cacao in Ecuador

The species *Theobroma cacao* originated in the basin of the great Amazon River, evolving through various populations that are now the source of its known and yet-to-be-discovered genotypic variability. This variability is greater in the upper Amazon. All the varieties that currently contribute to the cacao supply represent a tiny portion of the global diversity of the species; less than 25% of the planting material in cacao-producing countries has benefited from some level of genetic improvement, reflecting the limited utilization of this natural resource. Understanding the path to current varieties and production systems requires an exploration of the origin, domestication, dissemination, and use of this crop.

Recent archeological evidence points to the natives of the Mayo-Chinchipe culture, who lived in what is now the southeastern part of the Ecuadorian Amazon, Zamora-Chinchipe province, as the first to process cacao beans for beverages used in events and ceremonial rituals over 5000 years ago. These findings contradict earlier reports that placed the origin, domestication, and use of cacao 2000 years later, in southern Mexico. This discovery prompted a project to explore the Zamora Chinchipe zone (the author of this note participated in the first exploration) from 2010 to 2020 to locate wild cacao trees that would reinforce the available evidence. Approximately 300 trees were found, and their leaves, branches, and pods were collected (Figure 2.1).

During the exploration, only pods with cotyledons of violet, pink, and white colors were observed. The comprehensive analysis of pod morphology, cotyledon color, and subsequent DNA analysis results revealed the proximity of some trees to Nacional cacao, the ancestral variety present at the beginning of our cacao production. Part of the collected planting material is preserved in INIAP's genetic banks at EETP and Domono Farm, the latter located on the eastern slope of the Andean Mountain range. Some of the rescued materials were also established in sites near the collection area (agricultural schools).

Figure 2.1 INIAP, CIRAD personnel, and natives of the Zamora Chinchipe area gathered to plan work (a) to venture into the eastern foothills of the Andean Mountain range, Zamora Chinchipe area (b, c), with the aim of sighting and collecting planting material from ancestral Nacional cocoa trees (d).

The results of this project strengthen the theory that the origin of cacao trees sighted by the Spanish in the coastal region of present-day Ecuador came from a genetic group identified in the wild in the Zamora Chinchipe zone. It was called the Nacional group, although other genetic groups were also identified. At some point in the pre-Hispanic period, natives of the Mayo-Chinchipe culture transported cacao pods to the coastal region, and later, natives of this region included cacao beans among the products they traded with people from other coastal areas further north, including Central America. This is how cacao could have reached southern Mexico, where it was also used as a beverage in religious rituals.

2.3 Highs and Lows in the Economic History of Cacao

The first European in contact with cacao turned into chocolate and offered by native Americans was Cristóbal Colón, who did not pay attention to it. It was Hernán Cortez who, in 1528, brought several tons of beans to Spain; there, chocolate was mixed with some type of sweetener to reduce bitterness, becoming a costly and coveted drink for royalty. The rest of Europe gradually tasted chocolate, spreading consumption; to bring the source of supply closer, France and England planted cacao on some Caribbean islands and later in their African colonies. In the 19th century, chocolate consumption in Europe became the engine of global cacao demand, a role it still maintains; 60% of Ecuador's exports and 40% of global production are shipped to that destination.

The first shipments in the 17th century were supplied by the harvest of spontaneous cacao plants in the lower basin of the Guayas River (Guayaquil District). The increase in demand drove intentional cultivation, giving rise to cacao agriculture. Over time, it continued to be planted upstream on riverbanks along the tributaries of the Guayas River, a circumstance that facilitated river transport and is the origin of the term 'cacao arriba' which also referred to Nacional cacao at that time. Typically, riverbank soils are loamy and fertile, and being close to river channels, the water table does not move far from the surface during the dry period, allowing plants to better withstand drought. However, the expansion of cacao cultivation was accompanied by plantings further away from watercourses, on hillside terrains with greater moisture restrictions during the dry season. Until today, after diseases cut 40% of potential annual production, drought is the second most limiting factor.

At the beginning of the 18th century, the Guayaquil District (now comprising the provinces of Guayas, Los Ríos, and El Oro) was already producing around 1000 tons of cacao. The industrial revolution in Europe in the 19th century popularized chocolate consumption, boosted demand, and stimulated production; the period from 1780 to 1820 was a time of great economic prosperity. In response to the growing demand, new plantations advanced to the Carrizal Chone River basin, in what is now the province of Manabí, the second most productive area in Ecuador, and later to the Esmeraldas River basin, gradually shaping the current cacao geography. The commercial planting of cacao in the provinces of the Amazon region began in the 70s along the communication routes that were opened due to the nascent oil economy.

In the mid-19th century, cacao activity stagnated due to the economic downturn in Europe and the American Civil War (1861–1865); the combination of both events restricted demand and discouraged production. On the other hand, the frequent political-military crises of the nascent Republic of Ecuador (independent since 1820) also contributed to the decline in crops; production began to recover in the last quarter of the 19th century, and by the end of the century, this exceeded 15,000 tons. In 1916, when Moniliasis disease was detected, production had already surpassed 40,000 tons. The presence of Witches' broom in 1922 joined Moniliasis to reduce the production to its lowest point at the end of 1930, around 10,000 tons. Before the collapse, Ecuador had already become the world's leading cacao producer and exporter; meanwhile, Africa was beginning to emerge as an exporting power.

The decline in international demand due to the economic depression in the United States (1929–1933) and World War II (1939–1945) exacerbated the losses in production caused by diseases in cocoa fields without any maintenance work. In 1950, with just over 18,000 tons, production had retreated to the levels of the early 20th century. A similar phenomenon occurred in

Brazil with the entry of witches' broom detected in 1989 in the Bahia region; the rapid spread of the disease drastically reduced cacao production, which is still recovering.

With difficulties, in response to changing climate and price volatility, in a market unable to adjust smoothly to the supply and demand dynamics due to the perennial nature of the crop, Ecuador's cacao economy began to grow in the second half of the 20th century. By the early 2000s, production had approached 100,000 tons in response to the increase in cultivated area; productivity remained low, at 0.28 tons/ha. The average production growth rate, equal to 1.5% annually since the 1950s, contrasts with the growth rate in the 21st century, at least 7% annually. Half a million Ecuadorians depend directly and indirectly on cacao, with a cultivated area that already exceeds 600,000 hectares. The planting of the high-yielding clonal variety CCN 51, a significant technological leap, is the main reason production has quadrupled, and productivity has more than doubled in the last two decades.

2.4 Looking for a Solution to Sanitary Problems

In the 1930s, the observation of trees with the absence or scarce presence of diseases in local orchards (referred to as 'Refractories') caught the interest of some producers. They planted seeds hoping that the new plants would replicate the disease tolerance observed in the parent trees, but they ended up getting sick too.

In the late 1930s and early 1940s, English scientists from the Imperial College of Trinidad, now the University of the West Indies, explored accessible areas of the upper Amazon by river, seeking wild cacao trees to collect planting material. The goal was to study their behavior against diseases. With the collected material, a High Amazon Cacao Genetic Bank was established on the island of Trinidad. Genotypes from this bank were later distributed to other countries like Ecuador and Costa Rica. Scavina, Pound, and Iquitos Mixed Calabacillo (IMC), among others, are names of some rescued and distributed genotypes.

Also, in the early 1940s, scientists from the same university visited commercial cacao fields on the island to select Trinitarian-type cacao trees that they cloned using a protocol perfected in the 1930s. Each selected genotype received the code ICS (for Imperial College Selection) followed by a number. Like the upper Amazon cacaos, some ICS clones, including ICS 95, widely known in the cacao world, were distributed to Ecuador and other countries. In the 1950s, scientists from the same university crossed High Amazon cacaos with different ICS clones. Research in Ecuador closely followed the steps of the work conducted in Trinidad, at least until the 1970s.

From 1944 to 1948, as part of a collaborative project between the Departments of Agriculture of Ecuador and the United States, hundreds of traditional orchards were visited to select trees with Nacional-type characteristics. The selected trees came from spontaneous crosses of Nacional x Trinitarian, though it was difficult to determine the level of proximity to the pure Nacional cacao variety. Each clone was identified with the code EET followed by a number before being part of the Nacional type cacao genetic Bank at the Pichilingue Station. The EET code and numbering were also assigned to other clones that did not necessarily belong to the Nacional type, including the ICS 95 clone (code EET 111), a situation that still causes some confusion.

Over time, the bank with Nacional-type selections enriched itself with clones from more trees identified on other farms. The clones EET 19, EET 48, EET 64, EET 95, EET 96, and EET 103 distributed by INIAP for commercial planting in the 1970s originated from this Germplasm Bank. For reference, the EET 103 clone comes from a tree selected on the Porvenir farm (Vinces, Los Ríos, Ecuador), the cacao heartland of the country at that time.

In the early 1940s, personnel from the United Fruit Company had also collected planting material from trees of interest, establishing a genetic bank on its Tenguel estate in the Balao area; it later became known as Centro de Cacao de Aroma Tenguel (CCAT). The germplasm bank on this site was built with seeds from selected trees showing commercially interesting characteristics, regardless of whether they were close to the Nacional variety or not. This bank had about 3000 trees, each assigned the code CCAT followed by a number for L (line) and H (row) coordinates guiding their distribution in the field (Figure 2.2). The importance of this genetic resource was revalued in the 1990s

Figure 2.2 CCAT tree over 80 years old at the Tenguel Aroma Cocoa Center.

when INIAP conducted a detailed study of about 200 plants, focusing on the most productive ones with pod morphology like that of the Nacional variety.

The best CCAT trees, less than 100 of the 200 studied, after applying sanitary and productivity criteria, were cloned, and planted in Station Pichilingue to start the CGN (National Genotype Collection). The clones EET 544, EET 558, EET 575, and EET 576 delivered by INIAP for commercial use in 2009 come from plants evaluated for several years, first in this collection and later in multi-location tests. The clones EET 544, EET 558, and EET 103 form the basis for the production of 600 hectares of fine cacao on the well-positioned La Victoria state (Guayas, Ecuador), visited by cacao producers and chocolatiers from around the world interested in knowing and tasting Nacional-type cacao liquor.

New explorations in the Ecuadorian Amazon in the 1970s and 1980s (financed with funds from the London Cacao Merchants Association) allowed the rescue of more wild cacao trees, which are now part of the Chalmers and Allen collections of upper Amazon cacao at the Pichilingue Station. Some genotypes from this bank were used in crosses as part of a project funded by the U.S. Department of Agriculture that started in 2003. Additionally, in the first decade of the 21st century, as part of a project funded by SENACYT (Secretaría Nacional de Educación Superior, Ciencia, Tecnología e Innovación del Ecuador), the north of the Esmeraldas province was explored to rescue Criollo-type cacao trees (introduced from southern Colombia a long time ago) on traditional farms; the rescued material also rests in a Criollo-type cacao germplasm bank at the Pichilingue Station (Figure 2.3).

From 1998 to 2013, as part of various international cooperation efforts between INIAP, IPGRI, BIOVERSITY, and USDA/MARS, projects were executed for the genetic improvement of cacao varieties applying criteria of

Figure 2.3 Criollo cocoa tree (a) and criollo cocoa clone at the South Coast Experimental Station (b) from the central-northern area of Esmeraldas.

high performance and disease resilience. Three commercial varieties (INIAP 800, INIAP 801, and INIAP 802) with a performance level comparable to CCN 51 are the result of these projects, not to mention the availability of a genetically improved population that can still be used as a foundation for further studies and genetic improvement.

2.5 How Traditional Production Systems Were Formed?

In the last decades of the 19th century, coinciding with a second cacao boom, owners of large cacao plantations, taking advantage of their visit to Venezuela on their return from Europe, introduced Trinitarian cacao to Ecuador to boost production in response to growing demand. In general, Trinitarian cacao are more vigorous and resilient to less favorable environmental conditions, such as hillside and non-riverbank terrain. Trinitarian cacao emerged in the mid-18th century on the island of Trinidad, from the spontaneous cross between Criollo and Lower Amazon (Amelonado) cacao trees, genetically distant from each other, and distant from Nacional cacao; that explains its vigor, but that is another story. Over time, Trinitarian cacao planted in Ecuador also crossed spontaneously with Nacional cacao; the seeds of this cross were used to expand the area planted with cacao.

The trees present today in our traditional production systems, whose population has been assigned the name of Nacional x Trinitarian complex, have their origin in the natural cross of both. Additionally, the distribution and planting of hybrids from crosses of Nacional x Upper Amazon genotypes (Scavina, Silecia, Pound, and IMC 67) conducted at the Pichilingue Station and delivered to producers in the 1960s and 1970s caused new introgressions that increased the genetic variability of commercial cacao fields. For example, it is now common to see as a result of these natural crosses pods with bright green color typical of the Scavina genotype, although the original genotype introduced to the country, resilient to witches' broom, has very small pods and beans (it is not commercial).

Due to these genetic introgressions, visiting traditional orchards is a unique opportunity to observe segregations through the shape, size, and color of the pods, and even the tone of the bean's cotyledons. In hybrid tree populations, no two trees have the same genetic load, unlike the homogeneity of clonal orchards. A study of fresh beans from pods from 40 hybrid plants in a commercial orchard revealed that the sensory profile of the pulp changed with each tree, showing a trait that reflects the genetic heterogeneity of plants. The fermentation of this mixture of beans imparts unique flavors and aromas to cacao paste. The main explanation for the low productivity of these traditional production systems, between 0.2 to 0.5 tons per hectare, lies in the productive structure of the tree population. 80% of the harvest comes from the 20% most productive trees, a circumstance that reflects the high level of inefficiency of this productive system (Figure 2.4).

Figure 2.4 Tree from a traditional cocoa orchard endowed with a large load of pods.

2.6 Origin of Local Clonal Varieties, Management Practices, and Post-Harvest Quality Technology

With the headquarters at the experimental center on the Pichilingue estate, the first formally planned cacao research took place between 1944 and 1948. Hundreds of cacao farms were visited, and trees of the Nacional type were selected, focusing on those with a good load of pods and acceptable levels of disease tolerance. The selected trees were cloned to establish a cacao germplasm bank in Pichilingue. This population is the source of the varieties EET 19, EET 48, EET 64, EET 95, EET 96, and EET 103 delivered for commercial planting in the 1970s. Recent DNA analyses show that the predominant genetic load in the composition of commercial EET clones comes from Nacional and Amelonado (Lower Amazon) cacao populations.

The introgression of Trinitarian cacao into the ancestral Nacional cacao, at the end of the 19th century, gives meaning to this result; Amelonado cacao is one of the parents of Trinitarian cacao. On the other hand, in the 1950s, controlled crossings were made between the best selections of Nacional-type cacao with Upper Amazon genotypes, particularly with Scavina, Silecia, and Pound, which exhibited lower witches' broom infection, to obtain hybrid progenies. In the 1960s and 1970s, the commercial distribution of seedlings and pods from those hybrids with better performance was recommended. Some performed well initially, but over time, they became very vulnerable to this disease; the production of others was affected by sexual incompatibility during fertilization. A few achieved acceptable performances in areas with less favorable environments for diseases. Until a few years ago, pods of these latter ones were still distributed by the Portoviejo experimental station, as a source of planting material in cacao-producing areas in the Manabí province; there, the drier environment mitigates the health problem.

In the 1970s and 1980s, as part of a Technical Cooperation Agreement between INIAP and the UK's International Development Agency, studies were conducted on pollination, epidemiology, physiology, and agroclimatology to improve the understanding of the factors shaping the productive behavior of traditional cacao orchards. The results were assembled into a set of recommendations (rejuvenation pruning, height trimming, periodic removal of diseased fruits, orchard replenishment, and shade management) to rehabilitate and recover productivity. The absence of demonstrative plots at the commercial level and the analysis of production costs and profitability margins related to these practices were obstacles to the adoption. Indeed, although traditional production systems respond with yield increases to productive rehabilitation, the issue of the economic return on investment financing this change is usually not addressed.

As a reference, the removal of diseased fruits must be done every 8 days at an annual cost per labor equivalent to the value of 0.25 tons of dry cacao per hectare; therefore, the practice is justified only if the orchards are already producing at least 0.6 tons per hectare before implementing it. Regarding a well-qualified annual pruning, the cost per hectare is covered with the sale of 0.30 tons of dry cacao; in any case, these are indicators that need further work to gain precision in estimates. On one occasion, a colleague spoke about the benefits of the periodic removal of moniliasis diseased fruits as the most effective solution to minimize harvest losses from this cause; during the question-and-answer session, I expressed my concern about the profitability of removal, and he humbly replied that he did not have an answer to that question. This experience illustrates one of the problems faced by the technological update of this crop.

On the other hand, the heterogeneity of the productive structure of trees in the traditional production systems explains why 80% of the production is contributed by the 20% most productive trees; in contrast, there are trees whose harvest is worth no more than 0.30 dollars per year, and it is usual that two out of every 100 trees do not produce and only occupy space; this description explains the low productivity of this system. Producers rarely work with indicators, but they are aware of the heterogeneity of their trees. In a certain area with high precipitation and a high incidence of diseases, there is already a previous path drawn to harvest; producers avoid passing through trees with low yield, thus avoiding waste of time, energy, and scarce resources. When they have capital to invest, even at the cost of staying without income for a while until the new planting produces, they choose to cut down the old cacao trees to replace them with the CCN 51 clone. For this reason, the area occupied by traditional orchards as a production system continues to shrink.

In the 1990s and until the first decade of the current century, with the economic and scientific support of international research actors (United States Department of Agriculture-USDA, Centre for Agricultural Bioscience International-CABI, and Nestle), the beneficial action of microorganisms, particularly of the Trichoderma genus, for the biological control of cacao diseases was investigated. The results confirmed their antagonistic capacity at the laboratory level; progress was made in formulating prototypes to validate their performance in commercial orchards, but no additional progress was recorded beyond this point.

Since the 1990s, the international market has become more demanding regarding the differences in the physical and sensory quality of cacao beans as an added value factor; in this context, claims about the mixtures of Nacional and CCN 51 cacao beans detected in export batches received at the destination were frequent. Against this backdrop, INIAP began implementing a small laboratory to provide services and conduct research on post-harvest practices, sensory profile descriptions, and estimation of the mixing level between cacao types; the initiative arose within the framework of an agreement and project between the government of Ecuador and the EU (through *Centre de Coopération Internationale en Recherche Agronomique pour le Développement*-CIRAD) to improve the production and quality of Ecuadorian cacao.

Over time, the laboratory (now known as the Laboratory for the integral quality of Cacao and Coffee) improved its capacities to work in the field of physical and sensory quality; the chemical dimension of quality is supported by the nutrition laboratory of the Santa Catalina Experimental Station, located near Quito. These capacities were reinforced and grew with the project 'Study on the physical, chemical, and organoleptic parameters to identify fine or aroma cacao from ordinary ones' sponsored by International Cocoa Organization-ICCO and executed in Ecuador, Trinidad and Tobago, Venezuela, and Papua New Guinea (the undersigned was in charge of the international coordination of this project) at the beginning of this century. The results strengthened the country's position as a producer of fine cacao.

A product of this laboratory is the development of a protocol to define the percentage of the blending of CCN 51 cacao beans and Nacional-type cacao using the visible light spectrometry technique (there is a publication on the subject); currently, this tool is applied to detect and quantify the proportion in which each variety is present in the mixture. The laboratory also provides research services, chocolate bar production, and training for operators in the cacao sector. In the framework of another collaborative project with CIRAD (France), the laboratory was used to differentiate between varieties and origins of cacao, using the near-infrared spectrometry (NIRS) technique. The first works to collect, prepare samples, and determine the presence of cadmium (Cd) in cacao beans from different areas of the country were also carried out in this laboratory in conjunction with the soil analysis laboratory at Pichilingue Station.

2.7 Environment, Varieties, and Potential Yields

Currently, the cacao surface area in Ecuador is estimated to be over 600,000 hectares. Less than a third of this area is still composed of traditional production systems. The rest is clonal cacao, mainly CCN 51, whose surface area continues to grow, replacing traditional cacao, as well as oil palm, corn, and pastures. Although the incidence of diseases and water scarcity during the dry season restrict cacao average productivity, it currently exceeds 0.60 tons/ha; until 2000, it was close to 0.3 tons/ha. The cultivated area with irrigation is small, presenting a clear opportunity to increase production with the widespread adoption of this practice.

The massive planting of the clonal variety CCN 51 at the beginning of the 21st century allowed for a significant leap in productivity, production, and exports in a brief period. Depending on the level of technology, CCN 51 yields vary between 0.9 and 3.0 tons/ha; CCN 51 production systems with small and medium surfaces are located at the lower end and middle of the indicated range, while in a few properties, large units of up to 1000 hectares with cutting-edge technology, productivity is at the upper end.

The clone ICS 95, originating from Trinidad, is one of the parents of CCN 51, and the other parent is the clone IMC 67 from a wild tree sighted in the Upper Amazon in the late 1930s. The code 'IMC' comes from the words Iquitos Mixed Calabacillo, referring to the rescue zone and traits of the pod. The clonal variety CCN 51, with high yield, was developed on a farm in the Costa region, Naranjal area, through work conducted in the 1950s and 1960s by Homero Castro, an Ecuadorian researcher. The origin of this variety is a plant selected in the progeny of the cross ICS 95 x IMC 67, and the first batch planted with CCN 51 dates back to 1967. A little-known fact is that the same cross had already been made earlier in Trinidad, but no high-yield plants were identified in the resulting progeny, or at least not with the productivity of CCN 51; it can only be said that the genetic dice rolled in favor of Ecuador.

The average productivity of CCN 51 without irrigation is three to four times that of the traditional production systems with Nacional-type cacao, which is in the range of 0.2 to 0.5 tons/ha. With irrigation, this system can produce up to 0.8 tons/ha. Without irrigation in a similar environmental setting, CCN 51 doubles triples the yield of the EET varieties delivered by INIAP in the 1970s. By benefiting from irrigation, the yield of EET clones can reach up to 1.5 tons/ha, but the productivity gap with CCN 51 remains wide. These figures explain the preference of producers to plant CCN 51, a variety that, when well-managed and irrigated, yields at least 2.5 tons/ha. In deep and fertile alluvial soils, it can break the 3 tons/ha barrier, while in lower-quality soils such as hillsides and shallow slopes, the yield is rarely less than 2 tons/ha with irrigation, while the average without irrigation approaches 1 ton/ha.

At present, no one is willing to plant varieties that produce less than CCN 51. In 2016, INIAP introduced two new high-yield clonal cacao varieties (INIAP 800 and INIAP 801) with productivity comparable to that of CCN 51, the reference standard. The planting of both varieties is gaining ground, although limited by the narrow supply of planting material. Against this group of varieties, the benchmark for the genetic improvement of new cacao varieties is high. On the other hand, INIAP completed the validation of the high-yield variety INIAP 802 recently released (2023) for commercial cultivation; this variety yields as much as CCN 51 and sometimes more but never less. An important detail is that the plant that gave rise to INIAP 802 was selected from a cross between two genotypes of wild Upper Amazon cacao, Amazon 14 and EBC (Caquetá Botanical Expedition) 10 (there is some uncertainty about the identity of the first parent, the genetic analysis of which places it close to CCN 51, an aspect still pending clarification). It is also important to mention that one parent of INIAP 800 and INIAP 801 is CCN 51, while the other is an EET of the Nacional type.

Meanwhile, the productivity of cacao in traditional orchards remains depressed for several reasons: excessive productive heterogeneity among trees, high incidence of diseases, few plants per hectare, water scarcity, and lack of maintenance, among others. Marginal profitability is pushing the gradual disappearance of this system, being replaced by CCN 51, INIAP 800, INIAP 801, and now INIAP 802. The harvest from thousands of hectares of traditional cacao, provided good post-harvest practices are followed, is still marketed at attractive prices in high-value niche markets, a circumstance that compensates to some extent for its conservation, although the surface area with this production system is decreasing.

This is the story of the current cacao scenario that positions Ecuador as the third-largest world producer with growing production and average productivity. In 2022, exports reached 415,000 tons, and revenues amounted to $1.1 billion; in 2023, revenues exceeded $1.3 billion; In addition, 5% of this production is used for local consumption, adding value to the chain.

Promoting a culture of continuous innovation must become the guiding objective of the cacao sector. It must be supported by the formation of scientific capital, ongoing technical education programs for producers, solid demonstrations of productive and economic performance on a commercial scale, contests to reward the production of more cacao per square meter of land, competitions to choose the producer who maximizes the return on invested dollars, and other productivity indicators. The support of coordinated public and private policies facilitates the innovative process and generates changes in productive efficiency. Identifying opportunities for innovation and analyzing the priority and economy of these opportunities should always be on the discussion table at any meeting of cacao value chain operators.

2.8 Cultivars in Ecuador and Other Research

In the last 25 years, the INIAP National Cacao Program, with the support of producers on farms and other institutions, has conducted genotype and variety trials at 16 sites in different cacao zones of the country, in the Costa and Amazon regions. The clones EET 544, EET 558, EET 575, EET 576, INIAP 800, INIAP 801, and INIAP 802 delivered by INIAP for production between 2008 and 2023 are the result of this work.

A test established a decade ago in the northern province of Esmeraldas, near the Colombian border, was recently resumed; it had been discontinued due to the remoteness of the site and logistical support issues committed by a local organization of producers. However, favorable comments were heard during a recent visit to the area, at least about the performance of one of the studied genotypes. For this reason, this test is now the subject of new observations that could lead to the development of a new clonal variety of cacao with disease tolerance for this rainy environment with fragile soils. The genotypes established in that location come from populations that gave rise to the commercial varieties INIAP 800 and INIAP 801.

In general, the surface area of the different tests ranged from 0.35 to 2 hectares, density between 833 and 1,111 plants per hectare, the number of genotypes compared between 3 and 11, and the number of replications between 2 and 3. The objective was the comparison of selections arising from the genetic improvement processes conducted by INIAP since the 1990s until now. There is a field trial on the Domono experimental farm in the eastern foothills of the Andes; it compares the performance of clones and wild cacao trees, ancestors of Nacional cacao, rescued in the Zamora Chinchipe area, southeast of the Ecuadorian Amazon region.

When genetic improvement processes begin to show promising plants, small plots are established at the experimental center level; later, based on the initial performance of the compared genotypes, multi-local trials are established. This has been the path followed for the development and delivery of clonal varieties delivered by INIAP since the late 1970s. The varieties EET 103 and CCN 51 are the usual controls for these comparative trials, as there is sufficient evidence of their productive performance and the degree of relationship between the yields of both; that's why they are useful to assess the differences between the genotypes under study and also to check the validity of the figures. In this context, the following is a reference to clarify this issue: INIAP 800 yields the same and sometimes slightly more than CCN 51, while CCN 51 doubles the yield of EET 103, and often the difference is more significant, under rainfed conditions; hence, any figure outside this reference framework raises alarms during the interpretative process of the results.

Between 2010 and 2020, CIRAD and INIAP started a collaborative project to rescue wild cacao trees, ancestors of Nacional cacao, in the southeastern Ecuadorian Amazon. The planting material (pods and cuttings) rescued from these trees was established in a germplasm bank at the Domono farm of the Central Amazonian Experimental Station, another part in the Pichilingue Experimental Station, and another in a couple of agricultural schools near the explored area. Some trees have beans with cotyledons of ivory-white color, pink color, and combinations between pink and white; the beans of other trees are dark violet, like CCN 51. In certain genotypes, well-marked differences in leaf color of young shoots have also been observed. Based on this finding, the hypothesis has been proposed that early differences in leaf color could be useful as indicators to predict the color of cotyledons and make an approach to the sensory profile of particular genotypes; the relationship between leaf and cotyledon was confirmed at least for one genotype that had a pod at the time of the visit I made to the bank a decade ago; the observation of plants is a very exploratory comparative test that may lead to pleasant surprises.

In addition to recording various phenotypic data, including yield components, the analyses at CIRAD continue to reveal the genetic, biochemical, and sensory nature of the collected wild genotypes. One of the results is the confirmation of the theory that the ancestors of Nacional cacao on the Costa originated in what is now the southeastern Amazon of Ecuador. In addition to selecting genotypes with possible commercial interest, genetic improvement by crossing wild ancestors of Nacional cacao with other cacaos for the development of new varieties is an appealing idea. The information produced by this project is regularly published in indexed scientific journals.

The research agenda for the development of good sanitary practices, highly active in the 1990s, has stalled, even though the main factor restricting production is diseases. On the other hand, in recent years and up to the present, research in connection with the problem of Cd content in cacao has been quite active; the Polytechnic School of Guayaquil (ESPOL) and INIAP are leading this work. Both conduct laboratory and field tests to measure the impact of diverse types of amendments on the dynamics of Cd in the soil and in the plant at the 'hotspots' detected in our cacao zones. Recent reports show that the decrease in Cd content in the beans, in response to the application of some amendments, is significant.

2.9 Evolution of Production and Cultivated Area

More than half of the cacao planted area is concentrated in the Guayas River watershed, the largest in the country with 34,000 km^2. In the 1980s, the cacao area was in the range of 250,000 to 300,000 hectares. The distribution of cacao along the coastal region in the 1990s with a production of 90,000 tons, the majority of which was centered in that basin. Currently, this region continues to account for 65% of the bean supplies. The cacao surface area continues to grow in the Costa and Amazon regions, but overall, the geographical distribution pattern of the crop has not changed much; nevertheless, the cultivated area already exceeds 600,000 ha, and production in 2022 was above 400,000 tons, with sales of over 1 billion and 1,384 million dollars in 2022 and 2023, 10% of the agricultural gross domestic product (GDP).

If we compare the 411,000 tons exported in 2023 with the 100,000 tons at the beginning of the 21st century, an increase of 311,000 tons in the last two decades is evident, translating into an average annual growth rate of 7%. Large cacao-producing countries in Africa have had rates of 1% in recent years; the annual growth rate of cacao production in Peru is around 7%.

The cacao area continues to grow at the expense of replacing traditional orchards and replacing oil palm, corn, and pastures, among others. This phenomenon can be visualized by touring the fields, and although there is no precision on this matter, it is estimated that there is an increase of 10,000 hectares of new plantings annually.

2.10 Research Financing

Local funding for cacao research in the last 30 years has been complemented by external funds, tied to international projects, within the framework of cooperation agreements. Several projects have been executed sequentially since 1993: INIAP/ICCO/CFC, INIAP/IPGRI, INIAP/Biodiversity, INIAP/Bioversity, and INIAP/USDA (MARS), among others implemented recently. It is estimated that the total amount of external support through these projects is about 3 million dollars.

2.11 Conclusions

In conclusion, Ecuador's cacao industry has undergone significant transformations, with over 600,000 hectares dedicated to cacao cultivation. The widespread adoption of the high-yielding clonal variety CCN 51, alongside gradual adoption of newer varieties like INIAP 800 and 801, has driven a remarkable increase in productivity and exports. In addition, the recent delivery of the commercial variety INIAP 802 adds high-yield options that will strengthen the future of the country's cocoa production. Traditional production systems of the Nacional-type cacao, facing challenges such as disease susceptibility and low productivity, are gradually being replaced.

The cacao sector's dynamism is evident in the continuous expansion of cultivated areas and the country's emergence as the world's third-largest cacao producer. Research initiatives, such as rescuing wild cacao trees in the Amazon, envision a new line of research for the genetic improvement of the crop. The industry's economic impact is substantial, with cacao exports generating over $1.3 billion in revenue in 2023.

However, challenges persist, including disease management and the need for sustainable practices. The push for innovation, backed by scientific education and coordinated policies, remains crucial for the sector's future. While clonal varieties dominate, there's recognition of the value of conserving traditional orchards, particularly to take advantage of niche markets.

In terms of research, projects evaluating genotypes and varieties, along with efforts to address issues like Cd content in cacao, underscore the commitment to enhancing the industry. The collaboration with international partners has significantly contributed to research funding, emphasizing the importance of global cooperation in advancing Ecuador's cacao sector.

Bibliography

Amores, F., Palacios, A., Jiménez, J., & Zhang, D. (2009). Entorno ambiental, genética, atributos de calidad y singularización del cacao en el Nororiente de la provincia de Esmeraldas. Estación Experimental Tropical Pichilingue. INIAP, SENACYT, APROCANE, USDA. Boletín técnico N° 135. Quevedo, EC. 99 p. http://repositorio.iniap.gob.ec/handle/41000/1565

Amores, F., Vasco, S., Eskes, A., Suarez, C., Quiroz, J., Loor, R., Jiménez, J., Zambrano, J., Bolaños, M., Reynel, V., Terán, M., and Quijano, G., 2011. Selección en campo y en la estación de nuevas variedades de cacao en Ecuador. En: Eskes, AB (ed.). *Enfoques colaborativos y participativos para la mejora de la variedad de cacao.* Informe final del proyecto CFC / ICCO / Bioversity International sobre la productividad del cacao y la mejora de la calidad: un enfoque participativo (2004-2010). Fondo Común para los Productos Básicos (CFC), Amsterdam, Países Bajos; Intl. Cacao Organization (ICCO), Londres, Reino Unido; Bioversity Intl., Roma, Italia, pp. 59–72.

INIAP. (1976–2023). Informe anual (INIAP/IA-CACAO-CAFÉ 1993–2023). Instituto Nacional de Investigaciones Agropecuarias, Quevedo (Ecuador). Estación Experimental Tropical Pichilingue. Programa de Cacao y Café. https://www.bibliotecasdelecuador.com/

Loor, R., Amores, F., Vasco, S., Quiroz, J., Casanova, T., Garzón, A., Sotomayor Jr., I., Jiménez-Barragán, J. C., Tarqui-Freire, O. M., Zamora, G. A. R., Quijano-Rivadeneira, G. C., Plaza-Avellán, L. F., Guerrero-Castillo, H. E., & Zambrano-Flores, F. G. (2019). INIAP-EETP 800 'Aroma Pichilingue', nueva variedad ecuatoriana de cacao fino de alto rendimiento. *Revista Fitotecnia Mexicana, 42*(2), 187–189.

Loor, R. G., Risterucci, A. M., Courtois, B., Fouet, O., Jeanneau, M., Rosenquist, E., Amores, F., Vasco, A., Medina, M., & Lanaudet, C. (2009). Tracing the native ancestors of the modern *Theobroma cacao* L. population in Ecuador. *Tree Genetics & Genomes, 5*, 421–433. https://doi.org/10.1007/s11295-008-0196-3

Loor Solorzano, R. G., Fouet, O., Lemainque, A., Pavek, S., Boccara, M., Argout, X., et al. (2013). Correction: Insight into the wild origin, migration and domestication history of the fine flavour Nacional *Theobroma cacao* L. variety from Ecuador. *PLoS ONE, 8*(2), https://doi.org/10.1371/annotation/2357f0f1-7dc3-4781-afb0-29a8ce56b3f0

Ramlachan, N., Agama, J., Amores, F., Quiroz, J., Vaca, D., Zamora, C., Moore, J. M., Brown, J. S., Schnell, R. J., & Motamayor, J. C. (2011). Regional selection of hybrid Nacional cocoa genotypes in coastal Ecuador. INGENIC Newsletter, 12, 10 p.

Part II

Agroforestry system

3 Cacao

Shade and nutrition effects under Amazonian conditions

Santiago C. Vásquez, Mirian Capa-Morocho, Marlene Molina-Müller, and Fernando Granja

3.1 Introduction

Cacao is one of the main perennial tree crops worldwide and it is native to South American tropical forests. The scientific name of this species of the Malvaceae family, *Theobroma cacao* L., was defined by Carolus Linnaeus and published in his book *Species Plantarum* in 1753; 'Theobroma' is a Greek word that can be translated as 'Food of Gods', while 'cacao' is a word derived from the Olmec and the subsequent Mayan languages (Kakaw) (Bhattacharjee, 2018; Dillinger *et al.*, 2000).

The first historic register of cacao consumption points to the Olmecs (1500–400 BC), originally in the form of a drink. According to Coe's theory, they crushed the cacao seeds, mixed them with water, and added spices, chilies, and herbs. They began cultivating cacao in equatorial Mexico. Over time, the Mayans (600 BC) and Aztecs (400 AD) developed successful methods for cultivating cacao as well (Coe and Coe, 2013).

The diffusion of cacao traits in the 16th century is due to the Spanish conqueror Hernando Cortés, who drank it with the Aztec Emperor Moctezuma and brought it to the Spanish court in 1528 along with the equipment needed to make the drink. However, cacao's importance was recognized when the first Spanish harvest arrived at the Iberian Peninsula from New Spain, in 1598, starting the cacao trade and resulting in the establishment of the first chocolate shops (Coe and Coe, 2013; Young, 1994).

Although these records point to Mesoamerica as the center of origin and domestication of this crop, the latest studies of genetic diversity indicate Ecuador as a cacao hotspot. The archeological evidence shows the use of cacao 5,300 years ago in Santa Ana, La Florida (Palanda), Zamora Chinchipe province in the southeast of Ecuador, the upper part of the Amazon. To this finding, the results of genetic diversity analysis are added, showing its great genetic diversity and ancient domestication in this area, which places it as the center of origin of cacao (Valdez, 2019; Zarrillo *et al.*, 2018). Due to the great genetic diversity present in Ecuador, it is necessary to rescue, conserve, and characterize these cacao genetic resources with the purpose of identifying materials with potential traits to be produced or used as a basis for plant

DOI: 10.4324/9781003381761-5

breeding in studies focused on pests and disease resistance, environmental and nutritional requirements, aroma quality, and cadmium accumulation in the beans.

Of the four cacao groups recognized worldwide (Criollo, Forastero, Trinitario, and Nacional) in Ecuador, the production of Nacional cacao stands out. It is also known as 'Arriba', because its cultivation began approximately in the year 1600 in the Ecuadorian district of 'Arriba', currently known as the province of Los Ríos. This cacao was recognized for its high quality and production in this territory, as compared to the other provinces, which was attributed to deep alluvial soil, better drainage due to its higher altitude above sea level, and low prevalence of strong winds and high rainfall (Jaimez *et al.*, 2022; Maiguashca, 2014). This cacao type stands out for its fine aroma, which is highly desired worldwide and nationally.

In America, cacao is cultivated from Mexico to Brazil, with Ecuador and Brazil being the ones who have the greater production in the region, surpassing 600,000 ha planted and 300,000 tons produced (FAOSTAT, 2023). The countries that follow in hectares established are Colombia (9%), the Dominican Republic (9%), Peru (6%), and Venezuela (4%) (Sánchez *et al.*, 2019), with a sustained increase in the number of hectares involved in the crop. In recent times, cacao plantations in the Amazon region of Ecuador have increased significantly. However, the yields are low (Figure 3.1), concerning international demand (ESPAC, 2023).

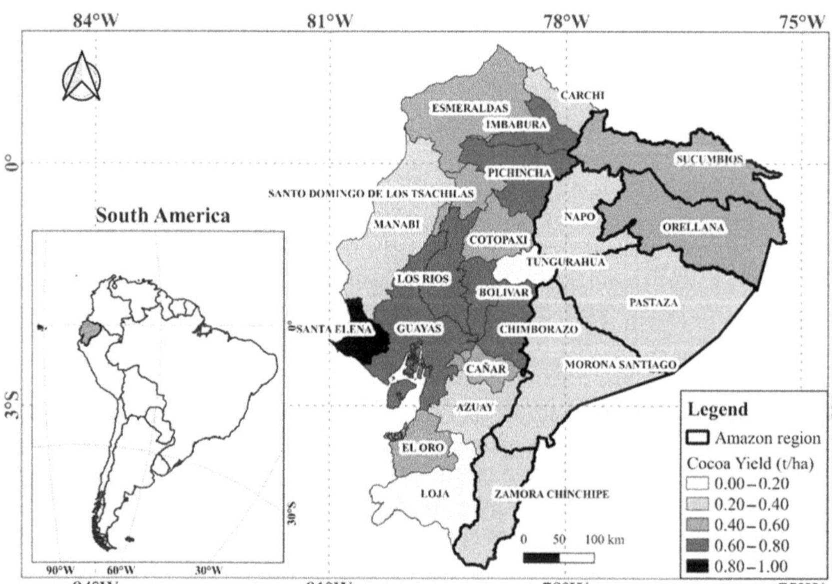

Figure 3.1 Mean cacao yield (t/ha) by province in Ecuador in the period 2017–2021.

Figure created with QGIS v.3.24.1 software, using information from ESPAC 2023.

The low yield in the area is because a large part of the production is carried out by small farmers, who, in many cases, need to be made aware of the basic parameters to be considered to obtain a higher yield. Therefore, even today, despite the studies carried out, it is necessary to deepen the crop needs in the Amazon region, concerning its nutritional and solar radiation requirements. Considering the above, this chapter provides the major findings of the effect of shade and nutrition in cacao cultivation under Amazonian conditions.

3.2 Nutrition

3.2.1 Crop responses to nitrogen

Nitrogen (N) is an essential nutrient for optimal crop growth and development of plants. The availability of nitrogen in the soil can significantly fluctuate in both space and time due to soil and climatic conditions such as precipitation, temperature, wind, soil type, and soil pH (Maathuis, 2009). Low availability of nitrogen reduces the growth and development of plants. Similarly, excessive nitrogen fertilization reduces yield (Cabello *et al.*, 2009; Erel *et al.*, 2008; Stefanelli *et al.*, 2011; Van Herwaarden *et al.*, 1998; Wang and Cheng, 2011). Excessive nitrogen also reduces quality characteristics such as organoleptic attributes, content of mineral nutrients, secondary metabolites, and nitrate content (Albornoz, 2016; Ehret *et al.*, 2014; Lairon, 2010; Nielsen *et al.*, 2009; Schreiner *et al.*, 2013).

Among the nitrogen-limiting factors, leaching is one of the main problems in rainy areas or over-irrigated areas; while in arid regions, the main problem is volatilization (Jones *et al.*, 2013; Leghari *et al.*, 2016; Lehmann and Schroth, 2003). Nitrogen uptake by plants depends on crop demand, root development within the soil, and the ability of the roots to extract nitrates from the soil.

3.2.1.1 Nitrogen sources

Approximately 78% of nitrogen is available in the atmosphere in inert structure (N_2), which cannot be absorbed directly by plants (Leghari *et al.*, 2016, Martinez-Dalmau *et al.*, 2021). In agricultural systems, nitrogen fertilizer input is an essential factor for incrementing the total nitrogen content in soil (Liu *et al.*, 2021). Nitrogen fertilizer requirements vary significantly from site to site (Greenwood et al., 1982). Long-term fertilization tends to increase the nitrogen content on the surface layer and accelerate nitrogen fixation and retention (Liu *et al.*, 2021).

There are different processes that can provide nitrogen to the plant, for example, industrial N fixation, atmospheric N fixation, biological N fixation, and organic sources (Leghari *et al.*, 2016). However, industrial N fertilizers (inorganic) are the most widely used by farmers. The most common industrial fertilizers are urea, nitric acid, ammonium nitrate, ammonium sulfate, and calcium and potassium nitrate.

Urea is the main nitrogen fertilizer used worldwide; urea accounts for 66% of global synthetic N fertilizer use (Mandal *et al.*, 2019; Zhang *et al.*, 2023) due to its high N content. Urea is produced by the reaction of ammonia with carbon dioxide. Anhydrous ammonia is concentrated, absorbed, and oxidized to produce nitric acid, while ammonium nitrate and ammonium sulfate are obtained by neutralizing nitric acid with ammonia and from the ammonization of sulfuric acid, respectively (Cheremisinoff and Rosenfeld, 2011).

Inorganic N fertilizer usage in agriculture has increased by 67 Tg N yr^{-1} between 1961 and 2016 (Elrys *et al.*, 2020). Moreover, N demand is expected to continue to rise while agricultural production is expected to increase from 50% to 100% by 2050 (Walling and Vaneeckhaute, 2022). In addition, increasing global energy prices have caused a substantial increment in fertilizer prices, leading to higher production costs. Therefore, nowadays, the efficient use of nutrients in plants has become a key factor.

The efficiency of N use in plants, despite being affected by soil and climatic factors, is also influenced by management practices such as application methods, timing, and N sources (Leghari *et al.*, 2016). The effects of different N sources on crop growth and nitrate accumulation have attracted great attention in recent years (Kaymak, 2013; Stagnari *et al.*, 2007; Wang, Li and Malhi, 2008). So, the following section reports on the effects of the N sources on cacao yield and quality.

3.2.1.2 *Effect of nitrogen sources on the yield and quality of cacao*

Nitrogen is primarily absorbed through fine roots as either ammonium or nitrate (Cheremisinoff and Rosenfeld, 2011). Previous research indicates that the N source applied to the crop influences its growth and production according to the N source applied (Alsam *et al.*, 2021, DeAndrade, *et al.*, 2009, Gardiazabal *et al.*, 2007, Granja and Covarrubias, 2018, Ma *et al.*, 2015, Molina and Covarrubias 2019, Yan *et al.*, 2020). To understand the effects of N sources on the yield and quality of cacao under Ecuadorian Amazon conditions, three different sources (ammonium sulfate, calcium nitrate, and urea) and control were evaluated (Capa-Morocho *et al.*, 2022).

In cacao crops, nitrogenous sources significantly influenced the productive variables and yield under Ecuadorian Amazon conditions. The number of fruits and production per plant may be higher when applying ammonium sulfate than when calcium nitrate, urea, and control are used (Table 3.1). The mean yield of cacao could increase by 48% with the application of nitrogen in the form of ammonium sulfate, in contrast to control. This may be due to energy losses in the absorption, assimilation, and transport metabolism of NH_4^+ and NO_3^-. In nitrate assimilation, the nitrogen from nitrate is converted into nitrite by nitrate reductase, then to ammonium, and finally to amide nitrogen in glutamine (Liu *et al.*, 2022). The nitrogen absorbed by the plant in the form of NH_4^+ allows energy optimization as compared to NO_3^- (Fageria *et al.*, 2003). The reduction of nitrate to ammonium implies to the plant an energy cost of 32 moles of photons per mole of NO_3^- assimilated and transported. By contrast, the cost for ammonia absorption, transport, reduction, and assimilation is about 9.45 moles of photons per mole of NH_4^+. Therefore, ammonium supplied to plants would save photoenergy compared to nitrate, about 10 ATPs per mole of nitrogen (Guo *et al.*, 2007, Raven, 1985, Salsac *et al.*, 1987). This energy could be used for metabolic processes and glucose synthesis. In that context, Baligar and Fageria (2017) indicate that nitrogen applied in the form of ammonium improves cacao growth compared to nitrate due to the lowest requirements for assimilation of NH_4^+ forms of nitrogen by plants, helping in producing higher dry matter.

On the other hand, the agronomic efficiency of the fertilizer responds to the nitrogen source. Although the agronomic efficiency for nitrogen in the Ecuadorian Amazon is low, ammonium sulfate presents better efficiency, producing 2.04 kg of dry cacao seed clone CCN-51 (Colección Castro Naranjal) for each kg of fertilizer unit used (Table 3.1). While, for each kg of calcium nitrate applied, the yield increases by 0.87 kg of dry seed. Ammonium sulfate has been shown to be very efficient because ammoniacal N is less likely to be lost due to leaching and denitrification (Barrios *et al.*, 2012). In addition, the efficiency of nitrogen sources depends on the affinity of plant species and other factors, such as pH, temperature, carbohydrate content at the root level, and osmotic potential (Marschner, 1995; Mengel and Kirkby, 2001). Li *et al.* (2013) indicate that many internal factors of plants, such as species, cultivar, and growth stages, determine the preference of plants for NH_4^+ or NO_3^-.

Other studies on N form preferences have shown that plants are adapted, or acclimated, to take up the most available nitrogen form to them, preferentially (Andersen and Turner, 2013; Boczulak *et al.*, 2014; Franklin *et al.*, 2017; Miller *et al.*, 2007). Plants adapted to acid soils may prefer ammonium (Ismunadji and Dijkshoorn, 1971). Prabhakaran (2010) and Zhang and Motilal (2016)

Table 3.1 Effects of nitrogen sources on productive variables of cacao CCN-51 (*Theobroma cacao* L) in the province of Zamora Chinchipe

Productive variables	*Nitrogen sources*				*SEM*[1]	P-value
	Control	*Urea*	*Calcium nitrate*	*Ammonium sulfate*		
Fruit numbers (fruits/tree)	10.7[c]	12.3[bc]	13.3[b]	16.7[a]	0.667	*0.001*
Production (kg/plant)	6.02[b]	6.97[b]	7.18[b]	9.70[a]	0.423	*0.002*
Dry seed yield (t/ha)	0.539[c]	0.748[ab]	0.650[b]	0.799[a]	0.027	*0.001*
Yield (t/ha)	0.539[c]	0.748[ab]	0.650[b]	0.799[a]	0.027	*0.001*
Agronomic efficiency (AE)		1.64[ab]	0.87[b]	2.04[a]	0.241	*0.046*

[1] SEM: Standard error mean, n = 10.

(Retrieved from Capa-Morocho *et al.*, 2022).

reported that cacao crops in the Amazon region have been grown in acidic and poor soils. However, under southern Ecuadorian amazon conditions, cacao soils have medium fertility and high organic matter content. Therefore, cacao reacts better at acidic pH, such as that provided by ammonium sulfate. Taken together, these findings suggest that cacao could have adapted to environmental conditions.

In cacao, the fat content is essential because it contributes to the flavor and facilitates industrial processing, such as transporting the chocolate mass through pipes, pumps, and machines (Almeida-Andrade *et al.*, 2019). Additionally, low-fat seeds represent higher grinding costs; hence processing factories prefer fat content above 55% (FEDECACAO, 2004).

Under Ecuadorian Amazon conditions, the nitrogen source influences the fat content in cacao seeds but does not significantly affect the ash and nitrogen contents (Figure 3.2). The clone CCN-51 presents an average of 3.16 grams of ash and 2.35 grams of nitrogen per 100 grams of dry matter. The application of ammonium sulfate increased the fat content by 4 grams as compared to the application of urea, which has the lowest value with 50.3 grams. This may be due to the nature of urea, where the ammonification process produces N losses. Consequently, the amount of nutrients available to the plant is reduced. In this way, the nutrient content could favor the formation of fats in cacao seeds. These results are consistent with previous studies where it is concluded that the lipid composition varies according to the level of nitrogen the plant is exposed to and according to the organ studied (Liu *et al.*, 2017, Ruan *et al.*, 2007, Yang *et al.*, 2011).

To conclude, ammonium absorption represents energy savings in the plant as compared to nitrate; this allows ammonium sulfate to be more efficient through better nitrogen use (Fageria *et al.*, 2003). On the other hand, the fatty

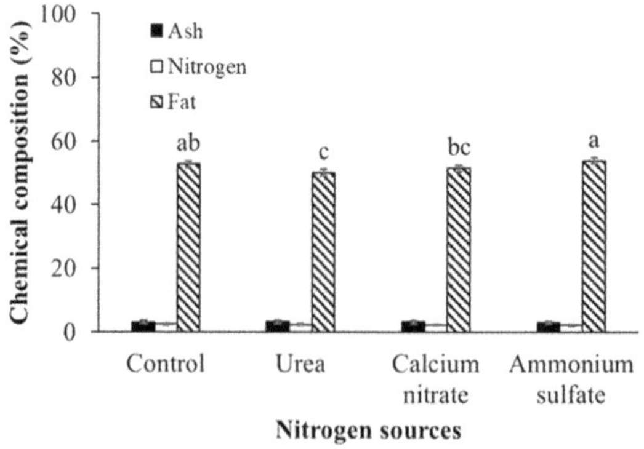

Figure 3.2 Chemical composition of CCN-51 cacao seeds per 100 g of dry matter. Different letters indicate significant differences between treatments ($p < 0.05$). Error bars are ± standard error.

(Adapted from Capa-Morocho *et al.*, 2022).

acid composition of cacao butter is influenced by genotype, climatic conditions, and harvest season (Spangenberg and Dionisi, 2001). Therefore, the cacao clone CCN-51 could have a greater preference for ammonium since, even in growth parameters, it has shown selectivity for this nitrogen source and not for nitrogen concentration (Baligar and Fageria, 2017).

3.2.2 Effect of nitrogen, phosphorus, and potassium on growth and quality of cacao

Extensive research has evaluated the effect of nutrients on cacao growth and production (Famuwagun *et al.*, 2018; Oyewole *et al.*, 2012; Sitompul *et al.*, 2014). However, there is no clarity on the specific effects of N, P, and K (NPK). There is even less information on the effects due to different environments, genotypes, and production systems. To understand the effects of NPK nutrients as independent factors and their interactions, the effects were evaluated in a cacao crop in production fertilized for 115 days, in the southern region of Ecuador.

The results showed that the growth of shoots (length and leaf area) was not affected by applying NPK nutrients (Herrera *et al.*, 2022). This result may be explained by the fact that the supply of nutrients from the soil was sufficient to achieve shoot growth during the evaluated period. Some areas of the Ecuadorian Amazon have soils with a high content of organic matter and a great biodiversity of soil fauna that can supply the nutritional crop needs (Huera-Lucero *et al.*, 2020). The shoots presented an average of 24.6 cm in length and 7763 cm^2 of leaf area. However, an influence of fertilization on the number of metamers per shoot was observed. The application of N (without P and K) presents the highest values of metamer per shoot. In contrast, the lowest values are presented by applying P (without N and K). P acts negatively on the number of metamers, obtaining an average of 7.33 metamers with the application of P and 9.3 without applying this element. In general, the number of metamers in all treatments is within the ranges reported in other studies (Almeida and Valle, 2007). Considering that, before applying fertilizers, the soil had a high level of P (22 mg kg^{-1}); the significant effect of this element could be due to excess. Novoa *et al.* (2018) report that excess phosphorus can affect the growth of vegetative organs. Puente-Páramo *et al.* (2016) indicate that high doses of phosphorus evaluated in different clones decreased the physiological efficiency of the use of this nutrient, finding that clone CCN-51 presented the lowest values. In addition, excess P can bind with calcium and magnesium, reducing their availability (Adams, 1980; Krouk and Kiba, 2020; Wang and Harrell, 2005), and therefore affecting plant growth, since these cations are part of the plant's structure.

Figure 3.3 shows the chemical composition of cacao fruits, clone CCN-51, in a physiological maturity state. In southern Ecuadorian Amazon conditions, triple fertilization with NPK improves the fat content of the exocarp. However, an increment of seed fat content was not observed, which is an essential

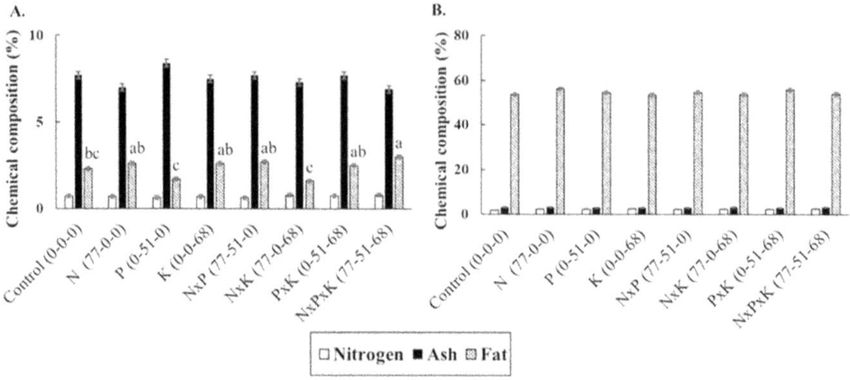

Figure 3.3 Chemical composition of CCN-51 cacao fruits per 100 g of dry matter: A. Exocarp and B. Seed. Different letters indicate significant differences between treatments ($p < 0.05$). Error bars are ± standard error.

(Adapted from Herrera *et al.*, 2022).

parameter for the food industry (Herrera *et al.*, 2022). The fat values in the exocarp agree with previous reports where values between 1 and 6% are indicated (Vriesmann *et al.*, 2011). Interestingly, all treatments found ash values were higher than 7%. This exceeds previously reported values that are between 2.79 and 6.70% (Veríssimo, 2012; Vriesmann *et al.*, 2011). In preliminary studies, the protein, fat, and ash content of the seeds were not affected by the NPK elements. This suggests that the quality of the fruit has low sensitivity to fertilization; however, more studies are required.

3.2.3 Movement of nutrients in the plant

Cacao exhibits a considerable genetic variability in morphological and physiological traits. When it is multiplied via seeds, it displays an orthotropic growth pattern, exhibiting cycles of leaf flushing in a rhythmic way, independent of climate, which indicates that the growth rhythm is under endogenous control (Almeida and Valle, 2007). The initially expanding leaves are pale green or pink due to anthocyanin content (Orchard *et al.*, 1980). Once expanded they are green, and in the final stage, they turn yellow with brown spots and finally fall from the tree (Abo-Hamed *et al.*, 1981). Buds of three to six pairs of leaves are produced in the flush; after a dormancy period of six to eight weeks, bud growth is reactivated and thus a new flush is produced. Leaf longevity varies with position within the canopy, incident irradiance, time of emergence, and height in relation to ground level. A flush usually coincides with the fall of old leaves, as the nutrient demand of the new flush is met by the translocation of nutrients and photoassimilates from old leaves (Almeida and Valle, 2007). The average lifespan of the leaves is estimated to be 450 days in shaded plants and 250 days in plants under full exposition to sun conditions (Müller *et al.*, 1992; Zuidema *et al.*, 2005).

Developing leaves have less photosynthetic capacity as they are sinks for photoassimilates, increasing the levels of fructose, glucose, and sucrose by imports from older leaves. At the end of the expansion, the glucose and fructose levels decrease and sucrose concentration increases, reflecting the end of photoassimilate importation and the beginning of endogenous sucrose synthesis (Baker and Hardwick, 1975). In the interflush, the stock of carbohydrates is reestablished and, with the return of a favorable balance, the appearance of a new flush can occur (Machado and Hardwick, 1988). Leaves show various stages of growth, reaching their maximum photosynthetic potential at full leaf expansion. In addition, the amount of photoassimilates is directly related to plant architecture and canopy volume. As the leaf area index increases, the photosynthetic capacity and the ability to process carbohydrates, produce fruits, and emit new leaves increase as well. (López *et al.*, 2015).

Nutrients are necessary for the plant in different quantities, so their deficit or excess can affect the physiological functioning of it. Nutrients are generally taken from the soil, but also by the leaves and other organs (Soria, 2008). Due to the influence of environmental conditions and aging, some nutrient concentrations increase while others decrease under similar conditions, depending on their mobility in the plant and the effects of dilution (Walworth and Sumner, 1987). Therefore, leaf nutrient content depends on leaf age, new leaf development, fruiting, light intensity, and seasonal effects (Van Vliet and Giller, 2017).

Studies conducted in the Amazon region on five-year-old CCN-51 cacao showed marked variations in leaf nutrient content as a function of their lifespan. This effect was tested using different fertilizer sources as well as without fertilization, and similar variations were obtained, indicating that the plant regulates these processes mainly endogenously as a function of leaf age rather than nutrient availability. The fact that the elements are not maintained in static quantities in leaves, but vary as age advances, may be useful to determine an optimal moment in which elements are found in relatively stable quantities. In this context, it is possible to choose an ideal moment to collect tissues samples, for instance for foliar analysis, because due to these variations in nutrient contents, a leaf sampled at the wrong moment, whether it is too young or too old, will result in misinterpretations of their contents. Hence, an optimal time to take samples for foliar analysis of full-light cacao crop in the Amazon region would be around four months after the flush of the shoot as shown in Figure 3.4.

Leaf nitrogen content between 88 and 199 days of shoot emergence present values around 2% that are considered a normal concentration in cacao leaves according to Van Vliet and Giller (2017). Nitrogen is a structural element of chlorophyll, where 75% of leaf N is directed to the chloroplast, consequently affecting leaf greenness and chlorophyll accumulation (Yamashita *et al.*, 2020). So, during leaf growth the process of cell division and expansion is continuous, requiring an accumulation of N in the chloroplasts, resulting in a constant increase in leaf N content until the leaf stops growing and the senescence stage begins.

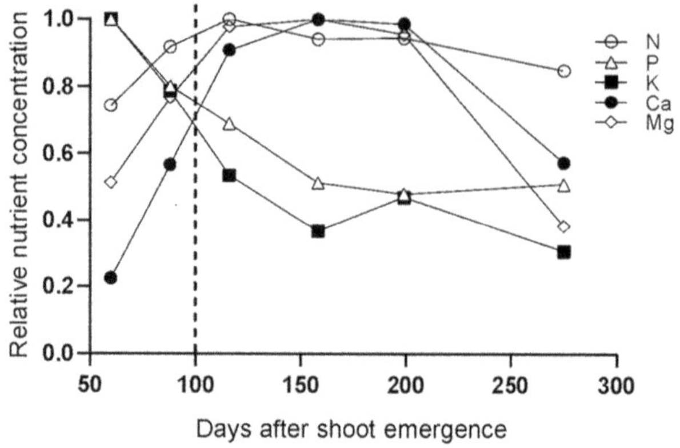

Figure 3.4 Macronutrients dynamics in leaves of CCN-51 cacao according to days after shoot emergence. The dashed line indicates the ideal time taken by samples for leaf analysis.

(Adapted from Romero *et al.*, 2022).

When analyzing the concentrations of P and K, a decrease is observed as the leaves age (Figure 3.5). Phosphorus is required for membranes and other cellular structures, suggesting a basic P cost for leaf 'infrastructure'. Another percentage of leaf P is used in metabolism, where it is rapidly cycled (Crous and Ellsworth, 2020). Therefore, as it is easily cycled the amount of P remaining in the leaf is low, as shown by leaf analysis, in addition to the decrease in P content as the leaf grows. Potassium is necessary for assimilating translocation from leaves to fruits and other sinks. As the leaf grows its photosynthetic

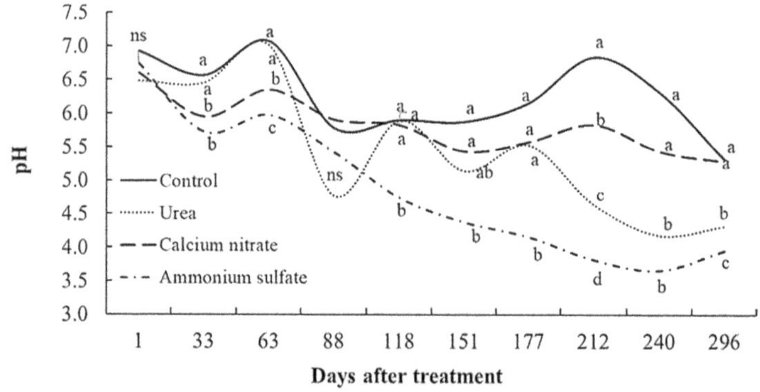

Figure 3.5 Effect of different sources of nitrogen fertilization on soil pH in cacao crop. Different letters for each day indicate significant mean differences between treatments according to Tukey's test ($p < 0.05$).

(Retrieved from Capa-Morocho *et al.*, 2022).

activity decreases; thus, the amount of K required to transport photoassimilates will be less, reflecting a low foliar K content (Usandivaras *et al.*, 2018).

For Ca and Mg, their concentrations increase and then begin to decrease as the leaf senescence. Calcium exhibits a dual function, as a structural component of cell walls and membranes and as an intracellular second messenger. Thus, Ca concentration in the cytosol is found in a low level and increased to generate a signal when the plant requires (Thor, 2019). This indicates that when the leaf arrives at the senescence stage, the leaf Ca content will be higher than at the beginning of leaf growth, due to its accumulation in the cell walls and in the cytosol in minor proportions. Mg is involved in the light-dependent reaction, the chlorophyll molecule is composed of a central Mg ion, and it also promotes the synthesis of adenosine triphosphate (ATP) and regulates the Calvin cycle by activating many enzymes (Wang *et al.*, 2018). When a leaf begins to grow and increases its photosynthetic activity, the leaf Mg content increases as it is required in the photosynthetic process; when the senescence stage begins, the content of this element decreases.

3.2.4 Nutrients in soils and fertilizer effects

Nutrient uptake from the soil in the cacao crop increases rapidly after the first five years of planting and then maintains a stable uptake rate for the rest of the plantation's life (García, 1993). The main element extraction, which is not replenished in the production system, is due to yield; primarily cacao's seeds; therefore, crops with higher yields will be more likely to demand more nutrients and the soil will require higher inputs. According to Furcal-Beriguete (2017), the extraction of elements from a cacao crop per 1000 kg of dry seeds are: 24 kg of N; 5.7 kg of P; 11.2 kg of K; 1.1 kg of Ca, and 3.8 kg of Mg. These extractions increase to 36 kg of N; 8 kg of P; 51 kg of K; 5 kg of Ca, and 8 kg of Mg if both seed and husks are considered as yield. It is noteworthy that although the seed does not require too much K, this element is necessary for the husks, which indicates that high K inputs are indispensable for cacao anyway. Another point to consider is that nutrient removal in seeds shows little variation, whereas large differences can be noticed for husks. This is caused by the fact that husks are affected more strongly by the environment and the type of fruits than seeds. Nutrients immobilized in the stem of cacao, and shaded trees are also considered lost for the system as they are excluded from nutrient cycling (Hartemink, 2005).

Cacao plantations without fertilizer application and in full sunlight, remove more nutrients from the system than are provided, although cacao established from virgin forest on fertile soils may not require fertilizer for many years. However, over time cacao production, especially without fertilizer, will deplete the soil of nutrients (Gockowski *et al.*, 2013). In conditions of the Ecuadorian Amazon in crops established in soils where secondary forest existed, no differences were found in crops that were fertilized as well as in crops that did not receive any fertilization during the first years of establishment. The same behavior was observed in different clones of Criollo type as well as for CCN-51. This indicates that in Amazonian soils at rest, fertilizer applications would not be so necessary when establishing a crop because the demand for nutrients is low and because jungle soils have adequate fertility. Therefore, nutrition strategies could preferably focus on amendments or stimulants that promote growth, rather than the edaphic contribution of elements. The association with other crops, such as in agroforestry systems, is still unclear because it may increase nutrient removal in early stages.

Cacao plants tolerate pH values between 5.0 and 7.5, according to Dostert *et al.*, (2012). In southern Ecuadorian Amazon, the zone of origin of cacao, acid soils predominate, which could limit the availability and absorption of N, K, Ca, Mg, and P, but at the same time could increase the availability of microelements such as Zn, Cu, Fe, and Mn. In some study areas sampled from soils of this region, pH values ranged from 4.7 to 5.6. This indicates quite acidic conditions for many crops; however, due to the fact that cacao is a crop that evolved in these environments, it has been noted to grow normally under these conditions.

In comparative studies, in cacao crop CCN-51 at productive stage, when testing different sources of N fertilization, it is observed that these have a strong effect on soil pH depending on fertilizer type as shown in Figure 3.5. Ammonium sulfate is a fertilizer with strong acidifying effects as well as urea in Amazonian soils; however, the acidification process is faster when ammonium sulfate is used because of its decalcifying effect. Another N source such as calcium nitrate tends to maintain a more stable pH in the soil similar to when no fertilization is applied. In these trials, no differences in growth, but yield increase were observed for the use of ammonium sulfate as mentioned earlier in this chapter. This would indicate the preference that cacao may have for acid soils and its affinity for ammonium. However, there is still a knowledge gap that requires more in-depth studies.

Due to the high amount of biomass that is generated by the humidity and temperature conditions, these soils usually have high organic matter contents that are between 4% and 11%. The cation exchange capacity may vary depending on the fertility and specific conditions of each zone, but normally there are no problems with salts or high electrical conductivity in this area. It has also been found that lime applications in its different forms constitute an efficient management alternative to raise the pH of this type of soils. To mention a case, it has been observed that applications of 1 kg per plant of lime, in the form of calcium hydroxide, raised the pH of a soil from 4.7 to an average value of 5.6 in the first month. However, this does not necessarily imply an increase in productive parameters or an improvement in the uptake of elements from soil by modifying the pH; there is still much controversial information about this topic. For instance, although according to Shamshuddin *et al.*, (2004), cacao crop is very sensitive to acidity, Wood (1985) argues that cacao is tolerant to acid conditions as long as the soil provides adequate nutrients.

A problem particularly associated with soil acidity is aluminum (Al) toxicity which is a major limiting factor in plant production on acid soils. Cacao is an

Al-sensitive crop; what is known as the 'soil acidity complex' is the combined impact of deficiencies of cations such as calcium and magnesium due to fixation of iron and Al and toxicity of Al. These combined effects can result in stunted growth of cacao seedlings (Shamshuddin *et al.*, 2004). The amounts of lime required to increase the pH of tropical clay soils could be prohibitive and can result in additional problems such as inducing zinc deficiency. Low soil pH per se is not toxic for plant growth. Kamprath (1970) recommends the use of smaller amounts of lime to correct the Al saturation of the cation exchange capacity to less than 15%. This is sufficient to overcome Al toxicity and provide calcium for crops (Van Vliet and Giller, 2017).

3.3 Shade

3.3.1 Shading and agroforestry systems

In Ecuador, the main cacao production systems are the monoculture system that predominates in the coastal region and the agroforestry systems in a smaller area in the Amazon region, although in the latter it is also produced as a monoculture (Abad *et al.*, 2020; Blare and Useche, 2019). The production of cacao under small-scale agroforestry systems is a traditional practice in the Amazon region of Ecuador. This system in indigenous communities is known as Chacra or Chakra and it has traditionally been considered as the space of women, although men participate in cutting and land clearing activities (Castellón, 2015). Cacao is widely cultivated under the Chacra system and is the leading fruit crop for commercial purposes in the Ecuadorian Amazon (Limongi Andrade et al., 2007); likewise, the use of agroforestry systems for the cacao crop has been reported as a traditional practice in several places in South America (De Almeida and Valle, 2010; Lobão *et al.*, 2007).

Agroforestry systems would provide different benefits, both in terms of subsistence through income and access to food, wood, and medicinal plants, as well as in terms of ecological sustainability, carbon sequestration, and habitat creation, avoiding soil degradation by improving its fertility (Tscharntke *et al.*, 2011). However, the evidence of the benefits of agroforestry systems is limited to a few crops such as coffee (Cerda *et al.*, 2020; Chaves *et al.*, 2008; Meylan *et al.*, 2017), while in cacao crop, several of this evidence is empirical or reports from farmers (Anglaaere *et al.*, 2011; Smith Dumont *et al.*, 2014; Vaast and Somarriba, 2014), with few studies showing the benefits on soil characteristics, growth, plant yield, and fruit quality in cacao crop (Blaser *et al.*, 2017).

Regarding soil fertility, the benefits of agroforestry systems are not conclusive, and increases in soil carbon (C), N, P, and K are reported (Ofori-Frimpong *et al.*, 2007). In another study, Blaser *et al.* (2017) report that in the superficial layer of the soil (first 15 cm) C and N increased; however, in the deeper layers (15–30 cm and 30–50 cm), no differences were found in these elements. Likewise, other studies did not find significant effects of agroforestry systems on soil nutrients (Jacobi *et al.*, 2014; Mohammed *et al.*, 2016). It should be considered that most fine roots in the cacao crop (80%) are found in the upper 25 cm of the soil (Niether *et al.*, 2019). Regarding other soil characteristics, no significant effects of agroforestry systems on pH, cation exchange capacity, and soil texture are reported (Blaser *et al.*, 2017).

With respect to the content of nutrients at the foliar level, in a study carried out by Blaser *et al.* (2017), no significant effects of the shade generated by the trees of the agroforestry system are reported. In contrast, Baligar *et al.* (2005) reported that excessive solar radiation in cacao reduced the absorption and entry of some nutrients; likewise, Isaac *et al.* (2007) state that there is an increase in nutrients at the foliar level in cacao when it grows under shade in agroforestry systems, which shows that more research is needed to understand the effect of shade on the movement of nutrients in the cacao plant grown under agroforestry systems.

Unclear results are also reported in experiments that studied the effect of shading by agroforestry systems on growth traits in cacao. Famuwagun *et al.* (2018) indicate that the elongation of the stem, the number of differentiated leaves, and the number of branches in young cacao plants are favored with the increase in sunlight, while these decrease when the intensity of shading increases. In another study, Schneider *et al.* (2017) did not find differences in plant height during the initial growth of cacao; however, the diameter of the stem increased significantly when the plants grew in monoculture in the absence of an agroforestry system.

Agroforestry systems are reported to frequently reduce cacao yields in the short term (Figure 3.6) (Armengot *et al.*, 2016; Tscharntke *et al.*, 2011). Likewise, Famuwagun *et al.* (2018) indicate that the number of flowers and fruits are reduced when the plants grow under shade within an agroforestry system. Similarly, it is reported that cacao yield was significantly higher in monoculture compared to agroforestry systems (Schneider *et al.*, 2017), which would be related to the greater possibility of capturing and using solar radiation, as well as, by the greater number of plants in monoculture systems, compared to the agroforestry system (Almeida and Gattward, 2018); in addition, the belowground competition reduces cacao growth and yield in agroforests (Figure 3.6) (Ahenkorah *et al.*, 1974, 1987; Niether *et al.*, 2019).

3.3.2 Solar radiation

Cacao is known to be native to the Upper Amazon basin in South America and domesticated approximately 5,300 years ago in southeast Ecuador (Fouet *et al.*, 2022; Valdez, 2019; Zarrillo *et al.*, 2018) and growing in understory environments with high rainfall and limited light availability (Baligar *et al.*, 2008; Raja Harun and Kamariah, 1983; Wood and Lass, 2001), exhibiting saturated rates of net carbon assimilation at photosynthetically active radiation (PAR) levels between 200 and 750 μmol m^{-2} s^{-1} (Almeida *et al.*, 2014; Avila-Lovera *et al.*, 2016). cacao would have a low light saturation: 95% of maximum photosynthesis occurs at 200 μmol m^{-2} s^{-1} (Baligar *et al.*, 2008), which would explain

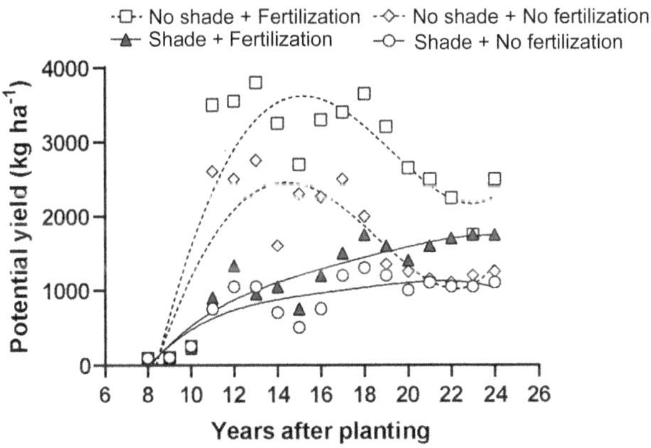

Figure 3.6 Cacao yields in response to shade under agroforestry systems with or without fertilizers. Curve fitting is polynomial.

(Adapted from Ahenkorah *et al.*, 1974).

the recommendations for cacao crops under shade in agroforestry systems (Tezara *et al.*, 2015). However, recent studies by Suárez *et al.* (2021) evaluated the physiological performance of cacao genotypes under different radiation intensities in an Amazonian environment, showing that cacao increases the net carbon assimilation rate when it is subjected to average PAR values on the day of 700 μmol m^{-2} s^{-1}, reaching average values of 1300 μmol m^{-2} s^{-1} at noon. This indicates that this genotype would present photoprotection mechanisms that avoid photoinhibition, increase the non-photochemical extinction coefficient, and present biochemical and morphological adjustments such as low content of chlorophyll, higher content of carotenoids, and low specific leaf area.

Likewise, it is reported that several Ecuadorian cacao clones grown in the coastal region of Ecuador respond positively to increased solar radiation, even with high values of 1000 μmol m^{-2} s^{-1}; adapting the physiological and morphological characters, without evidence of photoinhibition, all clones showed an increase of 35% on average in the net photosynthetic rate and optimal performance at high levels of radiation (Jaimez *et al.*, 2018; Tezara *et al.*, 2015). In another study Niether *et al.* (2018) measured the light transmission through the canopy of various cacao crop systems, they found that, above the canopy, under clear skies, PAR between 1580 and 2028 μmol m^{-2} s^{-1} was recorded, which suggests that the crop can grow in environments with high radiation. Conditions where there is high cloud density and low evaporative demand would not have photosynthetic limitations for cacao crops in full sun (Jaimez *et al.*, 2018).

Baligar *et al.* (2005) reported that high light levels (1050 μmol m^{-2} s^{-1}) were detrimental to the growth of shoots, roots, and leaves of seedlings, and reduced the uptake and entry of some nutrients. However, this cannot be generalized, at least until now there is no conclusive data, since apparently there is a differential response conditioned by the adaptability of the genotype, as well as by the climatic conditions where cacao is grown. The Amazonian region of Ecuador is characterized by high rainfall and cloudiness, which may be different from other geographic locations where there is less cloudiness and the amount of radiation received by a crop may be greater.

3.3.3 Response of genotypes to shading

Recent studies in the southern Amazon region of Ecuador in Yantzaza (El Padmi Experimental Station) were carried out with the objective of evaluating cacao growth traits in response to shading, in a young orchard. To generate shade, individual wooden structures were built in the shape of a cabin and were covered in the upper part with shade mesh that intercepted 35% and 80%, and control without shade; the structures were placed on each plant, and 0.3 m was always kept free between the shade mesh and the plant (Guamán *et al.*, 2022). Three genotypes were used: CCN-51, which has high yield potential and disease resistance, and is the most cultivated variety in Ecuador (Boza *et al.*, 2014; Jaimez *et al.*, 2022), two genotypes released by Instituto Nacional de Investigaciones Agropecuarias (INIAP, Ecuador), EETP-800 and EETP-801 (Estación Experimental Tropical Pichiligue), which would have superior quality characteristics, being classified as fine and aroma (Loor Solórzano et al., 2018, 2019).

Regarding plant growth, plant height, shoot length, and elongation rate, the number and length of metameres were evaluated. No significant effects of shade were found for all growth variables; however, significant differences were observed between the genotypes, highlighting EETP-800 with the highest values in all variables (Table 3.2). These results indicate that the decrease in solar

Table 3.2 Growth variables in young cacao plants under different shade levels

Genotype	*Shade (%)*	*Plant height (cm)*	*Shoot length (cm)*	*Metamere length (cm)*	*Metamere number*
CCN-51	0	90.7	20.5	3.0	8.3
	35	76.7	22.5	3.7	7.8
	80	74.5	22.8	3.1	7.3
EETP-800	0	121.8	55.7	0.4	16.6
	35	132.4	46.9	0.3	14.0
	80	144.3	51.1	0.3	15.1
EETP-801	0	90.9	26.6	2.1	12.2
	35	97.1	27.6	2.0	13.7
	80	93.1	22.3	1.6	14.0
E.E.		9.3	7.1	0.0	2.3
Genotype (G)		***	***	***	**
Shade (S)		ns	ns	ns	ns
G x S		ns	ns	ns	ns

Asterisks indicate significant differences between treatments at $**P < 0.01$ and $***P < 0.001$, while ns = not significant ($P > 0.05$), according to ANOVA.

radiation does not affect cacao growth, at least in the early stages of growth and development, but shows that the genotypes have different adaptations to the environment. In studies where the shade mediated by trees in agroforestry systems decreases cacao growth (Agele *et al.*, 2016; Famuwagun *et al.*, 2018) and yield (Esche *et al.*, 2023), it could be explained that in these systems, in addition to the decrease in radiation, there are other interactions and ecological conditions that are difficult to quantify, including interspecific competition for light, water, and nutrients.

Nutrient contents at the foliar level were analyzed; however, no significant effects of shade or genotype were found for macronutrients (N, P, K, Ca, and Mg), which were at adequate levels for young plants in a vegetative state (Snoeck *et al.*, 2016), although low levels of P were found, this is possibly due to the fact that the movement of this element decreases early compared to the other macronutrients, as shown by Romero *et al.* (2022) (Table 3.3). Likely, since it is a young crop, the demand for the crop was covered by the nutritional supply of the soil, probably, in higher stages such as in the reproductive stage or in the growth of fruits, the demand for nutrients in cacao plants will be greater, which changes the nutritional content at the foliar level (Bahia *et al.*, 2021). In addition, the nutritional content of the plant would respond mainly to soil fertility conditions (Moyin and Otoyosoye, 2002; Puentes-Páramo *et al.*, 2016).

Stomatal density was significantly affected by genotype and shade; CCN 51 reached the highest value (1262.56 stomata mm^{-2}) compared to EETP-800 (884.27 stomata mm^{-2}) and EETP-801 (970.33 stomata mm^{-2}) in controls without shade, with values similar to those reported by Daymond *et al.* (2011). With shade treatments, it decreased by 27.16%, 8.0%, and 0.20% in CCN 51, EETP-800, and EETP-801, respectively. However, no significant effects on stomatal conductance were found, although the 80% reduction of solar radiation caused a clear decrease in stomatal conductance as compared to 30% of shade and control without shade (Table 3.4). This suggests a possible acclimatization to high solar radiation increased stomatal density, due to the modification of the leaf area (Daymond *et al.*, 2011; Galyuon *et al.*, 1996; Harrison *et al.*, 2020; Suárez *et al.*, 2018). In conditions of high levels of solar radiation, there would be a greater supply of ATP and NADPH produced by the light reactions of

Table 3.3 Nutrients in the leaves of cacao plants and leaf optimum nutrient thresholds

Nutrient	*Lower limit*[a]	*Upper limit*[a]	*CCN-51*[b]	*EETP-800*[b]	*EETP-8001*[b]
N	1.8	2.5	2.51	2.79	2.64
P	0.17	0.25	0.15	0.13	0.15
K	1.2	2.4	1.32	1.56	1.37
Ca	0.3	1.5	1.27	1.46	1.17
Mg	0.2	0.8	0.30	0.29	0.31

[a] Indicates the values reported by Snoeck *et al.* (2016).

[b] Average values measured in fully expanded leaves of three cacao genotypes (CCN-51, EETP-800, and EETP-8001) 2.5 years old grown in Ecuadorian Amazon.

Table 3.4 PAR, stomatal conductance, temperature, and soil in three cacao genotypes under shade in Zamora Chinchipe

Variable	*Shade (%)*		
	0	*30*	*80*
PAR (μmol m^{-2} s^{-1})	19645.0	1029.8	288.5
PAR reduction relative to control (%)	0.0	46.9	85.5
Stomatal conductance (μmol m^{-2} s^{-1})	361.8	399.4	303.9
Leaf temperature (°C)	34.4	34.6	34.3
Soil temperature (°C)	32.13	32.45	31.55
Soil water content θ (%)	31	30	32

The measurements were made in full sun, each value is the average of three genotypes (CCN-51, EETP-800 and EETP-801) with five repetitions.

photosynthesis; in these conditions, the stomatal density generally increases (Lake *et al.*, 2001; Pellerin et al., 2000; Schoch *et al.*, 1980). Nevertheless, sensitivity to shading also depends on the genotype (Lewis *et al.*, 2021). In our studies, we found differences only due to genotype, that is, ETTP-801 387.4 μmol m^{-2} s^{-1}, ETTP-800 315.05 μmol m^{-2} s^{-1}, and CCN-51: 266.34 μmol m^{-2} s^{-1}.

3.4 Conclusions

The growth of cacao seems to be favored under Amazonian conditions, where there are conditions of great water availability, good soil fertility, and not excessive solar radiation. Our findings regarding the use of radiation and nutrient resources show that despite the clear acidifying effect on the soil, the application of ammonium sulfate improves cacao beans' yield and fat content, which can be very relevant for the chocolate industry. The nutritional content of the leaves varies during leaf growth; this is relevant to consider for an adequate leaf diagnosis; therefore, for a leaf analysis of the elements N, P, K, Ca, and Mg, the leaf sample should be collected around 100 days after the appearance of shoots, because very early or late sampling could lead to an erroneous diagnosis. Overall, our studies show an apparent genotypic effect on growth, as well as on morphophysiological variables; however, under the conditions of the Ecuadorian Amazon, the shading effect was not significant, possibly due to the cloudiness of the environment. Consequently, the amount of solar radiation that crops receive is not excessive, and on the other hand, there could be apparent acclimatization of cacao to the environment.

References

Abad, A., Acuña, C., and Naranjo, E. (2020), "El cacao en la Costa ecuatoriana: estudio de su dimensión cultural y económica", *Estudios de la Gestión: Revista Internacional de Administración*, 7, 59–83.

Abo-Hamed, S., Collin, H., and Hardwick, K. (1981), "Biochemical and physiological aspects of leaf development in cacao (*Theobroma cacao*) VI. Hormonal interaction between mature leaves and the shoot apex", *New Phytologist*, 89(2), 191–200.
Adams, F. (1980), "Interactions of phosphorus with other elements in soils and in plants", In: Khasawneh, F., Sample, E., and Kamprath, E. (eds), *The Role of Phosphorus in Agriculture*. American Society of Agronomy: Crop Science Society of America, and Soil Science Society of America. Book Series. ASA, CSSA, and SSSA Books, pp. 655–680. https://doi.org/10.2134/1980.roleofphosphorus.c24
Ahenkorah, Y., Akrofi, G. S., and Adri, A. K. (1974), "The end of the first cacao shade and manurial experiment at the cacao Research Institute of Ghana", *Journal of Horticultural Science*, 49(1), 43–51.
Ahenkorah, Y., Halm, B. J., Appiah, M. R., Akrofi, G. S., and Yirenkyi, J. E. K. (1987), "Twenty years' results from a shade and fertilizer trial on Amazon cacao (*Theobroma cacao*) in Ghana", *Experimental Agriculture*, 23(1), 31–39.
Agele, S., Famuwagun, B., and Ogunleye, A. (2016), "Effects of shade on microclimate, canopy characteristics and light integrals in dry season field-grown cacao (*Theobroma cacao* L.) seedlings", *Journal of Horticultural Sciences*, 11(1), 47–56.
Albornoz, F. (2016), "Crop responses to nitrogen overfertilization: a review", *Scientia Horticulturae*, 205, 79–83. https://doi.org/10.1016/j.scienta.2016.04.026
Almeida, A., and Valle, R. (2007), "Ecophysiology of the cacao tree", *Brazilian Journal of Plant Physiology*, 19(4), 425–448.
Almeida, A. A., Gomes, F. P., Araujo, R. P., Santos, R. C., and Valle, R. R. (2014), "Leaf gas exchange in species of the Theobroma genus", *Photosynthetica*, 52(1), 16–21.
Almeida, A., and Gattward, J. (2018), "Resposta do cacaueiro à variações da intensidade luminosa", In: Souza Júnior, J. O. (ed.), *Cacau: cultivo, pesquisa e inovação*. Editus, pp. 35–58. https://doi.org/10.7476/9786586213188.0002
Almeida-Andrade, J., Rivera-García, J., Chire-Fajardo, G., and Ureña-Peralta, M. (2019), "Propiedades físicas y químicas de cultivares de cacao (*Theobroma cacao* L.) de Ecuador y Perú", *Enfoque UTE*, 10(4), 1–12.
Alsam, M., Chattha, M., Khan, I., Maqbool, R., Chattha, M., Hussan, F., Nawaz, M., Hanif, M., Ayub, M., Aslam, M., Khan, M., and Hassan, M. (2021), "Effect of different organic mulches and nitrogen sources on the productivity of wheat crop grown in semi-arid area", *Pakistan Journal of Agricultural Research*, 34(3), 648–655. https://doi.org/10.17582/journal.pjar/2021/34.3.648.655
Andersen, K. M., and Turner, B. L. (2013), "Preferences or plasticity in nitrogen acquisition by understorey palms in a tropical montane forest", *Journal of Ecology*, 101, 819–825.
Anglaaere, L. C., Cobbina, J., Sinclair, F. L., and McDonald, M. A. (2011), "The effect of land use systems on tree diversity: farmer preference and species composition of cacao-based agroecosystems in Ghana", *Agroforestry Systems*, 81(3), 249–265.
Armengot, L., Barbieri, P., Andres, C., Milz, J., and Schneider, M. (2016), "Cacao agroforestry systems have higher return on labor compared to full-sun monocultures", *Agronomy for Sustainable Development*, 36, 1–10.
Avila-Lovera, E., Coronel, I., Jaimez, R., Urich, R., Pereyra, G., Araque, O., … and Tezara, W. (2016), "Ecophysiological traits of adult trees of Criollo cacao cultivars (*Theobroma cacao* L.) from a germplasm bank in Venezuela", *Experimental Agriculture*, 52(1), 137–153.
Bahia, B. L., Souza-Junior, J. O., Fernandes, L. V., and Neves, J. C. (2021), "Reference values and diagnostic ranges to assess the degree of nutritional balance for cacao plants", *Spanish Journal of Agricultural Research*, 19(1), e0801.
Baker, N., and Hardwick, K. (1975), "Biochemical and physiological aspects of leaf development in cacao (*Theobroma cacao*). III. Changes in soluble sugar content and sucrose synthesizing capacity", *New Phytologist*, 75(3), 519–524.
Baligar, V. C., Bunce, J. A., Bailey, B. A., Machado, R. C., and Pomella, A. W. V. (2005), "Carbon dioxide and photosynthetic photon flux density effects on growth and mineral uptake of cacao", *Journal of Food, Agriculture and Environment*, 3(2), 142–147.

Baligar, V., Bunce, J., Machado, R., and Elson, M. (2008), "Photosynthetic photon flux density, carbon dioxide concentration, and vapor pressure deficit effects on photosynthesis in cacao seedlings", *Photosynthetica*, 46(2), 216–221. https://doi.org/10.1007/s11099-008-0035-7

Baligar, V., and Fageria, N. (2017), "Influence of nitrogen forms and levels on the growth and nutrition of cacao", *Journal of Plant Nutrition*, 40(5), 709–718. https://doi.org/10.1080/01904167.2016.1262401

Barrios, M., García, J., and Basso, C. (2012), "Efecto de la fertilización nitrogenada sobre el contenido de nitrato y amonio en el suelo y la planta de maíz", *Bioagro*, 24(3), 213–220.

Bhattacharjee, R. (2018), "Taxonomy and classification of cacao", In: Umaharan, P. (ed.), *Achieving Sustainable Cultivation of Cacao*, 1st ed. Burleigh Dodds Science Publishing, pp. 588.

Blare, T., and Useche, P. (2019), "Differences in women's and men's conservation of cacao agroforests in coastal Ecuador", *Environmental Conservation*, 46(4), 302–309.

Blaser, W. J., Oppong, J., Yeboah, E., and Six, J. (2017), "Shade trees have limited benefits for soil fertility in cacao agroforests", *Agriculture, Ecosystems & Environment*, 243, 83–91.

Boczulak, S. A., Hawkins, B. J., and Roy, R. (2014), "Temperature effects on nitrogen form uptake by seedling roots of three contrasting conifers", *Tree Physiology*, 34, 513–523.

Boza, E. J., Motamayor, J. C., Amores, F. M., Cedeno-Amador, S., Tondo, C. L., Livingstone, D. S., ... and Gutiérrez, O. A. (2014), "Genetic characterization of the cacao cultivar CCN 51: its impact and significance on global cacao improvement and production", *Journal of the American Society for Horticultural Science*, 139(2), 219–229.

Cabello, M. J., Castellanos, M. T., Romojaro, F., Martinez-Madrid, C., and Ribas, F. (2009), "Yield and quality of melon grown under different irrigation and nitrogen rates", *Agricultural Water Management*, 96, 866–874.

Capa-Morocho, M., Romero-Maza, A., Romero, M., Molina-Muller, M., Vásquez, S., and Granja, F. (2022), "Efectos de las fuentes nitrogenadas en la morfofisiología, producción y calidad de cacao (*Theobroma cacao* L.) CCN-51 en el sur de la Amazonía ecuatoriana", *Tropical and Subtropical Agroecosystems*, 25, 119. https://doi.org/10.56369/tsaes.4316

Castellón Paco, M. (2015), "Género y producción de cacao en sistema chacra en la Reserva de Biósfera Sumaco en la provincia de Napo-Ecuador" (Master's thesis, Quito, Ecuador: Flacso Ecuador).

Cerda, R., Avelino, J., Harvey, C. A., Gary, C., Tixier, P., and Allinne, C. (2020), "Coffee agroforestry systems capable of reducing disease-induced yield and economic losses while providing multiple ecosystem services", *Crop Protection*, 134, 105149.

Chaves, A. R., Ten-Caten, A., Pinheiro, H. A., Ribeiro, A., and DaMatta, F. M. (2008), "Seasonal changes in photoprotective mechanisms of leaves from shaded and unshaded field-grown coffee (*Coffea arabica* L.) trees", *Trees*, 22(3), 351–361.

Cheremisinoff, N., and Rosenfeld P. (2011), "Industry and products", In: Cheremisinoff, N. P., and Rosenfeld, P. E. (eds), *Handbook of Pollution Prevention and Cleaner Production: Best Practices in the Agrochemical Industry*. William Andrew Publishing, pp. 1–24. https://doi.org/10.1016/B978-1-4377-7825-0.00001-7

Coe, S., and Coe, M. (2013), *True History of Chocolate*, 3rd ed. Thames & Hudson Publishing, pp. 288.

Crous, K., and Ellsworth, D. (2020), "Probing the inner sanctum of leaf phosphorus: measuring the fractions of leaf P", *Plant and Soil*, 454(1), 77–85. https://doi.org/10.1007/s11104-020-04657-3

Daymond, A., Tricker, P., and Hadley, P. (2011), "Genotypic variation in photosynthesis in cacao is correlated with stomatal conductance and leaf nitrogen", *Biologia Plantarum*, 55(1), 99–104. https://doi.org/10.1007/s10535-011-0013-y

De Almeida, A. A., and Valle, R. R. (2010), "Cacao: ecophysiology of growth and production", In: *Ecophysiology of Tropical Tree Crops*, pp. 37–70.

De Andrade, F., Buzetti, S., Andreotti, M., Arf, O., de Sá, M., and daCosta, J. (2009), "Sources and times of nitrogen application on irrigated corn crop", *Semina: Ciências Agrárias*, 30(2), 275–284. https://doi.org/10.5433/1679-0359.2009v30n2p275

Dillinger, T. L., Barriga, P., Escárcega, S., Jimenez, M., Lowe, D. S., and Grivetti, L. E. (2000), "Food of the gods: cure for humanity? A cultural history of the medicinal and ritual use of chocolate", *The Journal of Nutrition*, 130(8), 20575.

Dostert, N., Roque, J., Cano, A., La Torre, M., and Weigend, M. (2012), "Hoja botánica: cacao. *Theobroma cacao* L.", Giacomotti Comunicación Gráfica S.A.C., Lima, Peru. https://doi.org/10.13140/RG.2.2.31228.44165

Ehret, D. L., Frey, B., Forge, T., Helmer, T., Bryla, D. R., and Zebarth, B. J. (2014), "Effects of nitrogen rate and application method on early production and fruit quality in highbush blueberry", *Canadian Journal of Plant Science*, 94, 1165–1179.

Elrys, A. S., Metwally, M. S., Raza, S., Alnaimy, M. A., Shaheen, S. M., Chen, Z., and Zhou, J. (2020), "How much nitrogen does Africa need to feed itself by 2050?", *Journal of Environmental Management*, 268, 110488. https://doi.org/10.1016/j.jenvman.2020.110488

Erel, R., Dag, A., Ben-Gal, A., Schwartz, A., and Yermiyahu, U. (2008), "Flowering and fruit set of olive trees in response to nitrogen, phosphorus, and potassium", *Journal of the American Society for Horticultural Science*, 133, 639–647.

Esche, L., Schneider, M., Milz, J., and Armengot, L. (2023), "The role of shade tree pruning in cacao agroforestry systems: agronomic and economic benefits", *Agroforestry Systems*, 97, 175–185.

ESPAC. (2023), "Encuesta de superficie y producción agropecuaria", Instituto Nacional de Estadísticas y Censos, Quito-Ecuador, 42. https://shre.ink/HRhd (accessed January 22, 2023).

Fageria, N., Stone, L., and Santos, A. (2003), "Fertility management of irrigated rice (Manejo da fertilidade do solo para o arrozirrigado)", Santo Antonio de Goiás, Brazil, EMBRAPA Center for Rice and Beans.

Famuwagun, I. B., Agele, S. O., and Aiyelari, O. P. (2018), "Shade effects on growth and development of cacao following two years of continuous dry season irrigation", *International Journal of Fruit Science*, 18(2), 153–176.

FAOSTAT. (2023), "Data. Countries by commodity". https://www.fao.org/faostat/en/#rankings/countries_by_commodity_exports (accessed February 13, 2023).

FEDECACAO. (2004), "El beneficio y características físico químicas del cacao (*Theobroma cacao* L.)", Programa de comercialización. Produmedios, Bogotá, Colombia, 42. https://repository.agrosavia.co/bitstream/handle/20.500.12324/18060/42772_46877.pdf?sequence=1&isAllowed=y (accessed September 22, 2022).

Fouet, O., Loor Solorzano, R. G., Rhoné, B., Subía, C., Calderón, D., Fernández, F., ... and Lanaud, C. (2022), "Collection of native *Theobroma cacao* L. accessions from the Ecuadorian Amazon highlights a hotspot of cacao diversity", *Plants, People, Planet*, 4(6), 605–617.

Franklin, O., Cambui, C. A., Gruffman, L., Palmroth, S., Oren, R., and Näsholm, T. (2017), "The carbon bonus of organic nitrogen enhances nitrogen use efficiency of plants", *Plant, Cell & Environment*, 40(1), 25–35. https://doi.org/10.1111/pce.12772

Furcal-Beriguete, P. (2017), "Extracción de nutrientes por los frutos de cacao en dos localidades en Costa Rica", *Agronomía Mesoamericana*, 28(1), 113–129.

Galyuon, I. K. A., McDavid, C. R., Lopez, F. B., and Spence, J. A. (1996), "The effect of irradiance level on cacao (*Theobroma cacao* L.): I. Growth and leaf adaptations", *Tropical Agriculture*, 73(1), 23–28.

Gardiazabal, F., Mena, F., and Magdahl, C. (2007), "Efecto de la fertilización con inhibidores de la nitrificación (Entec® Solub 21) en paltos (*Persea americana* Mill) cv. Hass", in Actas VI Congreso Mundial del Aguacate, Viña del Mar, Chile.

García, A. (1993), "Sintomatología de las deficiencias nutricionales en cacao (*Theobroma cacao* L.)", *Boletín Técnico-Instituto Colombiano Agropecuario-ICA*, Palmira, Colombia. https://hdl.handle.net/20.500.12324/1156 (accessed September 22, 2022).

Gockowski, J., Afari-Sefa, V., Bruce, D., Osei-Asare, Y., and Agyeman, F. (2013), "Improving the productivity and income of Ghanaian cacao farmers while maintaining environmental services: what role for certification?", *International Journal of Agricultural Sustainability*, 11(4), 331–346. https://doi.org/10.1080/14735903.2013.772714

Granja, F., and Covarrubias, J. (2018), "Evaluation of acidifying nitrogen fertilizers in avocado trees with iron deficiency symptoms", *Journal of Soil Science and Plant Nutrition*, 18(1), 157–172. https://doi.org/10.4067/S0718-95162018005000702

Greenwood, D. J., Floate, M. J. S., Nelder, J. A., and Frissel, M. J. (1982), "Modelling of crop response to nitrogen fertilizer [and discussion]", *Philosophical Transactions of the Royal Society of London. Series B, Biological Sciences*, 296(1082), 351–362. https://www.jstor.org/stable/2395689

Guamán, A. O., Vásquez, S. C., Granja, J. F., Molina, M. L., and Capa, M. (2022), "Influencia del sombreo en las características morfofisiológicas y nutricionales en árboles jóvenes de cacao en la Amazonía sur Oriental de Ecuador", *Proceedings of the 8th International Research Congress REDU*. Ambato, Ecuador. https://doi.org/10.5867/Medwave.2022.S1.CI35

Guo, S., Zhou, Y., Shen, Q., and Zhang, F. (2007), "Effect of ammonium and nitrate nutrition on some physiological processes in higher plants – Growth, photosynthesis, photorespiration, and water relations", *Plant Biology*, 9, 21–29. https://doi.org/10.1055/s-2006-924541

Harrison, E., Arce, L., Gray, J., and Hepworth, C. (2020), "The influence of stomatal morphology and distribution on photosynthetic gas exchange", *Plant Journal*, 101(4), 768–779. https://doi.org/10.1111/tpj.14560

Hartemink, A. (2005), "Nutrient stocks, nutrient cycling, and soil changes in cacao ecosystems: a review", *Advances in Agronomy*, 86, 227–253. https://doi.org/10.1016/S0065-2113(05)86005-5

Herrera, R., Vásquez, S. C., Granja, F., Molina-Müller, M., Capa-Morocho, M., and Guamán, A. (2022), "Interacción de N, P y K sobre características del suelo, crecimiento y calidad de frutos de cacao en la Amazonía Ecuatoriana", *Bioagro*, 34(3), 277–288. https://doi.org/10.51372/bioagro343.7

Huera-Lucero, T., Labrador-Moreno, J., Blanco-Salas, J., and Ruiz-Téllez, T. (2020), "A framework to incorporate biological soil quality indicators into assessing the sustainability of territories in the Ecuadorian Amazon", *Sustainability*, 12(7), 3007. https://doi.org/10.3390/su12073007

Isaac, M. E., Timmer, V. R., and Quashie-Sam, S. J. (2007), "Shade tree effects in an 8-year-old cacao agroforestry system: biomass and nutrient diagnosis of *Theobroma cacao* by vector analysis", *Nutrient Cycling in Agroecosystems*, 78(2), 155–165.

Ismunadji, M., and Dijkshoorn. (1971), "Nitrogen nutrition of rice plant measured by growth and nutrient content in pot experiments. Ionic balance and selective uptake", *Netherland Journal Agricultural Science*, 19(4), 223–236. https://doi.org/10.18174/njas.v19i4.17303

Jacobi, J., Andres, C., Schneider, M., Pillco, M., Calizaya, P., and Rist, S. (2014), "Carbon stocks, tree diversity, and the role of organic certification in different cacao production systems in Alto Beni, Bolivia", *Agroforestry Systems*, 88(6), 1117–1132.

Jaimez, R. E., Amores Puyutaxi, F., Vasco, A., Gastón Loor, R., Tarqui, O., Quijano, G., … and Tezara, W. (2018), "Photosynthetic response to low and high light of cacao growing without shade in an area of low evaporative demand", *Acta Biológica Colombiana*, 23(1), 95–103.

Jaimez, R. E., Barragan, L., Fernández-Niño, M., Wessjohann, L. A., Cedeño-Garcia, G., Sotomayor Cantos, I., and Arteaga, F. (2022), "*Theobroma cacao* L. cultivar CCN 51: a comprehensive review on origin, genetics, sensory properties, production dynamics, and physiological aspects", *PeerJ*, 10, e12676. https://doi.org/10.7717/peerj.12676

Jones, C., Brown, B. D., Engel, R., Horneck, D., and Olson-Rutz, K. (2013), "Factor affecting nitrogen fertilizer volatilization", Montana State University, Extensión. https://www.sare.org/wp-content/uploads/Factors-Affecting-Nitrogen-Fertilizer-Volatilization.pdf (accessed June 12, 2022).

Kamprath, E. (1970), "Exchangeable aluminum as a criterion for liming leached mineral soils", *Soil Science Society of America Journal*, 34(2), 252–254. https://doi.org/10.2136/sssaj1970.03615995003400020022x

Kaymak, H. (2013), "Effect of nitrogen forms on growth, yield and nitrate accumulation of cultivated purslane (*Portulaca oleracea* L.)", *Bulgarian Journal of Agricultural Science*, 19(3), 444–449.

Krouk, G., and Kiba, T. (2020), "Nitrogen and phosphorus interactions in plants: from agronomic to physiological and molecular insights", *Current Opinion in Plant Biology*, 57, 104–109. https://doi.org/10.1016/j.pbi.2020.07.002

Lairon, D. (2010), "Nutritional quality and safety of organic food. A review", *Agronomy for Sustainable Development*, 30, 33–41.

Lake, J. A., Quick, W. P., Beerling, D. J., and Woodward, F. I. (2001), "Signals from mature to new leaves", *Nature*, 411(6834), 154–154.

Leghari, S. J., Wahocho, N. A., Laghari, G. M., Laghari, A. H., Bhabhan, G. M., Talpur, K. H., Bhutto, T. A., Wahocho, S. A., and Lashari, A. A. (2016), "Role of nitrogen for plant growth and development: a review", *Advances in Environmental Biology*, 10(9), 209–218.

Lehmann, J., and Schroth, G. (2003), "Nutrient leaching", In: Schroth, G., and Sinclair, F. L. (eds), *Trees, Crops and Soil Fertility*. CAB International, pp. 151–166. https://www.css.cornell.edu/faculty/lehmann/publ/Lehmann%20et%20al.,%202003,%20Leaching%20CABI%20book.pdf (accessed September 10, 2022).

Lewis, V., Farrell, A., Umaharan, P., and Lennon, A. (2021), "Genetic variation in high light responses of *Theobroma cacao* L. accessions", *Heliyon*, 7(6), e07404. https://doi.org/10.1016/j.heliyon.2021.e07404

Li, S., Wang, Z., and Stewart, B. (2013), "Response of crop plant to ammonium and nitrate N", *Advances in Agronomy*, 118, 205–397. https://doi.org/10.1016/B978-0-12-405942-9.00005-0

Limongi Andrade, R., Arévalo Vizcaino, V., Quiroz Vera, J., and Grijalva Olmedo, J. (2007), "CHACRAS Y MEDIOS DE VIDAS: Implicaciones para su manejo agroforestal", Instituto de Investigaciones Agropecuarias de Ecuador, Tena, Napo Ecuador.

Liu, M., Burgos, A., Ma, L., Zhang, Q., Tang, D., and Ruan, J. (2017), "Lipidomics analysis unravels the effect of nitrogen fertilization on lipid metabolism in tea plant (*Camellia sinensis* L.)", *BMC Plant Biology*, 17, 165. https://doi.org/10.1186/s12870-017-1111-6

Liu, S., Qin, T., Dong, B., Shi, X., Lv, Z., and Zhang, G. (2021), "The influence of climate, soil properties and vegetation on soil nitrogen in sloping farmland", *Sustainability*, 13, 1480. https://doi.org/10.3390/su13031480

Liu, X., Hu, B., and Chu, Ch. (2022), "Nitrogen assimilation in plants: current status and future prospects", *Journal of Genetics and Genomics*, 49(5), 394–404. https://doi.org/10.1016/j.jgg.2021.12.006

Lobão, D., Setenta W., Lobão, E., Curvelo, K., and Valle, R. (2007), "Cacau cabruca: sistemaagrossilvicultural tropical", In: Valle, R. R. (ed.), *Ciência, tecnologia e manejo docacaueiro*. Gráfica e Editora Vital Ltda., pp. 290–323.

Loor Solórzano, R. G., Amores-Puyutaxi, F. M., Vasco-Medina, S. A., Quiroz-Vera, J. G., Casanova-Mendoza, T. D. J., Garzón-Catota, A. I., ... and Zambrano-Flores, F. (2019), "INIAP-EETP-800 'Aroma Pichilingue', Nueva variedad ecuatoriana de cacao fino de alto rendimiento", *Revista Fitotecnia Mexicana*, 42(2), 187–189.

Loor Solórzano, R. G., Sotomayor Cantos, I. A., Jiménez Barragán, J. C., Tarqui Freire, O. M., Rodríguez Zamora, G. A., Casanova Mendoza, T. D. J., and Quijano Rivadeneira, G. C. (2018), "INIAP-EETP-800 e INIAP-EETP-801 nuevos clones de cacao fino y de aroma con alto rendimiento". https://repositorio.iniap.gob.ec/handle/41000/5240

López, O., Ramirez, S., Espinosa, S., Moreno, J., Ruiz, C., Villarreal, J., and Rojas, J. (2015), "Manejo Agroecológico del Cultivo de Cacao", In: *Manejo agroecológico de la nutrición en el cultivo del cacao*. Colección Universitaria Letras sin Papel, pp. 128.

Ma, Q., Wang, X., Li, H., Zhang, F., Rengel Z., and Shen J. (2015), "Comparing localized application of different N fertilizer species on maize grain yield and agronomic N-use efficiency on a calcareous soil", *Field Crops Research*, 180, 72–79. https://doi.org/10.1016/j.fcr.2015.05.011

Maathuis, F. J. M. (2009), "Physiological functions of mineral macronutrients", *Current Opinion in Plant Biology*, 12(3), 250–258. https://doi.org/10.1016/j.pbi.2009.04.003

Machado, R., and Hardwick, K. (1988), *Does carbohydrate availability control flush growth in cacao*, Santo Domingo, Dominican Republic, In Proc. 10th Int.

Maiguashca, J. (2014), La incorporación del cacao ecuatoriano al mercado mundial entre 1840 y 1925, según los informes consulares. *Procesos. Revista Ecuatoriana de Historia* 1, 67–97.

Mandal, S., Donner, E., Smith, E., Sarkar, B., and Lombi, E. (2019), "Biochar with near-neutral pH reduces ammonia volatilization and improves plant growth in a soil-plant system: a closed chamber experiment", *Science of the Total Environment*, 697, 134114. https://doi.org/10.1016/j.scitotenv.2019.134114

Marschner, H. (1995), *Mineral Nutrition of Higher Plants*, 2nd ed. Academic Press.

Martinez-Dalmau J., Berbel J., and Ordoñez-Fernández, R. (2021), "Nitrogen fertilization. A review of the risks associated with the inefficiency of its use and policy responses", *Sustainability*, 13, 5625. https://doi.org/10.3390/su13105625

Mengel, K., and Kirkby, E. (2001), *Principles of Plant Nutrition*, 5th ed., Kluwer Academic Publishers.

Meylan, L., Gary, C., Allinne, C., Ortiz, J., Jackson, L., and Rapidel, B. (2017), "Evaluating the effect of shade trees on provision of ecosystem services in intensively managed coffee plantations", *Agriculture, Ecosystems & Environment*, 245, 32–42.

Miller, A. E., Bowman, W. D., and Suding, K. N. (2007), "Plant uptake of inorganic and organic nitrogen: neighbor identity matters", *Ecology*, 88, 1832–1840.

Mohammed, A. M., Robinson, J. S., Midmore, D., and Verhoef, A. (2016), "Carbon storage in Ghanaian cacao ecosystems", *Carbon Balance and Management*, 11(1), 1–8.

Molina, J., and Covarrubias, J. (2019), "Influence of nitrogen on physiological responses to bicarbonate in a grapevine rootstock", *Journal of Soil Science and Plant Nutrition*, 19(2), 305–312. https://doi.org/10.1007/s42729-019-00030-1

Moyin, E., and Otoyosoye, B. (2002), "Utilization of agricultural wastes for the growth, leaf and soil chemical composition of cacao seedlings in the nursery", *Pertanika Journal of Tropical Agricultural Science*, 25(1), 53–62.

Müller, M., Serrano-Minar, P., and Biehl, B. (1992), "Photosynthetic characteristics during development of leaves from *Theobroma cacao*. L.", *Physiologia. Plantarum*, 853(A105), 599.

Nielsen, G. H., Nielsen, D., and Herbert, L. (2009), "Nitrogen fertigation concentration and timing of application affect nitrogen nutrition, yield, firmness, and color of apples grown at high density", *HortScience*, 44, 1425–1431.

Niether, W., Armengot, L., Andres, C., Schneider, M., and Gerold, G. (2018), "Shade trees and tree pruning alter throughfall and microclimate in cacao (*Theobroma cacao* L.) production systems", *Annals of Forest Science*, 75(2), 1–16.

Niether, W., Schneidewind, U., Fuchs, M., Schneider, M., and Armengot, L. (2019), "Below-and aboveground production in cacao monocultures and agroforestry systems", *Science of the Total Environment*, 657, 558–567.

Novoa, M., Miranda, D., and Melgarejo, L. M. (2018), "Efecto de las deficiencias y excesos de fósforo, potasio y boro en la fisiología y el crecimiento de plantas de aguacate (*Persea americana*, cv. Hass)", *Revista Colombiana de Ciencias Hortícolas*, 12(2), 293–307.

Ofori-Frimpong, K., Asase, A., Mason, J., and Danku, L. (2007), "Shaded versus unshaded cacao: implications on litter fall, decomposition, soil fertility and cacao pod development" In: S*ymposium on Multistrata Agroforestry Systems with Perennial Crops*, CATIE Turrialba, Costa Rica, 1721.

Orchard, J., Collin, H., and Hardwick, K. (1980), "Biochemical and physiological aspects of leaf development in cacao (*Theobroma cacao*). IV. Changes in growth inhibitors", *Plant Science Letters*, 18(3), 299–305.

Oyewole, O. S., Ajayi, I. O. J., and Rotimi, R. I. (2012), "Growth of cocoa (Theobroma cacao L.) seedlings on old cocoa soils amended with organic and inorganic fertilizers", *African Journal of Agricultural Research*, 7(24), 3604–3608.

Pellerin, M., Serrano, P., and Biehl, B. (2000), "Photosynthetic characteristics during development of leaves from *Theobroma cacao* L.", *Physiology Plant*, 853, 105–599.

Prabhakaran, K. (2010), "Cacao (*Theobroma cacao* L.)", In: Prabhakaran, K. P. (ed.), *The Agronomy and Economy of Important Tree Crops of the Developing World*. Elsevier, pp. 131–180.

Puentes-Páramo, Y., Menjivar-Flores, J., and Ortíz, A. (2016), "Eficiencia fisiológica de uso de NPK en clones autoincompatible y autocompatible de cacao (*Theobroma cacao* L.) en Colombia", *Revista de investigación agraria y ambiental*, 7(1), 17–24. https://doi.org/10.22490/21456453.1536

Raja Harun, R. M., and Kamariah, H. (1983), "The effects of shading regimes on the growth of cacao seedlings (*Theobroma cacao* L.)" *Pertunika*, 6, 1–5.

Raven, J. (1985), "Regulation of pH and generation of osmolarity in vascular plants: a cost-benefit analysis in relation to efficiency of use of energy, nitrogen and water", *New Phytologist*, 101(1), 25–77. https://doi.org/10.1111/j.1469-8137.1985.tb02816.x

Romero, M. A., Vásquez, S. C., Romero, A. E., Molina-Müller, M. L., Capa-Morocho, M. I., and Granja, F. (2022), "Nutrient dynamic in cacao leaves under different nitrogen sources: a reference tool for foliar analysis", *Revista Brasileira de Fruticultura*, 44. https://doi.org/10.1590/0100-29452022035

Ruan, J., Gerendas, J., Hardter, R., and Sattelmacher, B. (2007), "Effect of root zone pH and form and concentration of nitrogen on accumulation of quality-related components in green tea", *Journal of the Science of Food Agriculture*, 87(8), 1505–1516. https://doi.org/10.1002/jsfa.2875

Salsac, L., Chaillou, S., Morot-Gaudry, J., Lesaint, C., and Jolivet, E. (1987), "Nitrate and ammonium nutrition in plants", *Plant Physiology and Biochemistry*, 25, 805–812.

Sánchez et al. (2019), *La Cadena de Valor del Cacao en América Latina y El Caribe*. 104 pp.

Schneider, M., Andres, C., Trujillo, G., Alcon, F., Amurrio, P., Perez, E., Weibel, F., and Milz, J. (2017), "Cacao and total system yields of organic and conventional agroforestry vs. monoculture systems in a long-term field trial in Bolivia", *Experimental Agriculture*, 53(3), 351–374. https://doi.org/10.1017/S0014479716000417

Schoch, P. G., Zinsou, C., and Sibi, M. (1980), "Dependence of the stomatal index on environmental factors during stomatal differentiation in leaves of *Vigna sinensis* L. 1. Effect of light intensity", *Journal of Experimental Botany*, 31(5), 1211–1216.

Schreiner, M., Korn, M., Stenger, M., Holzgreve, L., and Altmann, M. (2013), "Current understanding and use of quality characteristics of horticultural products", *Scientia Horticulturae*, 163, 63–69.

Shamshuddin, J., Muhrizal, S., Fauziah, I., and Husni, M. (2004), "Effects of adding organic materials to an acid sulfate soil on the growth of cacao (*Theobroma cacao* L.) seedlings", *Science of the Total Environment*, 323(1), 33–45.

Sitompul, H. F., Simanungkalit, T., and Mawarni, L. L. (2014), "Respons Pertumbuhan Bibit Kakao (Theobroma Cacao L.) Terhadappemberianpupuk Kandang Kelinci Dan Pupuk Npk (16: 16: 16)", *Jurnal Agroekoteknologi Universitas Sumatera Utara*, *2*(3), 99703.

Smith Dumont, E., Gnahoua, G. M., Ohouo, L., Sinclair, F. L., and Vaast, P. (2014), "Farmers in Côte d'Ivoire value integrating tree diversity in cacao for the provision of ecosystem services", *Agroforestry Systems*, 88(6), 1047–1066.

Snoeck, D., Koko, L., Joffre, J., Bastide, P., and Jagoret, P. (2016), "Cacao nutrition and fertilization", In: Lichtfouse, E. (ed.), *Sustainable Agriculture Reviews*. Springer, pp. 155–202. https://doi.org/10.1007/978-3-319-26777-7_4

Soria, N. (2008), "Nutrición foliar y defensa natural", In *XI congreso ecuatoriano de la ciencia del suelo*.. Quito, Ecuador.

Spangenberg, J., and Dionisi, F. (2001), "Characterization of cacao butter and cacao butter equivalents by bulk and molecular carbon isotope analyses: implications for vegetable fat quantification in chocolate", *Journal of Agricultural and Food Chemistry*, 49(9), 4271–4277. https://doi.org/10.1021/jf001509g

Stagnari, F., Di Bitetto, V., and Pisante, M. (2007), "Effects of N fertilizers and rates on yield, safety and nutrients in processing spinach genotypes", *Scientia Horticulturae*, 114(4), 225–233. https://doi.org/10.1016/j.scienta.2007.06.016

Stefanelli, D., Winkler, S., and Jones, R. (2011), "Reduced nitrogen availability during growth improves quality in red oak lettuce leaves by minimizing nitrate content, and increasing antioxidant capacity and leaf mineral content", *Agricultural Science*, 2, 477–486.

Suárez, J. C., Melgarejo, L. M., Casanoves, F., Di Rienzo, J. A., DaMatta, F. M., and Armas, C. (2018), "Photosynthesis limitations in cacao leaves under different agroforestry systems in the Colombian Amazon", *Plos One*, 13(11), e0206149.

Suarez, J. C., Gelpud, C., Noriega, J. E., and Ortiz-Morea, F. A. (2021), "How do different cacao genotypes deal with increased radiation? An analysis of water relation, diffusive and biochemical components at the leaf level", *Agronomy*, 11(7), 1422. https://doi.org/10.3390/agronomy11071422

Tezara, W., Almeida, A., Valencia, E., Cortes, J., and Bolaños, M. (2015), "Actividad fotoquímica de clones élites de cacao (*Theobroma cacao* L.) ecuatoriano en el norte de la provincia Esmeraldas", *Investigación y Saberes*, 4(3), 563–574.

Thor, K. (2019), "Calcium—nutrient and messenger", *Frontiers in Plant Science*, 10, 440. https://doi.org/10.3389/fpls.2019.00440

Tscharntke, T., Clough, Y., Bhagwat, S. A., Buchori, D., Faust, H., Hertel, D., … and Wanger, T. C. (2011), "Multifunctional shade-tree management in tropical agroforestry landscapes–a review", *Journal of Applied Ecology*, 48(3), 619–629.

Usandivaras, L., Gutiérrez-Boem, F., and Salvagiotti, F. (2018), "Contrasting effects of phosphorus and potassium deficiencies on leaf area development in maize", *Crop Science*, 58(5), 2099–2109. https://doi.org/10.2135/cropsci2018.02.0092

Vaast, P., and Somarriba, E. (2014), "Trade-offs between crop intensification and ecosystem services: the role of agroforestry in cacao cultivation", *Agroforestry Systems*, 88(6), 947–956.

Valdez, F. (2019), "Evidencias arqueológicas del uso social del cacao en la Alta Amazonía", *Revista de Historia, Patrimonio, Arqueología y Antropología Americana*, 1, 117–134.

Van Herwaarden, A. F., Farquhar, G. D., Angus, J. F., Richards, R. A., and Howe, G. N. (1998), "Haying-off', the negative grain yield response of dryland wheat to nitrogen

fertilizer. I. Biomass, grain yield, and water use", *Australian Journal of Agricultural Research*, 49, 1067–1081.
Van Vliet, J., and Giller, K. (2017), "Mineral nutrition of cacao: a review", *Advances in Agronomy*, 141, 185–270. https://doi.org/10.1016/bs.agron.2016.10.017
Veríssimo, A. (2012), "Efeito da origem do cacauna sua qualidade comercial, funcional e sensorial. O caso do cacau catongo de Sao Tome e Principe e do Brasil", Tesis de grado. ISA, Universidad Técnica de Lisboa, Brasil, 87 p.
Vriesmann, L., Amboni, R., and Petkowicz, C. (2011), "Cacao pod husks (*Theobroma cacao L.*): composition and hot-water-soluble pectins", *Industrial Crops and Products*, 34(1), 1173–1181.
Walworth, J., and Sumner, M. (1987), "The diagnosis and recommendation integrated system (DRIS)", *Advances in Soil Science*, 6, 149–188.
Walling, E., and Vaneeckhaute, C. (2022), *Nitrate Handbook: Nitrogen Fertilizer and the Environment*. 1sted. CRC Press, 33 p.
Wang, H., and Cheng, L. (2011), "Differential effects of nitrogen supply on skin pigmentation and flesh starch breakdown of 'Gala' apple", *HortScience*, 46, 1116–1120.
Wang, J., and Harrell, D. (2005), "Effect of ammonium, potassium, and sodium cations and phosphate, nitrate, and chloride anions on zinc sorption and lability in selected acid and calcareous soils", *Soil Science Society of America Journal*, 69(4), 1036–1046. https://doi.org/10.2136/sssaj2004.0148
Wang, Z., Li, L., and Malhi, S. (2008), "Effects of fertilization and other agronomic measures on nutritional quality of crops", *Journal of the Science of Food and Agriculture*, 88, 7–23.
Wang, J., Wen, X., Zhang, X., Li, S., and Zhang, D. (2018), "Co-regulation of photosynthetic capacity by nitrogen, phosphorus and magnesium in a subtropical Karst forest in China", *Scientific Reports*, 8(1), 7406. https://doi.org/10.1038/s41598-018-25839-1
Wood, G. (1985), *Cacao*. Longman.
Wood, G. A. R., and Lass, R. A. (2001), *Cacao*, 4th ed. Blackwell Science.
Yamashita, H., Sonobe, R., Hirono, Y., Morita, A., and Ikka, T. (2020), "Dissection of hyperspectral reflectance to estimate nitrogen and chlorophyll contents in tea leaves based on machine learning algorithms", *Scientific Reports*, 10(1), 1–11. https://doi.org/10.1038/s41598-020-73745-2
Yan, M., Pan, G., Lavallee, J., and Conant, R. (2020), "Rethinking sources of nitrogen to cereal crops", *Global Change Biology*, 26(1), 191–199. https://doi.org/10.1111/gcb.14908
Yang, Y., Yu, X., Song, L., and An, C. (2011), "ABI4 activates DGAT1 expression in Arabidopsis seedlings during nitrogen deficiency", *Plant Physiology*, 156(2), 873–883. https://doi.org/10.1104/pp.111.175950
Young, A. (1994), *The Chocolate Tree: A Natural History of Cacao*, 1st ed. Smithsonian Institution Press, pp. 200.
Zarrillo, S., Gaikwad, N., Lanaud, C., Powis, T., Viot, C., Lesur, I., … and Valdez, F. (2018), "The use and domestication of *Theobroma cacao* during the mid-Holocene in the upper Amazon", *Nature Ecology & Evolution*, 2(12), 1879–1888.
Zhang, D., and Motilal, L. (2016), "Origin, dispersal, and current global distribution of cacao genetic diversity", In: Bailey, B. and Meinhardt, L. (eds), *Cacao Diseases*. Springer, pp. 29. https://doi.org/10.1007/978-3-319-24789-2_1
Zhang, Y., Wang, W., & Yao, H. (2023), "Urea-based nitrogen fertilization in agriculture: a key source of N_2O emissions and recent development in mitigating strategies", *Archives of Agronomy and Soil Science*, 69(5), 663–678. https://doi.org/10.1080/03650340.2022.2025588
Zuidema, P., Leffelaar, P., Gerritsma, W., Mommer, L., and Anten, N. (2005), "A physiological production model for cacao (*Theobroma cacao*): model presentation, validation and application", *Agricultural Systems*, 2(84), 195–225. https://doi.org/10.1016/j.agsy.2004.06.015

4 Social benefits and human development implications of organic cacao producers in Calceta – Ecuador

Marlon Arturo Cedeño Álava, David Giband, and Luz C. García Cruzatty

4.1 Introduction

The domestication of cacao, a native species in Ecuador, dates back more than 5,000 years ago in Palanda town, over the Ecuadorian Amazon [1]. It has global socioeconomic importance since its cultivation involves between 5 and 6 million farmers in many countries, who produce 4.7 million tons of cacao annually. Ecuador is the third largest producer in the world, after Ivory Coast and Ghana, producing 0.32 million tons in 2022 [2]. According to the Ministry of Agriculture and Livestock [3], in Ecuador there are 626,962 hectares of crops, with a production of 330,872 tons and an average yield of 0.61 tons of seeds per hectare ($t\ ha^{-1}$), with the main producing provinces being Los Ríos (21.85%), Manabí (20.74%), Guayas (18.41%), and Esmeraldas (17.17%).

Sustainable agriculture has become the main reason for agricultural policies in most countries. This concept requires, in particular, designing new agricultural practices that respect the environment and analyzing the study of agricultural systems with a systemic approach that integrates agronomic, environmental, economic, and social dimensions. While sustainable agriculture techniques (e.g., soil conservation techniques) have an older tradition and are becoming more widely known, concerns about the social dimensions also seek to integrate. Thus, the concepts of sustainable agriculture remain much more recent. Therefore, they deserve to be discussed and debated.

The particularities of this context seem to be mainly due to the fact that the social issues of sustainable agriculture give the impression of creating a division between, on one hand, those who pursue the integration of the environmental dimension in agriculture into a socioliberal economy and, on the other hand, those who oppose the integration of these social issues into the economic model.

The political acceptance in terms of the concept of sustainable agriculture is still recent. This concept therefore leaves room for many definitions, often contextual and sometimes contradictory. In these circumstances, it is important to deepen the concept, while agricultural policies refer more frequently to this category. After briefly describing the context in which the concept of

DOI: 10.4324/9781003381761-6

sustainable agriculture emerged in the world, we discuss researchers who focus more on economics and more critically those who integrate social aspects.

The current research seeks to evaluate the social benefits that sustainable agriculture offers to farmers and communities. More precisely, this investigation holds the statement that social implications must be integrated with the economic benefits that this production approach instills. Therefore, we will be led to better identify the main social problems raised by the concept of sustainable agriculture through an analysis based on the Human Development Index (HDI) issued by the United Nations.

This investigation is located in the town of Calceta, in the province of Manabí, Ecuador. The farmers settled in this area are especially dedicated to the cultivation of agricultural species, among which cacao is the most important because it is the basis of their agroforestry systems, where they cultivate in association woodland ~~forest~~ and fruit species such as orange, banana, mango, papaya, etc. It should be stated that these cultivation systems are not managed adequately due to the scarcity of economic resources to invest in irrigation, organic fertilizers, and agrochemicals. Most of these producers have small farms of 1 to 5 hectares and live in old houses built with local materials such as cane, wood, and toquilla (Figure 4.1).

In this rural area, the living conditions are precarious, since there is no wastewater disposal system, educational centers, and hospitals; and food supply markets are not accessible to all producers. A situation that should be addressed through appropriate public policies, which is why it is important to obtain updated information that allows us to know the agro-socioeconomic and environmental reality of different rural areas.

Figure 4.1 Living conditions of cacao farmers in Calceta.

4.2 Theoretical framework

The role organic agriculture is playing around the world goes further beyond the classic key goals pursued. The European Common Agricultural Policy focuses on income, supply, and quality putting aside the negative environmental impacts of agriculture but the last aspect is increasingly taking relevance in this scenario, clearly stating that without it, the key initial objectives will not be truthfully fulfilled [4].

The social revitalization that organic agriculture is promoting in rural areas is significant for farmers but still remains a loose end for governments as they have to see the pressure it generates as a priority issue to be addressed. The data about social benefits is in some cases not conclusive but it also confirms that with more sustainable production systems the economic benefits and the social interaction adopt a positive position; in most situations, for the need to hire more labor, higher demand for domestic products, and greater civic participation [5].

The transformation of society from an agrarian to an urban industrial economy makes the decline of rural areas in the world seem inevitable; that is why resilient strategies must be applied so people living in these zones can adapt to and survive in these changes. 'Three conditions for sustainable rural development in the knowledge economy are discussed: 1) development of new economic activities that can respond to potential urban demand; 2) local entrepreneurship that can establish and expand these new activities; and 3) social capital that can support the entrepreneurship in new activities with access to credits, labor, human capital, external markets and external knowledge for learning and innovation' [6].

4.2.1 Environmental performance

The negative externality caused by agricultural production has been a debatable topic that is always on the table. Waibel and Fleischer (1998) established a cost of negative effects for Germany of 125 million Euro per year corresponding to the main effects of the use of pesticides on human health, the quality of water, the chemical residuals found in food, and the damage caused to flora and fauna.

Pretty *et al.,* (2000) also found the environmental cost of the United Kingdom was even higher than previously estimated, reaching 190 million euros per year. Once again, the presence of pesticides justifies the shift from conventional to organic production. All these findings make the positive environmental effects, the main reason to look for an ecological system of production. This was also confirmed by a more extensive study where organic farming has less harmful effects than conventional farming on a per hectare basis [9]. However, not all researchers favor the idea of organic production as they will place the quantity of output over the marginal environmental effects.

The use of agrochemicals can affect human health in various ways. For example, direct exposure or contact with these chemicals can cause health problems, such as skin, eye, and respiratory irritation, nausea, dizziness, and even more serious illnesses if exposure is prolonged or in large quantities. The use of chemicals is widely documented in cacao farming. It has been confirmed that farmers need adequate practical training and support on their use to reduce associated risks, protect the environment, and ensure sustainable cacao production.

In Ghana, the second largest cacao-producing country in the world, a study revealed that approximately 70% of the insecticides used by farmers are approved by the Ghana Cacao Board, especially using neonicotinoids and pyrethroids. Although these farmers claimed to have adequate knowledge about pesticides, most showed inadequate storage, application, and disposal of pesticides. These inadequate practices and the inevitable exposure to chemicals would be related to the health problems experienced by farmers [10].

Cacao production is increasing and entails a substantial environmental cost due to changes in production systems, which is why it is important to move toward sustainable production systems, thus improving the quantification of environmental impacts. As an international methodology, it has been recommended to introduce a systematic tool to model and evaluate the environmental burdens associated with cacao throughout its life cycle [11].

4.2.2 *Food quality*

After the pandemic in 2020, the world started to think seriously about the quality of food consumed, and the perception of organic products by clients sharply rose in the main markets of the world. Woese *et al.*, (1995) found no evidence of the quality of food organically produced, but this is mainly due to the lack of comparative research between conventional and organic farming. However, the risk of the contamination of food when using pesticides and nitrites still prevails.

Agrochemical residues can remain in the food we consume. Although there are regulations to control allowable residue levels, constant exposure to small amounts of these chemicals through the diet can pose long-term health risks, such as developmental problems, cancer, or other chronic effects.

Specifically in cacao, it has been documented that the accumulation of Cadmium (Cd) threatens global food security, human health, and the future of chocolate shops. The increase in the bioavailability of Cd is a problem fostered by poverty, incorrect agronomic management, and the lack of knowledge about the propensity of cacao to assimilate Cd. The progressive accumulation of Cd, with a half-life of 10 to 30 years, in the human body, even at ultra-trace levels, can cause serious health complications. If Cd accumulates in the food chain through cacao products, the consequences on children, who are the main consumers of chocolate, include morbidity and mortality that could result in a significant demographic transition by 2050 [13].

4.2.3 *Farm income*

Without any doubt, the main concern of any agricultural policy is to make sure that the families of farmers generate sufficient income that allows them to keep on farming. Offermann and Nieberg (2000) found that the profits obtained by organic farmers were similar to the conventional ones. Despite the difference in output, the high price premia equilibrate this income difference but this finding shows equality in profits if the organic farm is seen on a diary basis and not as a per hectare production area.

This strong price position makes it possible for organic farming to obtain higher profit contributions than arable farms in several European countries making it possible to compete and be attractive for conventional farmers. Statistic time series studies often show similar results between organic and conventional farms concerning margins of profits. These findings may suggest that organic farms will have the same issues as conventional ones in adapting to changing external conditions, farm growth, and rationalization [15].

4.2.4 Rural development

Maybe this is the key objective for every government in the world; create economic development, increase income for households, and decrease the rates of unemployment are the main policies to generate common well-being and increase wealth for nations. It has been often said that the bigger use of the labor force and also the nonuse of chemicals and pesticides in organic farming are its main advantages and strengths but compared with the size of this subsector the implications tend to be minimized [16].

Finally, there are two main aspects that foresee a better future for organic farming. First, it is the marketing and processing of its products that are developed mainly as a small business but totally and legally independent giving the opportunity to families to obtain extra income and in some cases, those families can work cooperatively as organizations creating employment for locals as it is in the case of Fortaleza del Valle.

Second, with organic farming, the image of rural areas changes positively, not only because of its benefits to the environment but also because of the finances of young people who will choose to persevere in those regions instead of migrating to urban areas looking for better opportunities which in most of the cases will not happen [17].

4.2.5 Social implications of cacao farming

Cacao production, especially when done through organic methods, can generate a series of significant social benefits. These benefits include improvements in the living conditions of local communities, the generation of employment, the strengthening of local institutions, and the promotion of gender equality. Furthermore, organic agriculture tends to encourage sustainable practices that can preserve the natural environment, offering healthier and more nutritious products to local communities and the global market. These social benefits extend to improving the quality of life of farmers, promoting the health and well-being of communities, and contributing to the economic development of cacao-producing regions.

Cacao supply companies are increasingly committed to improving the sustainability of the crop, which is why more and more research is being carried out that addresses the multiple dimensions of sustainability of cacao production or the interrelationships between these dimensions. In this sense, a study was carried out in Ecuador and Uganda using a sample of 395 cacao farms to explore the sustainability of the farms and identify the social and economic benefits. The authors identified important sustainability issues related to the social and governance dimensions, which, compared to the environmental and economic dimensions of sustainability, receive much less attention from public institutions and private companies.

However, important social and economic benefits of sustainable cacao production were also identified, consisting of benefits for human health derived from the reduction of the use of pesticides and the conservation of resources for the long-term productivity of the farms, although at the expense of investments and their profitability [18].

Although some research has shown that trade-related interventions have a positive impact on quality of life by guaranteeing higher and more stable incomes, most research does not compare these incomes with the cost of living and/or the poverty line. So, contributions to Sustainable Development Goals and poverty reduction are unknown. Incorporating such comparisons would allow us to evaluate whether the cacao trade can improve the living conditions of farmers in terms of covering all living costs and supporting poverty reduction [19].

In this sense, Van Vliet *et al.,* (2021) showed that around half of cacao producers in Ivory Coast and Ghana earn a gross income that is below the World Bank extreme poverty line and the majority of them do not earn a sustainable income, indicating that the cacao trade alone may not be able to support poverty reduction and economic development. Additional research is needed to understand what factors influence the ability of households to appropriate the economic benefits associated with cacao production; for example, by identifying resource availability and opportunities for alternative income generation.

In Ghana, the second largest cacao-producing country in the world, studies have been carried out to compare organic and conventional cacao production systems, showing that there is higher environmental performance in organic cacao growing systems in terms of water extraction (+29%), species diversity (+26%), soil degradation (+24%), genetic diversity (+24%), and greenhouse gases (+22%). In addition, the better performance of organic farming systems was proven compared to conventional ones in terms of profitability (+20%) due to premium market prices, gender equity (+27%), and commitment to sustainability issues (+25%) [21].

In Ecuador, the third country in terms of cacao production, Cedeño *et al.,* (2021) documented greater economic sustainability of organic cacao production systems compared to conventional systems. This greater profitability of organic agriculture is due to the lower requirement for labor and a greater market appreciation for organic products. The greater profitability of organic agriculture and the use of environmentally friendly inputs in the production process make farms competitive.

García-Briones *et al.,* (2021) analyzed the characteristics of the cacao production and marketing chain and stated that the plans for the competitive improvement of Ecuadorian cacao have not considered the farmers' production environment. Analyzing the factors involved in the cacao chain, they conclude that the resilience of Ecuadorian cacao production depends on the articulation of five actions based on: (1) improving the rural environment of the producer, (2) promoting associativity and cooperativism, (3) improve the productivity and consumption of the national market, (4) consolidate the international market, and (5) research aimed at improvement.

In Indonesia, Leksono *et al.,* (2021) conducted an analysis of farmers' perceptions of the limitations of the organic farming system and the development of strategies for organic cacao cultivation based on local capacity. The results indicated several limitations in cacao cultivation, including pest control, low productivity, and marketing. Through an analysis of strengths, weaknesses, opportunities, and threats, the authors recommend: 1) expanding land used for organic agriculture to produce large quantities of products for export; 2) strengthening farmers' institutions to collaborate with industry and research institutions; and 3) improving the implementation of the organic cacao cultivation system in the countryside to satisfy the local community demand for healthy organic products.

Although the social benefits of cacao and organic production are known, in Ivory Coast, a country that in recent years has been the leading cacao producer in the world, several organizations through NGOs carry out actions to improve the empowerment of women and the education and professional training of young people, with the aim of improving the living conditions of cacao-producing populations.

To analyze the results of these actions, Hugues (2020) conducted 2,291 interviews to evaluate the indicators associated with the living conditions of cacao farmers through education, women empowerment, and professional training. The main findings were the following: 1) Living conditions in the communities similar to the national situation. The poverty rate is 42%, with a higher incidence among large households; 2) Children's French (in our case Spanish) and math skills are average and adult women's skills are low; 3) There is child labor. In these communities, three out of every five children dedicate themselves completely to their studies, while one out of every six works; 4) Women are involved in production activities, especially in the cultivation of cacao, food crops and other income-generating activities. However, the empowerment of women is not relevant.

It is possible that the situation in other countries is similar; so, it is important to carry out surveys and studies and document the reality of farmers, essential information for local authorities and governments in power to generate public policies to improve the economy of peasants and therefore the quality of life of cacao producers.

4.3 Methodology

This research delimitated the study area, covering the parishes of Valle del Carrizal, La Fortaleza, Quiroga, and Rio Grande from Canuto, which belong mainly to the districts of Bolívar. It also covers Tosagua, Chone, Junín, and Portoviejo for further data collection through the use of questionnaires and interviews made with the local farmers. The population of the whole region covers around 1200 farmers. The data was processed using the HDI but adjusted to the sample of the population of the region.

The HDI is a composite index calculated each year by the United Nations Development Programme (UNDP) to assess the level of development of countries based not on strictly economic data, but on the quality of life of their nationals. The HDI incorporates three factors:

Life expectancy at birth, because it is significant for people's future living conditions (food, housing, drinking water, etc.) and their access to medicines; the level of education, which determines the professional and social autonomy of the individual; gross national income (GNI) per capita, which reveals people's standard of living and, therefore, their access to culture, goods and services, transportation, and so forth (Figure 4.2).

The HDI comes in the form of a number between zero and one, the latter number symbolizing the highest level. This indicator, created in 1990, is now preferred to per capita income, which today seems too simplistic to assess the level of development. The importance given to the HDI is based on the idea that the freedom of men and women depends on human development. As shown in Table 4.1 there are countries with the best standards of living around the world [26].

4.4 Current situation

4.4.1 Ecuador compared with South America

The situation for Ecuador over the years has varied in so many ways, some positive and some negative, but these phenomena are mainly caused by external shocks converging directly in political crises and socioeconomic

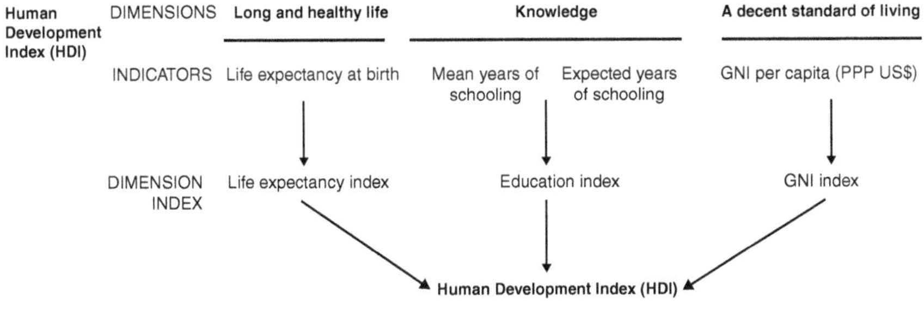

Figure 4.2 Human Development Index composite.

Table 4.1 Top five Human Development Index rankings

Rank	*Country*	*HDI Score*
1	Switzerland	0.962
2	Norway	0.961
3	Iceland	0.959
4	Hong Kong	0.952
5	Australia	0.951

Source: Human Development Index reports: https://hdr.undp.org/

development policies. Ecuador appears in seventh place ahead of countries like Venezuela, Bolivia, and Guyana where people live in extreme poverty or under the poverty threshold which gives an idea of the not-so-good living conditions for Ecuadorians in particular those who work in the countryside. (Table 4.2)

4.4.2 HDI trends in Ecuador

Despite Ecuador being in the 95th place in the HDI, the situation is a bit different from what it looks like; since the 90s, the country belongs to the medium class in the HDI context which includes developing countries. For 15 years, it was immersed in a range between 0.65 and 0.7 showing an upward tendency, despite the economic crisis in 1998 when the financial bubble blew up and people were forced to leave the country (Figure 4.3).

It was during the next ten years when this uprising trend consolidated overcoming the barrier of medium nations and placing Ecuador among the countries with a high HDI reaching a peak of 0.765 [26]. Some economic analysts

Table 4.2 Ecuador's Human Development Index compared with South American countries

Rank	*Country*	*HDI Value*	*Change from 2020*
47	Argentina	0.842	0.002
87	Brazil	0.754	-0.004
118	Bolivia (Plurinational State of)	0.692	-0.002
88	Colombia	0.752	-0.004
42	Chile	0.855	0.003
95	**Ecuador**	**0.740**	**0.009**
108	Guyana	0.714	-0.007
105	Paraguay	0.717	-0.013
84	Peru	0.762	0.000
58	Uruguay	0.809	-0.012
120	Venezuela	0.691	-0.004

Source: Human Development Index reports: https://hdr.undp.org/

Note: Data presented here were used in the preparation of the 2021/2022 Human Development Report 'Uncertain Times, Unsettled Lives: Shaping our Future in a Transforming World', released on 8 September 2022.

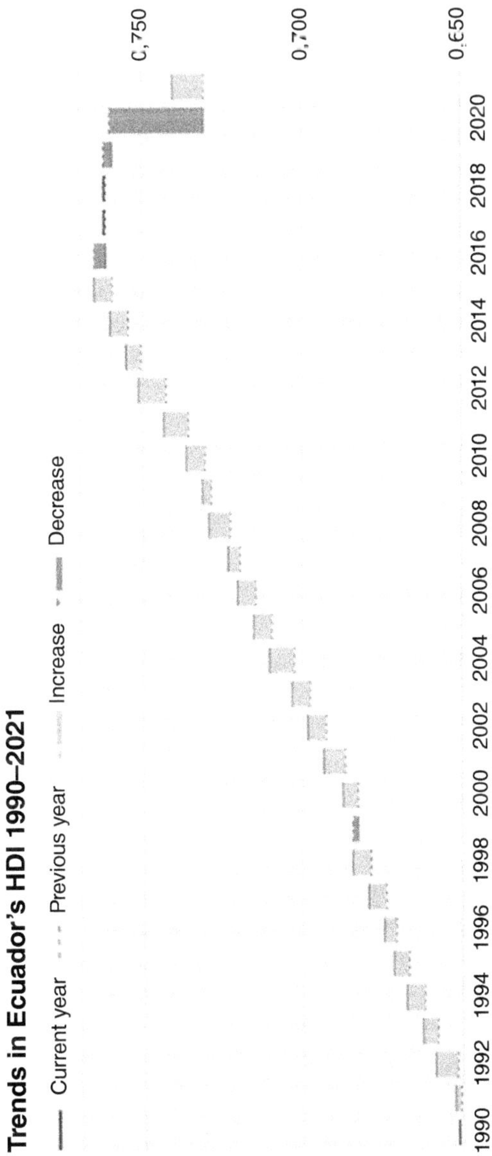

Figure 4.3 Human Development Index trend in Ecuador since the 90s.

explain this singularity to the development policies promoted by President Correa during his ten years in command but some others think that the fiscal policy expansion led to an increase in GDP which affected directly the GNI per capita, an essential part of the composite of HDI.

Today's rank of Ecuador still places it as a developed nation with a 0.74 HDI value. The downward trend shown in the last years has its explanation in the oil price crisis suffered by producer countries but what is really important to observe is that despite the HDI, the standard living conditions of farmers are not what the index says which implies the limitations of it that will be analyzed forward.

4.5 Results

After processing the data collected, the first step was to examine the long and healthy life dimension with its indicator being the life expectancy at birth that will result in the life expectancy index (LEI). To better understand this index, it is necessary to know how it is computed; there are two fixed values—the minimum and maximum expectancy life in years—the former is 20 and the latter 85. The life expectancy value in Ecuador is obtained from the available data found at INEC[1] and the current one is 73.7. As a result, the LEI is 0.826. The following formula is used to obtain the LEI:

$$Life\ Expentancy\ Index(LEI) = \frac{LE - 20}{85 - 20}$$

$$\text{LE} = \text{Life expectancy at birth}$$

The second dimension to go through is education with its indicators, expected years of schooling index (EYSI)and mean years of schooling index (MYSI). There are no minimum values in this section, just maximums, 18 and 15 that correspond to the previously mentioned indicators. The current values of the indicators are found in the same government source that carries out domestic census every ten years approximately depending on the situation of the country. The formula to compute the index is a bit more complex than the previous one, and the present Education Index of Ecuador is 0.703:

$$Education\ Index(EI) = \frac{MYSI - EYSI}{2}$$

MYSI:

$$Mean\ Years\ of\ Schooling\ Index = \frac{MYS(Mean\ Years\ of\ Schooling)}{15}$$

EYSI:

$$Expected\ Years\ of\ Schooling\ Index = \frac{MYS(Mean\ Years\ of\ Schooling)}{18}$$

Finally, there is the decent standard of living dimension with its indicators, the GNI at Purchasing Power Parity (PPP) per capita. This is a common metric used to measure the income of people. It also has a roof and a bottom, the former 75000 which represents 1 and the latter 100 which is 0. The Income Index in Ecuador is 0.70 and the formula to obtain it is as follows:

$$Income\ Index(II) = \frac{ln\,ln(GNIpc) - ln\,ln(100)}{ln\,ln(75000) - ln\,ln(100)}$$

The HDI is equal to $^3\sqrt{LEI * EI * II}$ giving a final result of 0.740. The following tables will present the current dimensions and values of Ecuador's HDI and the indicators for every dimension (Tables 4.3 and 4.4)

With all this data we proceed to the next step which is to adjust the methodology to the population (1200 farmers) of Fortaleza del Valle. It can be seen that the results are not similar at all; it is not even close and what is even worse is the index is almost lower than 0.55 which will place the region of Fortaleza del Valle in the low human development section if that figure is reached. These results have a relation with the situation seen at the moment of collecting the data.

Table 4.3 Ecuador's Human Development Index dimensions and values

Dimension	*Indicator*	*Value*	*Minimum*	*Maximum*
Health	Life expectancy (in years)	73.7	20	85
Education	Years of schooling expectancy	14.6	0	18
	Years of schooling mean	8.8	0	15
Income	GDP per capita (2021 PPP $)	10.132	100	75000

Source: Human Development Index reports: https://hdr.undp.org/

Table 4.4 Ecuador's Human Development Index indicators

Life expectancy index	**0.83**
Education index	**0.70**
Expected years of schooling	0.81
Mean yarns of schooling	0.59
Income index	**0.70**
HDI	**0.740**

Table 4.5 Fortaleza del Valle's Human Development Index dimensions and values

Dimension	*Indicator*	*Value*	*Minimum*	*Maximum*
Health	Life expectancy (in years)	59.4	20	85
Education	Years of schooling expectancy	10.9	0	18
	Years of schooling mean	6	0	15
Income	GDP per capita (2021 PPP $)	4800	100	75000

Table 4.6 Fortaleza del Valle's Human Development Index indicators

Human Development Index Indicators	
Life expectancy index	**0.,61**
Education index	**0.50**
Expected years of schooling	0.60
Mean yarns of schooling	0.40
Income index	**0.59**
HDI	**0.563**

Source: Cedeño (2023)

The situation most of the farmers live in is not the most suitable according to decent standards. This is shown in Figure 4.1 where the houses are in bad condition (Tables 4.5 and 4.6).

4.6 Conclusions

Due to the various definitions it generates, the concept of sustainable agriculture can be seen as an effort to integrate the environmental dimension into the activity that constitutes agriculture in the national economy, as a questioning of the structures of the complex food industry marked by a market economy whose externalities it generates contribute to the crisis of agriculture, in particular to the social crisis in general. In this last perspective, ecology does not offer magical solutions to difficult problems. Agriculture based more on ecological principles could increase the efficiency of production and decrease the pressure on the environment, but the use of resources will always set limits to sustainable production. Difficult decisions about population control, energy conservation, land use, and other dimensions of a society will need to address sustainability (Oison, 1992). For this author, ecology can help guide these decisions, but it will not provide the means to escape from it. This domain belongs to politics and civil society. Therefore, the concept of sustainable agriculture leaves room for interpretation and alternative social visions. What are they? What do they imply as changes? Structural? What social change strategies do you prefer?

Note

1 Instituto Nacional de Estadísticas y Censos.

References

[1] S. Zarrillo, N. Gaikwad, C. Lanaud, T. Powis, C. Viot, I. Lesur, O. Fouet, X. Argout, E. Guichoux, F. Salin, R. Solorzano, O. Bouchez, H. Vignes, P. Severts, J. Hurtado, A. Yepez, L. Grivetti, M. Blake and F. Valdez, "The use and domestication of *Theobroma cacao* during the mid-Holocene in the upper Amazon," *Nature Ecology and Evolution*, vol. 2, no. 12, pp. 1879–1888, 2018.

[2] L. C. García, J. B. García and F. Sánchez, "Variability and combinatorial ability in fruit characters of new *Theobroma cacao* hybrids in Ecuador," *Revista Brasileira de Ciencias Agrarias*, vol. 18, no. 3, p. e3265, 2023.

[3] M. "Ministerio de Agricultura y Ganadería," Informe de Rendimientos Objetivos de cacao, 2021. [Online]. Available: http://sipa.agricultura.gob.ec/index.php/informe-de-rendimientos-objetivos/rendimiento-de-cacao. [Accessed: 15 March 2024].

[4] A. Häring, S. Dabbert, F. Offermann and H. Nieberg, "Benefits of organic farming for society," in *European Conference – Organic Food and Farming*, Copenhagen, Denmark, 2001.

[5] R. J. MacRae, B. Frick and R. C. Martin, n.d."Economic and social impacts of organic production," *Canadian Journal of Plant Science*, 2088, vol. 87, no 5, pp. 1037–1044.

[6] Y. Lia, H. Westlundb and Y. Liua, "Why some rural areas decline while some others not: An overview of rural," *Journal of Rural Studies*, 2019, vol. 68, pp. 135–143.

[7] H. Waibel and G. Fleischer, Kosten und Nutzen des chemischen Pflanzenschutzes in der deutschen Landwirtschaft aus gesamtwirtschaftlicher Sicht, Wissenschaftsverlag Vauk KG, 1998.

[8] J. N. Pretty, C. Brett, D. Gee, R. E. Hine, C. F. Mason, J. I. Morison, H. Raven and M. D. Rayment, "An assessment of the total external costs of UK agriculture," *Agricultural Systems*, no. 65 no 2, pp. 113–136, 2000.

[9] M. Stolze, A. Piorr, A. Häring and S. Dabbert, "The environmental impacts of organic farming in Europe," in *Organic Farming in Europe: Economics and Policy*, e. by Dr Stephan Dabbert, Dr Nicolas Lampkin, Dr Johannes Michelsen, Dr Hiltrud Nieberg, Dr Raffaele Zanoli. Stuttgart: University of Hohenheim, 2000.

[10] K. O. Boateng, D. Enock, K. Isaac, K. A. Godfred, A. Emmanuel and D. Godfred, "Knowledge, perception, and pesticide application practices among smallholder cacao farmers in four Ghanaian cacao-growing regions," *Toxicology Reports*, vol. 10, pp. 46–55, 2023.

[11] H. Hosseinzadeh-Bandbafha and M. Kiehbadroudinezhad, "Environmental impacts of chocolate production and consumption," in Galanakis, C. M. (ed.), *Trends in Sustainable Chocolate Production*, pp. 229–258, Springer, Lisbon, 2022.

[12] K. Woese, D. Lang, C. Boess and K. W. Bögl, "Ökologisch und konventionell erzeugte Lebensmittel im Vergleich – Eine Literaturstudie. BgVV (Bundesministerium für gesundheitlichen Verbraucherschutz und Vetrinärmedizin)," in *Hefte, 4*, Dahlem, 1995.

[13] N. R. Maddela, K. Dhatri, L. C. García, S. Chakraborty, K. Venkateswarlu and M. Megharaj, "Cacao-laden cadmium threatens human health and cacao economy: A critical view," *Science of the Total Environment*, vol. 720, no. 137645, 2020.

[14] F. Offermann and H. Nieberg, "Economic performance of organic farms in Europe," in *Organic Farming in Europe: Economics and Policy*, University of Hohenheim, Stuttgart, 2000.

[15] A. Häring, "Impact assessment of different policy options on organic farming: Farm level case studies," in *EU-Project FAIR3-CT96-1794: Organic Farming and the CAP*, Technical Deliverable Rix, 2001.

[16] S. Dabbert and R. Zanoli, "Blueprint for organic farming policy in Europe. EU-Project FAIR3-CT96-1794," in *Organic Farming and the CAP*, Draft of Technical Deliverable Rxiv, 30.05.2001, 2001.

[17] B. Tress, "Landwirt schafft Landschaft – Umstellungspotential und landschaftliche Konsequenzen der ökologischen Landwirtschaft in Dänemark," in *Forschungsrapport 121: Roskilde*, Universität Roskilde, 2000.
[18] L. Tennhardt, G. Lazzarini, R. Weisshaid and C. Schader, "Do environmentally-friendly cacao farms yield social and economic co-benefits?," *Ecological Economics*, vol. 197, article 107428, 2022.
[19] I. Dreoni, M. Schaafsma and Z. Matthews, *The Social Impacts of cacao Production: A Systematic Review*, UKRI GCRF TRADE Hub, 2021.
[20] J. A. Van Vliet, M. A. Slingerland and Y. R. Waarts, "A living income for cacao producers in Côte d'Ivoire and Ghana?," *Frontiers in Sustainable Food Systems*, vol. 5, pp. 732831, 2021.
[21] J. Bandanaa, I. K. Asante, I. S. Egyir, C. Schader, T. Y. Annang, J. Blockeel, I. Kadzere and A. Heidenreich, "Sustainability performance of organic and conventional cacao farming systems in Atwima Mponua District of Ghana," *Environmental and Sustainability Indicators*, vol. 11, no. 100121, 2021.
[22] M. Cedeño, D. Giband and L. C. García, "Sustainability and rural development through the production of organic cacao in Calceta – Ecuador: Economic analysis and assessment of the implications on the living standards of farmers," in *Proceedings of the 1st International Conference on Water Energy Food and Sustainability (ICoWEFS 2021)*, Portugal, Springer, 2021.
[23] A. R. García-Briones, B. F. Pico-Pico and R. Jaimez, "La cadena de producción del Cacao en Ecuador: Resiliencia en los diferentes acto-res de la producción," *Revista Digital Novasinergia*, vol. 4, no. 2, pp. 152–172, 2021.
[24] A. S. Leksono, I. Mustafa and Z. P. Gama, "Organic cacao farming in Indonesia: Constraints and development strategies," *Organic Agriculture*, vol. 11, pp. 445–455, 2021.
[25] K. K. Hugues, "Sustainability in the cacao production chain in Côte d'Ivoire: Schooling, women empowerment and household living condition," *Open Journal of Social Sciences*, vol. 8, pp. 191–208, 2020.
[26] Human Development Index (HDI), n.d."Human Development Index (HDI)," United Nations Development Programme, 2023. [Online]. Available: https://hdr.undp.org/data-center/human-development-index#/indicies/HDI [accessed 25 07 2023].

5 Analyzing cacao agroforestry complexity in Manabí, a biodiverse coastal province of Ecuador

Romina Fossati, Anne-Gaël Bilhaut, Ezequiel Zamora-Ledezma, Andrés González-González, Armelle Mazé, and Juan Fernández-Manjarrés

5.1 Introduction

One of the main problems with the production of tropical commodities such as coffee, avocados, oil palm, and cacao, among others, is that they are located in biodiversity-rich countries with high rates of habitat destruction (Bager et al., 2020). In addition, there are different types of production of these commodities, ranging from intensive single-species plantations, which are poor harbors of uncultivated biodiversity, to biodiverse and mixed agroforestry systems, which are often used as buffer zones around protected areas. As part of the complexity of these agroforestry systems, there is a multitude of farmers, associations, and communities striving for a way of life in which agroforestry provides not only income but also food security to their households.

This study focuses on the coastal province of Manabí, Ecuador (South America, Figure 5.1). Our choice of study area is based on the premise that the processes described here are common to many tropical areas of high biodiversity where agroforestry is practiced. Indeed, the province of Manabí in Ecuador is part of the Tumbes–Choco–Magdalena biodiversity hotspot, which contains a high number of species and endemism due to the concentration of a wide variety of ecosystems in a relatively small area (Sierra & Al., 1999).

Furthermore, Manabí is one of the most important cacao-producing provinces in Ecuador due to its favorable natural conditions in terms of soil and climate (Sanchez, 2019). Due to its location in the coastal region of the country, the province of Manabí is particularly affected by the climatic variability caused by the *El Niño* and *La Niña* phenomena.

In this chapter, we first define agroforestry and why it is important in the fight against climate change. We then describe in detail the biodiversity of the area and the trends in habitat destruction over the last two decades. To understand the social aspects of agroforestry systems, we present an analysis of the actors involved in research and development in the area, and finally, we conclude on how different approaches to agroforestry can affect the short- and long-term sustainability of these systems.

DOI: 10.4324/9781003381761-7

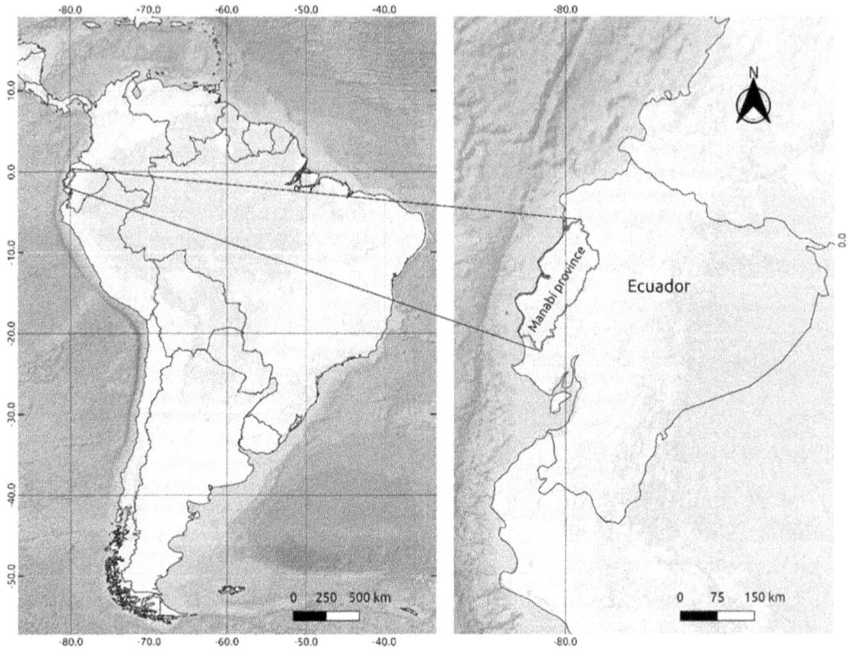

Figure 5.1 Position of Ecuador in South America and of the province of Manabí.

5.1.1 What is agroforestry?

Agroforestry can be defined as a dynamic ecological land-use system in which woody perennials (trees, shrubs, palms, bamboo, etc.) share the same space with agricultural crops and/or animals, with the spatial and temporal interactions that result (FAO, 2022; Ospina-Ante, 2009). Woody perennials can take tree- and shrub-like forms, but they also include grasses, palms, and ferns. For the World Agroforestry Centre (ICRAF), an agricultural practice to be identified as agroforestry must meet two requirements: woody perennials must be deliberately grown on the same unit of land as agricultural crops and there must be significant interactions between the woody and non-woody components of the system (Ollinaho & Kröger, 2021). As a land-use management system, it does not only allow diversification and sustainable production, but it is also structurally and functionally more complex than monocultures (van Noordwijk et al., 2019).

Agroforestry is well-known for its crucial role in biodiversity conservation and has many benefits: it plays a prominent role in providing habitats for different species, helps preserve their germplasm, helps reduce the conversion of natural habitats into agricultural landscapes, and creates corridors between habitat remnants and key areas for flora and fauna (Haggar et al., 2019). In addition, the diversity of species contributes to the stability and resilience of agricultural lands (P.R. Nair, 2007).

There are three basic types of agroforestry systems: agrisilviculture (mixing crops and trees), silvopasture (combining pasture/livestock and trees), and agrosilvopasture (including crops, pasture/livestock, and trees) (P. K. R. Nair, 1991). However, this basic typology can be further subdivided as the type and composition of individual agroforestry systems vary across the globe in response to the biophysical, socio-economic, and climatic characteristics of different regions (Kumar et al., 2014).

In the tropics, agroforestry is generally considered to be a subsistence practice (P. R. Nair, 2007), and consequently most fruit-based systems are found here, as fruit trees are an indispensable component of agroforestry systems (Lauri et al., 2020; P. K. R. Nair, 1991). Fruits, nuts, and fodder lead to product diversification, generating stable economic benefits for households (Ortiz Angela Somarriba Eduardo, 2008), so that agroforestry systems can be considered socio-ecological systems (Castañeda-Ccori et al. 2020).

Most commodity crops, such as coffee or cacao, are produced in agroforests commonly associated with tropical and subtropical areas identified as biodiversity hotspots (Robbins et al., 2015). In particular, cacao (*Theobroma cacao* L.) is one of the world's most important commodities, being the third most important agricultural export product worldwide and the second most valuable commercial crop in the tropics. It is estimated that more than 80% is produced by 7–8 million small family farms in about 50 countries worldwide (Díaz-Montenegro et al., 2018).

5.1.2 The role of agroforestry in the fight against climate change

Climate change is one of the most concerning threats to the environment, with negative impacts on human health, natural resources, food security, economy, and others. Its main cause is found in the increase in the concentration of greenhouse gases (GHG) in the atmosphere, especially CO_2. A significant proportion of CO_2 emissions is due to land use change, deforestation, and the use of fossil fuels. Environmentally friendly, agroforestry systems can be part of the solution to climate change, becoming an important tool to face the current scenario (Ortiz Angela Somarriba Eduardo, 2008).

Trees, the most important component of agroforestry, are an outstanding tool. They create microclimates that reduce the ambient temperature and heat stress, as well as conserve soil moisture and produce nitrogen-rich forage. Moreover, they reduce storm runoff and recharge aquifers, and depending on the species and density of planting, they can evacuate groundwater. These properties of trees and woody perennial plants in general, change the risks and impacts of droughts and floods (Rosenstock et al., 2019).

Finally, trees on farms, and more generally in landscapes, help mitigate climate change by storing carbon in biomass and soils thereby reducing atmospheric CO_2 stock (Rosenstock et al., 2019). They can sequester and store between 12 and 288 t C ha^{-1} of carbon, which makes growing cacao in agroforestry systems an attractive option. In addition to these ecological capacities, trees play a crucial social and economic role in agroforestry. They can also produce products such as fruits and fuel, leading to diversification of production and generating stable economic benefits for households (Ortiz Angela Somarriba Eduardo, 2008).

5.2 Cacao agroforestry systems and their importance in the Manabí region

5.2.1 Some generalities of cacao production in Ecuador

Cacao is expanding rapidly in Latin America, driven by both an expanding consumer base and growing external markets (Borda et al., 2021). Its cultivation has doubled in the last 50 years in Ecuador (Castañeda-Ccori et al., 2020), where it is a traditional crop of great economic, social, and environmental importance (INIAP, 2018). It is the country's third export product and accounts for 20% of the total cultivated area.

Two main genotypes of cacao are grown in Ecuador, 'Fino de aroma' (or *Nacional*) and CCN-51 cacao. A half of the cacao cultivated surface corresponds to the *Nacional* type, also called 'fine-aroma cacao'. This variety accounts for 70% of Ecuador's cacao exports, making it the world's leading producer of fine aroma cacao (Díaz-Montenegro et al., 2018), although Ecuador is only the world's fifth largest cacao-producing country. The *Nacional* variety is mainly grown in agroforestry systems and its production is practically organic. It is characterized by its yellow color and unique aroma and taste.

In the last decade, however, the cultivation of the '*Nacional* Cacao Complex 51 (CCN-51)' variety has become more widespread, mainly due to its greater resistance to pests and diseases, as well as its higher yield. Contrary to the *Nacional* genotype, CCN-51 trees are grown in full sun, usually in monocultures with the addition of pesticides and fertilizers. The characteristic color of the fruit is red (Cartagena et al., 2020).

5.2.2 Current extent of cacao

According to the National Institute of Statistics (INEC), Manabí is one of the provinces with the highest agricultural production. Currently, the cacao crop accounts for 61.96% of the province's permanent cultivated area, with the remaining devoted to bananas (22.57%), palm oil (5.47%), and coffee (4.87%) (INEC, 2021). While 55.17% of cacao producers in Manabí use the *National* genotype, the remaining 36.51% rely on the CCN-51 genotype and 8.32% use *National* grafting (Barrera et al., 2019).

Of the 124,139 hectares of cacao cultivated in Manabí in 2021, 108,555 hectares were in monocultures, while 15.584 hectares corresponded to agroforestry systems, associated with other crops such as fruit trees and timber. While the area of cacao cultivated in monocultures is more important than that in agroforestry systems, Manabí is the province with the largest area where cacao is combined with other species in agroforestry systems (INEC, 2021). As we can see in Figure 5.2 **A**, most cacao plantations occur in the tropical dry forest ecosystems, with a portion in the piedmont of moist forests, while Figure 5.2 **B** shows the total area of cacao cultivated in each canton of Manabí. Of the 22 cantons into which Manabí is divided, only 18 cultivate this product, and those with the largest number of plantations are Chone with 40,483 ha, El Carmen with 34,369 ha, Pichincha with 17,824 ha, and Flavio Alfaro with 13,894 ha (SIPA, 2021).

5.2.3 Types of cacao systems in Manabí

Although there is no official documentation on the different types of agroforestry systems in Manabí, fieldwork by rural development experts has provided some clues (C. Marlin, personal communication). As shown in Figure 5.3, four types of agroforestry systems can be distinguished: first, there are traditional *Nacional* cacao plantations that are more than 40 years old, grown alongside fruit and timber trees, with very little pruning management and low yields **(Figure 5.3 A)**. This type of agroforestry is key in the food security and diversity of products that can be marketed for smallholders:

> *…Cacao cultivation here in Manabí continues as a matter of custom, rooted ancestrally. Where there is Nacional cacao, there are 50-60 associated species among trees, shrubs, fruit trees, the fauna of the place… all this makes the producer to be self-sufficient because they have other options such as cassava, banana, orange, tangerine, grapefruit, cherimoya…*
>
> (project manager of national non-governmental organization)

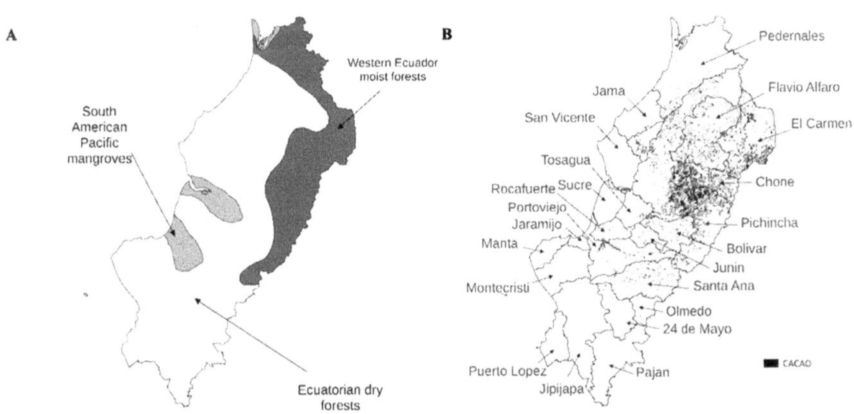

Figure 5.2 Most of the cacao plantations occur in the tropical dry forest ecosystems, with a portion in the piedmont of the moist forests. **A)** Ecoregions of the Manabí province following the classification of Olson (Olson, 2001). **B)** The surface area of cacao plantations for each canton.

Data from: https://datosabiertos.gob.ec ((SIPA, 2021), downloaded 27/03/2023).

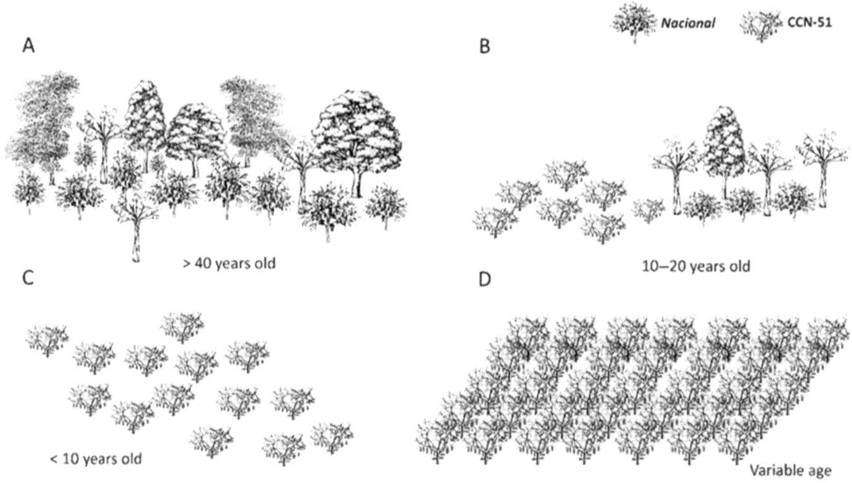

Figure 5.3 Cacao typology. A) Systems of Cacao Nacional of more than 40 years. B) Combinations of Nacional and CCN-51 plantations with other tree species. C) CCN-51 plantations of less than 10 years. D) CCN-51 plantations of more than 20 hectares.

Second, there are younger plantations between 10 and 20 years old, the majority of which are of CCN-51 in combination with the *Nacional* variety. Although there are still some associated timber and fruit trees, the priority in this case is still cacao. The performance is higher and, in terms of management, pruning is carried out, but excluding the use of pesticides or herbicides **(Figure 5.3 B)**. A third category of plantations, which have been growing recently, corresponds to those having between 0- and 10-year-old trees, the majority of which are managed as monocultures of CCN-51 and rely on the use of pesticides and herbicides **(Figure 5.3 C)**. Finally, there is the last category with management similar to the latter, that is monoculture with the use of pesticides and herbicides, but in this case in larger farms of more than 20 hectares and of variable age **(Figure 5.3 D)**.

The cultivation of CCN-51 has been controversial since the beginning. Not only the external markets but local actors also see this highly productive variety that requires no tree shade as a threat to the heritage production of fine-flavor cacao in Ecuador:

> *...CCN-51 is an early, highly productive material that has been adapting to different areas, but it can leave us as a country out of the competition. CCN-51 is grown in most countries that grow cacao with very high yields and less expensive labor. We must continue working on domestic cacao, which is what can make the difference...*
>
> (project manager of national non-governmental organization).

5.3 The ecology and conservation in Manabí

5.3.1 From mangroves to cloud forests

Manabí has one of the highest levels of diversity that can be found in Ecuador and corresponds to the area between the coast and the coastal mountain range, where different dry and humid ecosystems converge such as savannas, dry scrub, deciduous forests, semi-deciduous forests, premontane evergreen forests, montane low-cloud forests, and coastline forests, among others. In this area, up to nine different types of ecosystems have been found on a surface of only 8,000 hectares (Ministerio del Ambiente de Ecuador, 2013).

The province of Manabí encompasses three ecoregions **(see Figure 5.2 A)**, namely the western humid forests, the mangroves, and the dry forests of Ecuador (Olson, 2001). The western humid forests of Ecuador are one of the most threatened ecoregions in the world, as most of the original habitat has already been destroyed. The natural fragmentation of these humid forests led to the rapid evolution of several endemic species during past geologic times, especially bird species, many of which are currently threatened (WWF, 2021).

The mangroves of Manabí are small areas of great biodiversity and ecological importance. However, they have undergone major habitat changes and are critically endangered. Unfortunately, 38% of their total area is in critical condition and at risk due to high fragmentation. As a result, many of the species that inhabit these mangroves are threatened with local extinction (WWF, 2021).

The dry forests of Ecuador are considered to be an area of great biological importance due to the large number of species and the amount of endemism present. In fact, they are one of the last remnants of tropical dry forests, one of the most endangered ecosystems in South America (Portillo-Quintero & Sánchez-Azofeifa, 2010). The climate of this area is linked to the oceanic Humboldt Current and the warm current of El Niño. Surprisingly, this dry forest is surrounded by humid forests that receive up to 2000 mm of rain per year. Precipitations vary from 300 to 1500 mm per year, and in El Niño years rainfall can be 200 times higher than in very dry years (WWF, 2021). The flora and fauna of these forests are characterized by certain physical adaptations that allow the different species to survive extreme conditions of temperature and drought. Although its diversity is much lower than that of humid forests, this ecoregion is recognized for its high endemism (WWF, 2021). In fact, the strong evolutionary and ecological pressures created by this extreme climate have promoted high species richness, including several endemic species, but have also made dry forests more vulnerable to human intervention.

5.3.2 Biological diversity

Ecuador is well known as a megadiverse country. It has been classified as one of the 17 most megadiverse countries in the world. Its location in the Neotropics as well as the presence of the Andes Mountains and the influence of marine currents on its coasts can explain this peculiarity (see the preceding section). However, up to date, there have been no exhaustive studies that accurately describe the biodiversity of the flora and fauna of this highly diverse country (MAE, 2010). Nevertheless, global databases provide the best available description of biodiversity.

According to the Global Biodiversity Information Facility (GBIF) consulted on 16 February, 2021 (GBIF, 2021), the data obtained includes 128,356 spp., which were filtered to eliminate entries having no spatial coordinates, duplicate species, or taxa that did not include an assigned species. The data was compiled in a pivot table in Excel detailing the number of species separated by *Phyllum* and by Class, leaving 4,083 species in Manabí, with *Plantae* being the most abundant *Phylum* with 2,159 species, followed by *Animalia* with 1,848 species (Figure 5.4).

5.3.3 Endangered species, especially forest birds but not exclusively

In terms of endangered species, according to the official database of the International Union for Conservation of Nature (IUCN) (IUCN, 2021), there are more than 60 species in Manabí that are considered to be at some kind of risk. More than 67% of these species are vulnerable, 23% are endangered, and 10% are critically endangered (**Figure 5.5**).

Of the more than 60 species of conservation concern, most, if not all, are associated with forest ecosystems whose populations are threatened by deforestation and habitat fragmentation. Among bird species, we can mention the *Lilacine amazon*, a small parrot native to Ecuador, the purple quail-dove (*Geotrygon purpurata*), the Peruvian tern (*Sternula lorata*), and the grey-cheeked parakeet (*Brotogeris pyrrhoptera)*. In the class of mammals, we find the Ecuadorian capuchin (*Cebus aequatorialis*), the black-headed spider monkey *(Ateles fusciceps*), and the Hammond's rice rat (*Mindomys hammondi)*

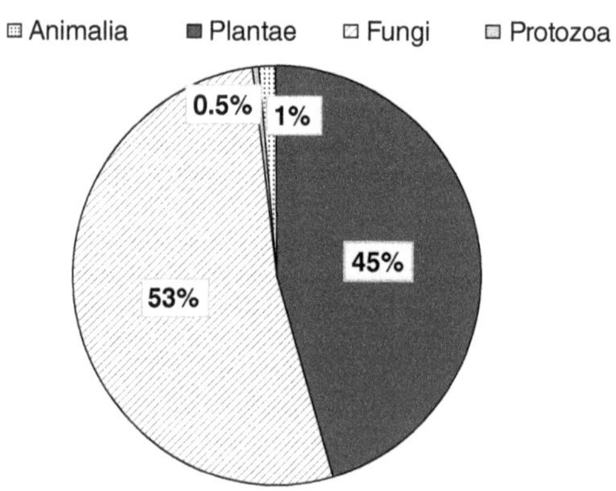

Figure 5.4 Pie chart representing the biodiversity of Manabí according to the Global Biodiversity Information Facility.

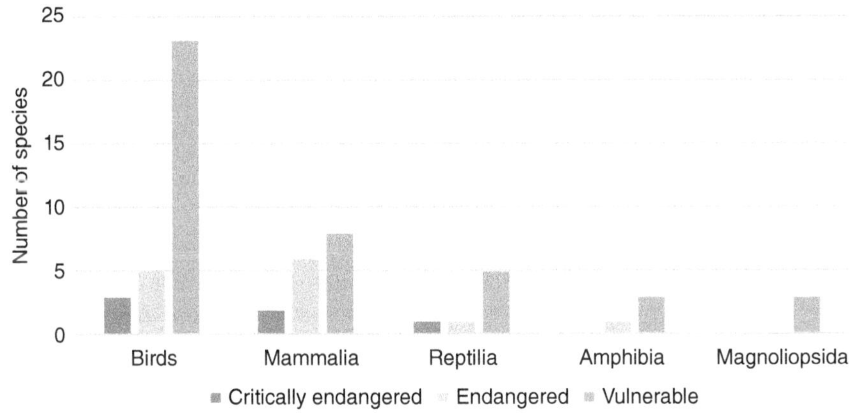

Figure 5.5 Number of endangered species in Manabí by taxonomic class according to the IUCN lists.

known only in Ecuador. In terms of amphibians and endangered reptiles, we find the *Dendropsophus gryllatus* frog also endemic to Ecuador. As for plants (Magnoliopsida class), we can mention two endemic species: *Macleania subsessilis* and *Thibaudia litensis*. Table 5.1 lists the critically endangered and endangered species.

Table 5.1 List of critically endangered and endangered species in Manabí according to the IUCN database

Threat	*Group*	*Scientific Name*
Critically Endangered	Birds	*Amazona lilacina*
		Phoebastria irrorata
		Pterodroma phaeopygia
	Mammals	*Ateles fusciceps* ssp. *fusciceps*
		Cebus aequatorialis
Endangered	Birds	*Brotogeris pyrrhoptera*
		Neomorphus radiolosus
		Pseudastur occidentalis
		Sternula lorata
		Geotrygon purpurata
		Mindomys hammondi
		Zalophus wollebaeki
	Mammals	*Ateles fusciceps*
		Balaenoptera borealis
		Balaenoptera musculus
		Cebus capucinus ssp. *capucinus*
	Amphibians	*Dendropsophus gryllatus*

5.4 Mapping the actors involved in Manabí's cacao agroforestry

5.4.1 The complexity of the actors' network

It is well known that most of the world's cacao (84%) is produced in monocultures and the rest of it is mixed with other species in agroforestry. The product diversification strategy adopted by these farms protects the households from fluctuations in the price of the main product, cacao. However, the low productivity of this product increases poverty and has a growing impact on land and biodiversity (Ravenhill, 2014). Ecuador is not an exception. The living conditions of most cacao smallholders are precarious; 99% of them have less than 10 hectares (Sanchez, 2019) and their farms are rather small and usually require the participation of the whole family.

Agricultural activities in developing countries are fragile as smallholders depend on external support to finance their activities, obtain improved varieties, control pests, and finally, market their crops. In order to obtain a panorama of the socio-ecological actors of cacao agroforestry in Manabí, we conducted anonymous semi-structured interviews with seven representatives of the cacao agroforestry sector. They took place virtually in February and March 2021 and the actors interviewed included: two researchers from the Technical University of Manabí (UTM), one researcher from the Escuela Politécnica del Litoral (ESPOL), a representative of the Fortaleza del Valle Corporation, one representative of the NGO Maquita, a representative of the provincial government of Manabí (GPM), and a consultant from the European Union (EU).

Furthermore, our focus was on actors supporting the institutional ecosystem around the smallholder cacao farmers, rather than the commercialization value chain. The information gathered from the interviews, indicating collaborative links (formal and informal), was used to create a network of actors using the software *Cytoscape* (The Cytoscape Consortium, 2022) as a means of representing the cartography. At least 28 different actors were identified, comprising eight different institutional types **(Figure 5.6)**. The questionnaire and transcriptions of the interviews are available on request from the first author.

In our opinion, the most striking result of our cartography is the large number of international agencies (IAs) and foreign private companies (PCs) that provide support in Manabí, especially economic support for conducting research **(Figure 5.6)**. The various IAs present in the region work primarily with producer associations (PAs), but also collaborate with local universities (U). However, the scale of collaboration is not the same. For instance, Spanish IAs have close links with nongovernmental organizations (NGOs) working at the national level (NGO2), while the local producers' association has no links with international organizations other than a local NGO (NGO1).

The local producers' association (called a 'corporation' in Ecuadorian legal parlance) is considered 'second-level' status and is made up of four 'first-level' associations, representing a total of 983 small producers of fine-flavored cacao

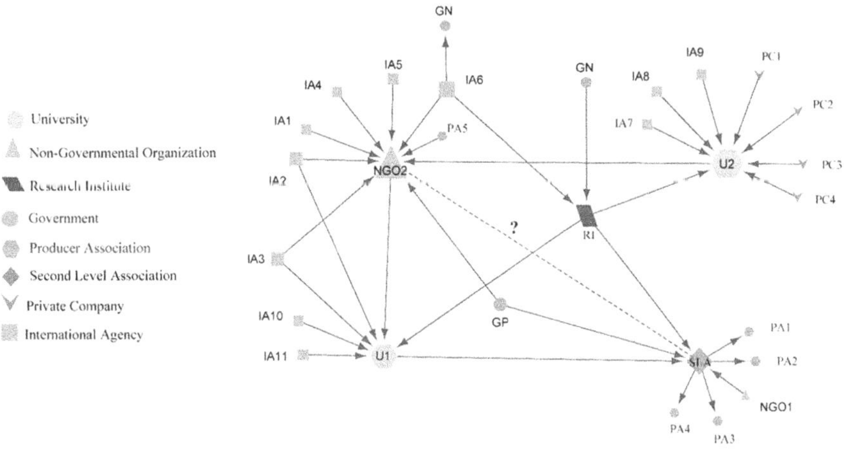

Figure 5.6 Network of actors in Manabí related to cacao agroforestry.

(represented as SLA in **Figure 5.6**). In other words, the 'corporation' is an association of associations. As for the national NGO, it interacts with 38 grassroots organizations in the province of Manabí and four second-level associations (not shown in **Figure 5.6** because the interviewee did not provide details). The overall mission of this NGO is to work with small producers, contribute to the improvement of productive units, promote the active participation of women and young people, and generally strengthen associativity. In contrast, the 'corporation' seems to focus mainly on the marketing of the products. The interviews did not reveal any explicit links between the national NGO and the local 'corporation'.

The provincial government (GP) also has links with the 'corporation' and a national NGO but it was not clear whether there was any effort or need to get them to work together and collaborate at any level. To our surprise, no NGO with an interest in biodiversity was found as part of this network, which is rather surprising given the importance of this megadiverse area. A complex network of collaboration was found around funding and marketing (previous sections) but not around protecting agroforestry areas as part of a biodiversity initiative that includes landscape management.

There are two universities with a strong presence in the region. Both universities have very active agronomy faculties that interact with producers at different levels, although the focus seemed to be mainly on the development of improved varieties, cacao physiology, and cultural practices. It is not known if the agronomy faculties interact with the biology programs regarding biodiversity conservation. Both universities have links with the National Research Institute (RI) for the development of new technologies in Quito, the capital city. This institute, in turn, works with the government (GN) to disseminate any technological improvements. The interviews did not reveal any relationship between these two universities, but this does not mean that they do not exist. An important difference between the two universities is that one of them currently has close relationships with European agencies and PCs that fund research, but again, this does not mean that the other university does not have any. The European agencies appear to have close links at several levels. One of the European agencies supports the GN, the RI, and the national NGO. At the local level, it also has links with a Quito-based NGO, although these are not as strong.

The various interviewees agreed with what is reported in the current literature, namely that cacao production has several problems. These include low production due to the age of the plantations and their poor management, limited access to water, the presence of pests, limited use of technology, poor post-harvest practices, the presence of heavy metals such as cadmium, limited access to credit, and poor technical support. These problems not only affect crop yields but also the producers' economy and quality of life. This increases migration to large cities due to lack of opportunities, thus weakening the international competitiveness of Ecuadorian products in the international markets (Cartagena et al., 2020; Vásquez & Ulloa, 1997).

5.4.2 The marketing bottlenecks

The financial and marketing situation of smallholder farmers was described by respondents as fragile, and the lack of association was cited as one of the main problems facing producers. Indeed, there is a perception that smallholders receive lower prices in return for their sales if they do not go through the associations. However, only less than 3% of the cacao marketed each year in Manabí comes from PAs (Barrera et al., 2019):

> "*...The lack of associativity in rural cooperatives has been present for years. This has been a limiting factor, because perhaps there was no strategy to promote associativity in order to grow, to defend their interests. In Manabí there were 2700 formally constituted associations but most of them are associations that were created for a project and after the project ended the associations dissolved and did not make further contact. The 2016 earthquake eroded the associative theme as well as the 2020 pandemic.*"
>
> (Local project manager)

The lack of association and organization among small producers is also responsible for the fact that the fine-flavor *Nacional* cacao, which is essentially organic because it is grown in plots mixed with other tree species, is marketed as conventional cacao.

Most of the cacao production is exported to the world market through export companies that buy undifferentiated types of cacao beans in bulk (organic and nonorganic are frequently mixed as well as *Nacional* with other genotypes), as this is less demanding for the producers, although is less profitable. These export companies have an indirect presence through the local traders (not shown in the network) who are responsible for collecting the cacao production from the smallholders. Not surprisingly, local traders take advantage of the socioeconomic conditions of the producers and the distance from the buying centers. The lack of a market that distinguishes between different types and qualities of cacao therefore leads to a large proportion of *Nacional* cacao plantations being replaced with CCN-51, which is more productive but which in turn poses various problems from an economic, social, and environmental point of view. On the other hand, there are large producers who trade directly with foreign companies that recognize the quality of their product and hence obtain a better price.

While the niche market of extra-high-quality cacao may be an attractive idea for smallholder farmers, several challenges remain to be solved: access to the specialty market is very restricted and maintaining certification is extremely expensive. The global cacao market has changed in recent years as more and more consumers have become aware of the costs of cacao production, not only in environmental terms but also in terms of social inequality. Participation in this new market requires a high level of certification committed to organic and fair production. These requirements would be particularly important for varieties marketed as '*cacao Fino de Aroma*'. However, given the fragile economic and social conditions of smallholder farmers, reaching this demanding market seems to be a long way off.

All the stakeholders interviewed agreed that since the *Nacional* cacao variety is considered to be part of Ecuador's identity, and taking into account that its production in agroforestry systems is practically organic, without the use of herbicides or pesticides, its cultivation and production should be encouraged with fair financial compensation to the producers, instead of the CCN-51 variety. This differentiation would motivate Manabí producers to promote the cultivation of high-quality cacao, using a mix of native trees, which can also have a positive impact on biodiversity too.

5.5 Conclusions

Manabí is one of the most important provinces in Ecuador, both for its demographic density and for its agricultural importance. In fact, Manabí is the third most populated province in the country and has one of the largest areas of agricultural exploitation. This is especially true for cacao cultivation, which occupies one of the first positions in this category (Coronel Ponce et al., 2019). In addition, it is also the province with the largest number of agroforestry systems, which, as discussed above, are an important tool for reducing the negative impacts of the greenhouse effect and the conservation of biodiversity.

Agroforestry should be understood through an interdisciplinary lens. There is a need for clear policies and frameworks to support the many different interest groups involved in its research and development. There are ecological and socioeconomic factors that influence the adoption of agroforestry; so, future research and policies should aim not only for the economic sustainability of agroforestry systems but also for a balance between environmental and socio-economic points of view. Studies are required to deepen our understanding of the importance of *Nacional* cacao production in agroforestry systems, compared to monocultures, because of its direct positive impact on biodiversity and household income.

Promoting association initiatives among smallholders for the commercialization of cacao is as important as the technical aspects of cacao production and should be a priority to increase household income in this competitive market. Associations need support to afford the certification process so that they can be paid accordingly for organically produced cacao.

Agroforestry systems are also fundamental tools for the conservation of biodiversity because they host a great diversity of species, providing habitat, resources, and food for a wide variety of animal and plant species, and could serve as a strategy to protect bird species that require trees in their habitats Commodity production is commonly associated with areas identified as biodiversity hotspots. So, it is extremely important to implement policies aimed at preserving and protecting primary forests too.

Acknowledgments

This work was made possible thanks to the support of the D-Critical project (ANR-20-CE32-0001) funded by the French National Agency. We would also like to thank the interviewees for their time and valuable information provided.

References

Bager, S., Persson, M., & Reis, T. (2020). Reducing commodity-driven tropical deforestation: Political feasibility and 'Theories of Change' for EU policy options. *SSRN Electronic Journal*, 1–39. https://doi.org/10.2139/ssrn.3624073

Barrera, V., Alwang, J., Casanova, T., Dominguez, J., Escudero, L., Loor, G., Peña, G., Párraga, J., Arévalo, J., Quiroz, J., Tarqui, O., Plaza, L., Sotomayor, I., Zambrano, F., Rodríguez, G., García, C., & Racines, M. (2019). La cadena de valor del cacao y el bienestar de los productores en la provincia de Manabí-Ecuador. *SSRN Electronic Journal*, *5*(564), 1–19.

Borda, A., Morales, O., Teegen, H., Rees, G. H., & Gonzalez-Perez, M. A. (2021). Addressing sustainable rural development with shared value: A Peruvian model from the cacao industry. *Sustainability (Switzerland)*, *13*(14), 1–17. https://doi.org/10.3390/su13148028

Cartagena, Y., Guanín, T., García, C., Parra, R., & Martínez, A. (2020). *Estación Experimental Santa Catalina Economía Agrícola y Cambio Climático Informe Anual 2019*. https://doi.org/10.1111/1477-9552.12329.2

Castañeda-Ccori, J., Bilhaut, A. G., Mazé, A., & Fernández-Manjarrés, J. (2020). Unveiling cacao agroforestry sustainability through the socio-ecological systems diagnostic framework: The case of four Amazonian rural communities in Ecuador. *Sustainability (Switzerland)*, *12*(15), 1–17. https://doi.org/10.3390/SU12155934

Coronel Ponce, D., Saenz Véliz, R., & Rosete Blandariz, S. (2019). Biodiversidad como indicador de sostenibilidad. *UNESUM-Ciencias: Revista Científica Multidiciplinaria*, 137–154.

Díaz-Montenegro, J., Varela, E., & Gil, J. M. (2018). Livelihood strategies of cacao producers in Ecuador: Effects of national policies to support cacao farmers and specialty cacao landraces. *Journal of Rural Studies*, *63*(December 2017), 141–156. https://doi.org/10.1016/j.jrurstud.2018.08.004
FAO. (2022). *About Agroforestry*. https://www.fao.org/forestry/agroforestry/80338/en/
GBIF. (2021). *GBIF*. https://www.gbif.org/
Haggar, J., Pons, D., Saenz, L., & Vides, M. (2019). Contribution of agroforestry systems to sustaining biodiversity in fragmented forest landscapes. *Agriculture, Ecosystems and Environment*, *283*, 106567. https://doi.org/10.1016/j.agee.2019.06.006
INEC. (2021). *Estadisticas Agropecuarias*. https://www.ecuadorencifras.gob.ec/estadisticas-agropecuarias-2/
INIAP. (2018). *Estacion experimental Portoviejo - Programa de Cacao y Café. Informe técnico anual 2018*. http://repositorio.iniap.gob.ec/handle/41000/5670
IUCN. (2021). *Red List of Threatened Species*. https://www.iucnredlist.org/
Kumar, P., Singh, R. P., Singh, A. K., & Kumar, V. (2014). Quantification and distribution of agroforestry systems and practices at global level. *HortFlora Research Spectrum*, *3*(1), 1–6.
Lauri, P., Meziere, D., Dufour, L., Gosme, M., Gary, C., Jagoret, P., Wéry, J., Dupraz, C., Lauri, P., Meziere, D., Dufour, L., Gosme, M., & Simon, S. (2020). Fruit-trees in agroforestry systems - Review and prospects for the temperate and Mediterranean zones. *3rd European Agroforestry Conference – Montpellier*, 23–25 May 2016
MAE. (2010). Cuarto Informe Nacional Para El Convenio Sobre La Diversidad Biológica. In *Ministerio De Ambiente: Vol. I*. https://www.cbd.int/doc/world/ec/ec-nr-04-es.pdf
MAE (Ministerio del Ambiente de Ecuador). (2013). Sistema de clasificacion de los Ecosistemas del Ecuador continental. Subsecretaria de Patrimonio Natural. Quito.
Nair, P. K. R. (1991). State-of-the-art of agroforestry systems. *Forest Ecology and Management*, *45*(1–4), 5–29. https://doi.org/10.1016/0378-1127(91)90203-8
Nair, P. R. (2007). The coming of age of agroforestry. *Journal of the Science of Food and Agriculture*, *87*(2), 1613–1619. https://doi.org/10.1002/jsfa
Ollinaho, O. I., & Kröger, M. (2021). Agroforestry transitions: The good, the bad and the ugly. *Journal of Rural Studies*, *82*, 210–221. https://doi.org/10.1016/j.jrurstud.2021.01.016
Olson, D. M., et al. (2001). Terrestrial ecoregions of the world: A new map of life on Earth. *Bioscience*, *51*, 933–938.
Ortiz Angela Somarriba Eduardo, R. L. (2008). Almacenamiento y tasas de fijación de biomasa y carbono en sistemas agroforestales de cacao (*Theobroma cacao*) y laurel (Cordia alliodora). *Agroforestería de Las Américas*, *2008*, 26–29.
Ospina-Ante, A. (2009). Preliminares a la agroforestería ecológica documentada de la región amazónica colombiana. *Momentos de Ciencia*, *6*(1), 49–54.
Portillo-Quintero, C. A., & Sánchez-Azofeifa, G. A. (2010). Extent and conservation of tropical dry forests in the Americas. *Biological Conservation*, *143*(1), 144–155. https://doi.org/10.1016/j.biocon.2009.09.020
Ravenhill, J. (2014). Global value chains and development. *Review of International Political Economy*, *21*(1):264–274. https://doi.org/10.1080/09692290.2013.858366
Robbins, P., Chhatre, A., & Karanth, K. (2015). Political ecology of commodity agroforests and tropical biodiversity. *Conservation Letters*, *8*(2), 77–85. https://doi.org/10.1111/conl.12169
Rosenstock, T. S., Wilkes, A., Jallo, C., Namoi, N., Bulusu, M., Suber, M., Mboi, D., Mulia, R., Simelton, E., Richards, M., Gurwick, N., & Wollenberg, E. (2019). Making trees count: Measurement and reporting of agroforestry in UNFCCC national communications of non-Annex I countries. *Agriculture, Ecosystems and Environment*, *284*, 106569. https://doi.org/10.1016/j.agee.2019.106569

Sanchez, V. H. (2019). Diagnóstico de la cadena de valor del Cacao en Ecuador. In Sánchez, V., Zambrano, J., y Iglesias, C. eds. *La cadena de valor del Cacao en América Latina y el Caribe* (pp. 49–62) INIAP, Quito. https://www.fontagro.org/new/uploads/adjuntos/Informe_CACAO_linea_base.pdf

Sierra, R., & Al, E. (1999). *Un Estudio Basado en la Biodiversidad de Ecosistemas y su Ornitofauna Por Rodrigo Sierra, Felipe Campos, y Jordan Chamberlin WILOLIFE CONSERVATION SOCIETY (WCS) ARIZONA STATE UNIVERSITY.*

SIPA. (2021). *Cifras Agroproductivas.* http://sipa.agricultura.gob.ec/index.php/cifras-agroproductivas

The Cytoscape Consortium. (2022). *Cytoscape 3.9.1 User Manual.* http://manual.cytoscape.org/en/stable/#cytoscape-version-user-manual

van Noordwijk, M., Rahayu, S., Gebrekirstos, A., Kindt, R., Tata, H. L., Muchugi, A., Ordonnez, J. C., & Xi, J. (2019). Tree diversity as basis of agroforestry. In: van Noordwijk M, ed. *Sustainable development through trees on farms: agroforestry in its fifth decade.* Bogor, Indonesia: World Agroforestry (ICRAF) Southeast Asia Regional Program. pp. 17–44.

Vásquez, M., & Ulloa, R. (1997). *Estrategia para la Conservación de la Biodiversidad Biológica en el Sector Forestal del Ecuador.* http://biblio.flacsoandes.edu.ec/catalog/resGet.php?resId=45263

WWF. (2021). *Terrestrial Ecoregions of the World.* WWF. https://www.worldwildlife.org/publications/terrestrial-ecoregions-of-the-world

6 Resilience of cacao-based agroforestry systems to climate change

Latin American experiences

Carlos A. Salas-Macías, Fernando Sánchez-Mora, Karime Montes Escobar, Javier de la Hoz-M, Ricardo Limongi-Andrade, Raul V. Mora-Yela, and Felipe R. Garcés-Fiallos

6.1 Introduction

Cacao (*Theobroma cacao* L.) is a very important plant species for farmers in tropical and subtropical regions of Latin America. The production focuses on fine-flavor cacao crops, especially in countries such as Ecuador, Peru, Brazil, Colombia, and Mexico, among the most important at the continental level (Antolinez-Sandoval et al., 2020). In recent years, the global demand for cacao has increased, leading to an increase in cultivated areas in tropical forest areas and the replacement of traditional agroforestry systems (AFS) with monocultures (Armengot et al., 2016).

For the purposes of this review, we used Latent Dirichlet Allocation (LDA) thematic modeling to discover thematic patterns in a collection of documents. We searched the Scopus database for research papers published in English between 1986 and 2023. The key words for the search were 'agroforestry AND cacao', and once the results were obtained, they were filtered by Latin American countries. The final sample consisted of 172 documents.

Data processing was performed using LDAShiny (de la Hoz-M et al., 2021), an open-source package for the R programming language. To increase the consistency of the topics, each summary was tokenized using bigraphs. The correct number of topics was selected by running simulations with k ranging from 5 to 50 and a Gibbs sampling process (Oh et al., 2019) with 500 iterations. A measure of topic coherence (Jones, 2019) was used to determine the quality of the LDA model. Subsequently, a naive labeling algorithm provided by the textmineR package (Xiong et al., 2019) based on bigrams was employed, and two agroforestry experts analyzed the topics using two sources of information: the most common (most likely) word lists and a sample of the titles.

DOI: 10.4324/9781003381761-8

Country	Articles
Brazil	45
Colombia	16
Mexico	15
Costa Rica	11
Usa	11
Germany	9
France	7
Peru	6
Switzerland	6
Ecuador	5
Bolivia	4
Canada	3
United Kingdom	3
Cameroon	2
Czech Republic	2
Spain	2
Antigua	1
Belgium	1
Finland	1
Italy	1

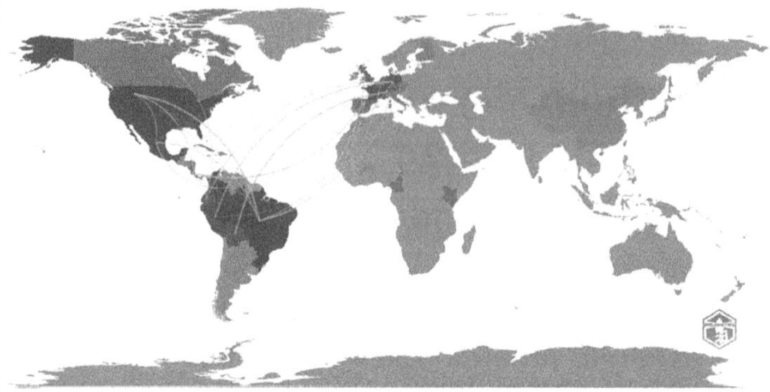

Figure 6.1 Analysis of the distribution of 172 cacao-based AFS research papers in Latin America from 1986 to 2023, as well as a visual depiction of the country collaboration network.

Figure 6.1 illustrates the countries in the Americas that have carried out the most research in the field of cacao-based AFS research. Brazil (45), Colombia (16), Mexico (15), Costa Rica (11), and Peru (6) stand out as the main contributors to high-impact research, as evidenced in the global scientific literature. On the other hand, the other countries present a lower frequency of publications and a lower-level impact. It also shows the network of international co-authorship relationships among 20 countries, the largest connected set. The analysis groups these 20 countries into four distinct clusters. There is a stronger cooperative relationship between countries in the same cluster than between countries in different clusters. Of course, this does not mean that there is no cooperation between countries in different clusters, but there may be some common research topics between countries in the same cluster, which makes their cooperation closer. More frequent connections (represented by the thickness of the line) can be observed between Brazil, Colombia, Mexico, and Costa Rica.

Table 6.1 shows the names of the themes generated from the words with the highest number of repetitions. The words were ranked by relevance. After searching within the research papers for each topic, prevalence rankings were generated from the 20 words with the highest number of repetitions. The top five were: t_3 (Soil Quality), t_6 (Climate-smart system), t_11 (Soil organic management), t_14 (Crop productivity model), and t_5 (Sustainable farmer practices).

Table 6.1 Topic frequencies

T	*D*	*P*	*Top_terms*	*Label*
3	11	5,171	soil, qualiti, content, chemic, fertil, high, indic, properti, physic, soil_qualiti, sampl, fruit, function, pod, bean	Soil Quality
6	11	5,564	product, climat, chang, system, potenti, coffe, environment, area, sustain, climat_chang, land, agricultur, local, produc, base	Climate-smart system
11	11	5,76	system, organ, soil, monocultur, matter, organ_matter, manag, crop, treatment, convent, evalu, depth, product, year, product_system	Soil organic management
14	10	5,603	plant, product, model, crop, system, harvest, densiti, econom, develop, yield, produc, perform, area, analysi, altern	Crop productivity model
15	10	5,496	farmer, practic, organ, market, livelihood, system, land, agricultur, farm, cultiv, commun, econom, social, sustain, knowledg	Sustainable farmer practices
5	9	4,486	ant, speci, plantat, tree, forest, epiphyt, bromeliad, rich, divers, shade, system, speci_rich, found, nativ, composit	Ant Plant interactions
19	9	4,84	leaf, plant, system, tc, condit, tree, clone, area, rate, valu, site, water, light, variabl, cultivar	Plant breeding
22	9	4,242	yield, tree, increas, level, tradit, crop, agroforest, farm, measur, prune, napu, palolo, shade, manag, higher	Agroforest yield enhancement
12	8	4,202	tree, shade, timber, yield, year, shade_tree, alliodora, speci, pod, plot, plantat, growth, cordia, product, system	Shade tree management
13	8	4,641	carbon, system, stock, year, soc, biomass, carbon_stock, tree, storag, store, soil, forest, estim, emiss, ag	Carbon storage
17	8	5,668	cabruca, forest, speci, conserv, bahia, atlant, atlant_forest, area, region, nativ, brazil, biodivers, southern, tree, shade	Forest conservation
4	7	4,215	nutrient, soil, litter, system, rubber, biomass, balanc, tree, product, stock, cycl, organ, phosphoru, forest, depth	nutrient cycling
8	7	4,045	timber, shade, tree, system, product, yield, incom, famili, servic, high, provid, contribut, farm, canopi, caf	Yield enhancement system
9	7	4,097	af, forest, secondari, product, system_af, nativ, secondari_forest, varieti, system, divers, tree, amazon, collect, ecosystem, method	Agroforestry systems
1	6	3,625	tree, speci, divers, tree_speci, afss, system, fallow, structur, sa, plant, level, effect, composit, florist, tree_divers	Tree species diversity
10	6	4,142	speci, system, rich, conserv, biodivers, speci_rich, monocultur, divers, intens, manag, commun, forest, number, dung, record	Biodiversity management

(*Continued*)

Table 6.1 (Continued)

T	*D*	*P*	*Top_terms*	*Label*
16	6	4,21	divers, speci, tree, cultiv, genet, popul, chiapa, mexico, crop, structur, region, genet_divers, system, conserv, municip	Genetic diversity
18	6	3,674	farm, grown, dog, drought, crop, concentr, helminth, system, effect, region, evalu, investig, caus, domest, live	Drought resilience
21	6	4,254	forest, landscap, agroforest, land, agricultur, disturb, bird, increas, local, effect, shade, abund, bat, tree, veget	Agroforest landscaping effects
23	6	3,607	structur, control, spatial, plant, system, banana, pest, high, complex, biolog, mirid, effect, base, talamanca, potenti	Spatial pest control
2	4	2,701	tamarin, speci, individu, lion, lion_tamarin, forest, tree, associ, predat, abandon, cabruca, structur, higher, number, basal	Predator-associated tree diversity
20	4	3,555	fruit, root, speci, pollin, fine, shade, product, system, cover, set, abund, studi, banana, inga, plantat	Shade fruit agroforestry
7	3	2,201	saf, en, la, forest, food, simpl, speci, system, habitat, diet, import, found, rang, famili	Forest food systems

T = Number of topics, D = Number or paper research, and P = prevalence.

6.2 Climate change and cacao production

Cacao production is closely linked to climatic conditions (Baffour-Ata et al., 2023; Suh & Molua, 2022; Blaser-Hart et al., 2021). As global temperatures increase and rainfall patterns become unpredictable, the suitability of certain regions for cacao cultivation may be affected, presenting significant challenges for cacao production. Lack of adequate rainfall can result in droughts negatively affecting plant growth (Nurhayati et al., 2022; Gateau-Rey et al., 2018) and flower and fruit formation. On the other hand, variability in rainfall and temperature is associated with the incidence of pests and diseases (Delgado-Ospina et al., 2021; Asante et al., 2017; Ndoumbè-Nkeng et al., 2009).

Climate variability affects not only the quantity but also the quality of harvests due to multiple factors (Daymond & Hadley, 2008; End et al., 1988). As with the direct effects of climate change on cacao cultivation, there are also indirect impacts that affect the entire supply chain (Avadí, 2023; Malek et al., 2022). Demand for high-quality cacao is steadily increasing, but fluctuations in production can lead to market shortages.

In the face of these challenges, it is crucial to develop adaptation and mitigation strategies to ensure the resilience of cacao production. Farmers and agroforestry experts are working on identifying and selecting cacao varieties that are more resilient to climate change, among other adaptation technologies (Wongnaa & Babu, 2020). Scientific research and collaboration between different actors, such as farmers, academic institutions, governments, and international organizations, play a key role in the search for sustainable solutions (Tschora & Cherubini, 2020).

In this regard, some initiatives in Latin America aim to generate information on cacao varieties with genetic characteristics that allow them to survive and produce more efficiently under adverse climatic conditions; among them in Ecuador, we can cite García-Cruzatty et al. (2023), Jaimez et al. (2021), Tezara et al. (2015), and Apshara et al. (2013); in Colombia, Apshara et al. (2013); in Bolivia, Niether et al. (2020); in Brazil, Apshara et al. (2013); in Venezuela, Tezara et al. (2020) and de Almeida et al. (2016); in Peru, Apshara et al. (2013); in Mexico, Vásquez et al (2022). Similarly, work is being done on the adoption of adaptation and improvement strategies throughout the cacao value chain, citing in Ecuador Middendorp et al. (2018); in Bolivia, Lori et al. (2022); in Costa Rica, Rousseau et al. (2012) and Schnell et al. (2007); in Brazil, Schroth et al. (2016, 2016), dos Santos et al. (2012), and Papalexandratou et al. (2011); in Peru, Lagneaux et al. (2021) and Wong et al. (2020); in Venezuela, Vera et al. (2000); in Colombia Núñez et al. (2023) and Ortiz-Rodríguez et al. (2016); in Mexico, Brito-Vega et al. (2018), Díaz-José et al. (2013), and Torres de la Cruz et al. (2011); and in Nicaragua, Suazo et al. (2014).

Similarly, it is mentioned that a predominant factor among sustainable practices in cacao cultivation is diversification in production systems (Danso-Abbeam et al., 2023; Hashmiu et al., 2022; dos Reis et al., 2019). Such diversification is achieved through the inclusion of shade trees and other complementary crops. In any case, although it is still necessary to deepen the understanding of technical, socioeconomic, and environmental aspects related to this integration (Quandt et al., 2023), its dissemination as an adaptation strategy that promotes better practices (Centeno-Alvarado et al., 2023; Wartenberg et al., 2018) and provides several ecosystem services that could substantially improve the productivity (Deheuvels et al., 2012) and quality of life of the actors in the production chain is of great importance.

6.2.1 Ecosystem services provided by cacao-based agroforestry systems

AFS have long been promulgated as sustainable land use practices that offer numerous benefits to farmers and the environment (Ahmad Dar et al., 2023; Cassamo et al., 2023; da Cruz Silva et al., 2023; Pérez-Neira et al., 2023; Centeno-Alvarado et al., 2023; Kheiri et al., 2023; Talipov & Mukimov, 2022). These integrated systems, which combine cacao cultivation with various shade trees and other crops, have proven to be a valuable strategy to address the challenges arising from climate change (Habib-ur-Rahman et al., 2022). The composition of tree species and their spacing are essential to obtain ecosystem benefits and improve the yield of almonds in crops such as cacao (Bai et al., 2017).

6.2.2 Microclimate

The cacao crop needs shade to survive and grow (Bai et al., 2017). The incorporation of shade trees plays a crucial role in creating favorable microclimates (Litschel et al., 2023), providing essential protection against temperature fluctuations and extreme weather conditions. These microclimates, in turn, provide optimal light conditions for cacao tree growth and development (Mortimer et al., 2018), while protecting against heat stress and providing natural moisture regulation. In this regard, access to climate data and early warning systems is critical for farmers to make preemptive decisions and adjust their practices accordingly.

6.2.3 Soil and water management

Some complementary crops planted between the rows of cacao trees function as cover crops, which act as a living layer that covers the soil, reducing direct exposure to rain and the impact of drops (Zhu et al., 2023) and erosive processes. Others can fix nitrogen (Domínguez-Núñez, 2022; de Sousa et al., 2017; Perin et al., 2004), enriching the soil with this essential nutrient for growth. Similarly, the leaves and remains of shade trees represent a valuable contribution of organic matter (Lalremsang et al., 2022). This contribution is of great importance, as it significantly increases nutrient content and improves their ability to retain water efficiently. This resulting nutritional balance greatly benefits both crops and the agricultural system as a whole, promoting healthy and sustainable growth.

In the same sense, some practices associated with ASF technologies, such as terracing and contour planting, help to avoid nutrient loss due to erosion and improve soil fertility (Biswas et al., 2022) and biota (McQueen & Treonis, 2020; Vaupel et al., 2023). In addition, the pore spaces created by deep tree roots provide improvements in water retention and soil structure, as well as hydraulic redistribution (Burgess et al., 1998).

6.2.4 Pollination

A unique feature of AFS is their tendency to create an environment appropriate to the habitation of beneficial organisms, thus aiding and boosting the pollination process. This phenomenon is mostly linked to the AFS framework's supply of shade trees (Staton et al., 2022; Toledo-Hernández et al., 2017). The diversity of plant species with different flowering times increases the chances of pollinators finding suitable resources throughout the year. By having a more diverse and resource-rich habitat, pollinators can establish more stable and healthy populations, which, in turn, increases pollination efficiency in the system and biological pest control (Boinot et al., 2019), thus strengthening the resilience of the system as a whole. This biological interaction decreases the need for chemical synthesis inputs and promotes an ecological balance in the plantation (Beer et al., 2003), which reduces the exposure of pollinators to toxic substances, thus contributing to their health and abundance.

6.2.5 Biodiversity

Biodiversity conservation occupies a central place in the context of the cacao-based AFS. This importance lies in their ability to promote habitat diversification by including multiple trees (Kassa et al., 2023). AFS are characterized by the presence of diverse plant strata, which provide a wide variety of ecological niches for numerous species (Harvey et al., 2007). This integration of biodiversity contributes significantly to the enrichment of wild populations, thus fostering a more balanced ecosystem and contributing to biodiversity conservation at different scales (Molua et al., 2023).

In addition to their impact on biodiversity, it is crucial to highlight that AFS also play a fundamental role in protecting habitat and promoting functional connectivity, two essential aspects for maintaining ecosystem health (Fremier et al., 2013). This protection and facilitation of functional connectivity has a positive effect on the ability of ecosystems to adapt to environmental changes, thus strengthening the resilience of the system (Zhang et al., 2023).

It is essential to keep in mind that cacao is an umbrophilous plant, which means that it requires shade for optimal development. Both monoculture systems and AFS usually start with short-cycle species in the first years of cultivation, such as those of the *Musa* genus, in order to take advantage of and keep the land free of weeds through temporary shade. However, the key distinction is that AFS are characterized by the incorporation of fruit or timber trees as a permanent source of shade, which allows them to generate economic income over time.

It is important to note that there are well-planned AFS, where shade species are carefully chosen and effectively managed to ensure the sustainable growth of cacao and other crops. On the other hand, there are less-planned AFS, where farmers allow plants already present in the field or new additions to develop more spontaneously. In both cases, the choice and management of shade species play a key role in the long-term success of cacao farming systems (de Lima et al., 2014).

In Latin America, numerous plant associations involving cacao cultivation are found, a prime example of which is Ecuador. In this country, farmers have adopted several cacao-based AFS associations in various regions as detailed in Table 6.2. In the Coastal Region, specifically in the Los Ríos Province, the cantons Quevedo and Valencia (Tapia-Vera et al., 2021; Prieto-Benavides et al., 2012), the cantons Mocache and Palenque (Torres et al., 2018), and the canton Montalvo (Alcívar et al., 2019) have witnessed these practices. Likewise, in

Table 6.2 Species grown in cacao-based AFS in Ecuador

Scientific name	*Common name*	*Use*	*Reference*
Albizia guachapele (Kunth) Dugand	Guachapelí	Forestal/ Frutal	Verdezoto Vargas et al., 2023; Prieto-Benavides et al., 2012
Anacardium occidentale L.	Marañon	Frutal	Torres et al., 2018
Ananas comosus (L.) Merr.	Piña	Frutal	Vargas Tierras et al., 2018
Artocarpus altilis (Parkinson) Fosberg	Fruta de pan	Forestal	Jadán et al., 2015

(*Continued*)

Table 6.2 (Continued)

Scientific name	*Common name*	*Use*	*Reference*
Bactris gasipaes Kunth	Chontilla	Frutal	Prieto-Benavides et al., 2012; Subía et al., 2017; Vargas Tierras et al., 2018
Carica papaya L.	Papaya	Frutal	Prieto-Benavides et al., 2012; Vargas Tierras et al., 2018
Cedrela odorata L.	Cedro	Forestal	Prieto-Benavides et al., 2012; Jadán et al., 2015; Torres et al., 2018; Verdezoto Vargas et al., 2023
Cedrelinga cateniformis (Ducke) Ducke	Chuncho	Forestal	Jadán et al., 2015; Pérez & Freile, 2017; Subía et al., 2017
Ceiba samauma (Mart.) K. Schum.	Huimba	Forestal	Jadán et al., 2015
Centrosema macrocarpum Benth.	Centrosema	Forrajera	Alcívar et al., 2019
Chimarrhis glabriflora Ducke	Mecha	Forestal	Jadán et al., 2015
Chlorophora tinctoria (L.) Gaudich. ex Benth.	Moral fino	Forestal	Prieto-Benavides et al., 2012
Chrysophyllum cainito L	Caimito	Forestal/ Frutal	Celi-Delgado and Aguirre-Mendoza, 2022
Chrysophyllum sanguinolentum (Pierre) Baehni	-	Forestal	Jadán et al., 2015
Citharexylum poeppigii Walp.	Picolete	Forestal	Jadán et al., 2015
Citrus spp.	Citrics	Frutal	Quiroz, 2010; Prieto-Benavides et al., 2012; Torres et al., 2018; Vargas Tierras et al., 2018; Celi-Delgado and Aguirre-Mendoza, 2022;
Colubrina arborescens (Mill.) Sarg	Caoba de montaña	Forestal	Tapia-Vera et al., 2021
Cordia alliodora (Ruiz & Pav.) Oken	Laurel	Forestal	Quiroz, 2010; Prieto-Benavides et al., 2012; Jadán et al., 2015; Torres et al., 2018; Celi-Delgado and Aguirre-Mendoza, 2022
Cordia macrantha Chodat	Laurel Prieto	Forestal	Tapia-Vera et al., 2021
Cremastosperma gracilipes R.E. Fr.	Jaya kara yura	Forestal	Jadán et al., 2015
Cybistax donnell-smithii (Rose) Seibert	Guayacán blanco	Forestal	Prieto-Benavides et al., 2012; Tapia-Vera et al., 2021
Erythrina glauca Willd.	Palo prieto	Forestal	Quiroz, 2010
Erythrina poeppigiana (Walp.) O.F. Cook	Bombón	Forestal	Quiroz, 2010, Torres et al., 2018
Erythrina spp.	Poró	Forestal	Subía et al., 2017
Gmelina arborea Roxb. ex Sm.	Melina	Forestal	Verdezoto Vargas et al., 2023
Guadua spp.	Guadua	Forestal	Alcívar et al., 2019
Guarea guidonia (L.) Sleumer	Cedrillo	Forestal	Jadán et al., 2015
Handroanthus spp.	Guayacán	Forestal	Torres et al., 2018; Verdezoto Vargas et al., 2023

(*Continued*)

Table 6.2 (Continued)

Scientific name	*Common name*	*Use*	*Reference*
Inga edulis Mart.	Guaba	Forestal/ Frutal	Quiroz, 2010; Pérez and Freile, 2017; Torres et al., 2018; Vargas Tierras et al., 2018; Alcívar et al., 2019
Inga laurina (Sw.) Willd.	Guabo	Forestal	Donato et al., 2016
Inga manabiensis T.D. Penn.	Guaba de mico	Frutal	Prieto-Benavides et al., 2012
Laurus nobilis L.	Laurel	Forestal	Pérez and Freile, 2017
Mammea americana L.	Mamey	Frutal	Prieto-Benavides et al., 2012
Mangifera indica L.	Mango	Frutal	Prieto-Benavides et al., 2012; Torres et al., 2018
Matisia cordata Bonpl.	Zapote	Frutal	Prieto-Benavides et al., 2012
Musa spp.	Plátano	Frutal	Prieto-Benavides et al., 2012; Pérez and Freile, 2017; Torres et al., 2018
Myrica spp.	Laurel	Forestal	Donato et al., 2016
Ocotea bofo Kunth	Jigua	Forestal	Verdezoto Vargas et al., 2023
Oryza sativa L.	Arroz	Ciclo corto	Torres et al., 2018
Passiflora edulis Sims	Maracuyá	Ciclo corto	Torres et al., 2018
Persea americana Mill.	Aguacate	Frutal	Prieto-Benavides et al., 2012; Torres et al., 2018; Vargas Tierras et al., 2018
Piptocoma discolor (Kunth) Pruski	Pigüe	Forestal	Jadán et al., 2015
Pouteria sapota (Jacq.) H.E. Moore & Stearn	Zapote	Frutal	Torres et al., 2018
Quercus humboldtii Bonpl.	Roble	Forestal	Verdezoto Vargas et al., 2023
Ruagea glabra Triana & Planch.	Cedrillo	Forestal	Jadán et al., 2015
Schizolobium parahyba (Vell.) S.F. Blake	Pachaco	Forestal	Prieto-Benavides et al., 2012; Donato et al., 2016; Verdezoto Vargas et al., 2023; Celi-Delgado and Aguirre-Mendoza, 2022
Swietenia macrophylla King	Moral fino	Forestal	Torres et al., 2018; Verdezoto Vargas et al., 2023; Alcívar et al., 2019
Tectona grandis L. f.	Teca	Forestal	Torres et al., 2018; Verdezoto Vargas et al., 2023
Terminalia amazonia (J. F. Gmel) Exell	Yumbingue	Forestal	Jadán et al., 2015
Triplaris guayaquilensis Wedd.	Fernán Sánchez	Forestal	Torres et al., 2018; Tapia-Vera et al., 2021; Verdezoto Vargas et al., 2023
Vochysia braceliniae Standl.	Tamburo	Forestal	Jadán et al., 2015
Zanthoxylum brisasanum (Cuatrec.) P.G. Waterman.	Lagartillo	Forestal	Jadán et al., 2015
Zea mays L.	Maíz	Ciclo corto	Torres et al., 2018

Guayas Province, the Yaguachi canton has also used these AFS associations (Verdezoto Vargas et al., 2023).

In the Ecuadorian Amazon Region, the Province of Napo, in particular the Arosemena Tola Canton (Pérez & Freile, 2017), as well as the cantons Archidona and Tena (Jadán et al., 2014), the Provinces of Sucumbíos and Orellana (Vargas Tierras et al., 2018), Zamora Chinchipe Province, and Chinchipe canton (Celi-Delgado and Aguirre-Mendoza, 2022), along with Orellana Province and La Joya de los Sachas canton (Subía et al., 2017), have also experienced the implementation of cacao-based AFS. These associations are similar to those reported in several Latin American countries (Table 6.3).

Table 6.3 Species grown in cacao-based AFS in Latin America

Reference	*Country*	*Species*
Alpizar, Fassbender, Heuveldop, Folster y Enríquez (Alpízar et al., 1986)	Costa Rica	*C. alliodora*
Beer (1988); Beer et al., (1990)	Costa Rica	*Erythrina poeppigiana* and *C alliodora*
Concha et al. (2007)	Perú	*Inga* sp., *Citrus nobilis* Lour., *Piptadenia* sp. *Sickingia williamsii* Standl., *Persea americana, Cecropia* sp, *C. aurantium, Calycophyllum spruceanum* (Benth.) Hook. f. ex K. Schum., *Mangifera indica* L., *C. odorata, Colubrina glandulosa* Perkins, *Coffea arabica* L., *S. williamsii*, and *C. papaya*
Delabie et al. (2007)	Brazil	*E. fusca* Lour.
Goñas et al., 2022	Perú	*Musa* spp, *Cordia* spp, *and Persea* spp.
Jacobi et al., 2014	Bolivia	*Schizolobium amazonicum* Huber ex Ducke, *C. ochroxylum, Anadenanthera* sp. *S. amazonicum, C. ochroxylum, Amburana cearensis* (Allemão) A.C. Sm., and *S. macrophylla*
Jadán et al., 2015	Ecuador	*Oenocarpus bataua* Mart., *Solanum sycophanta* Dunal, *Iriartea deltoidea* Ruiz & Pav., *Wettinia maynensis* Spruce, *C. odorata, C. cateniformis, C. alliodora, T. amazonia*, and *Myroxylon balsamum* (L.) Harms
Kummerow, Kummerow, y Da Silva (1982)	Brazil	*E. glauca.*
del Marín et al., 2016	Colombia	*A. guachapele, C. alliodora, C. odorata, Acacia glomerosa* Benth., *Croton* sp, *C. papaya*, and *Guarea trichilioides* L.
Middendorp et al., 2018	Ecuador	*Cecropia* sp., *Erythrina* sp., *Dussia* sp., *Psidium* sp., *Cupania* sp., *Pouteria* sp., *Carica* sp., *Citrus* sp., *Inga* sp., *Annona* sp., *Pourouma* sp., *Cedrela* sp., and *Cordia* sp.

(*Continued*)

Table 6.3 (Continued)

Reference	*Country*	*Species*
Monroe et al., 2016	Brazil	*E. glauca, Hevea brasiliensis* (Willd. ex A. Juss.) Müll. Arg., *G. sepium*, and *Musa* spp.
Morán-Villa et al., 2022	México	*Cedrela odorata* L, *E. americana* Mill., *Gliricidia sepium* (Jacq.) Walp., *Tabebuia rosea* (Bertol) DC., *Swietenia macrophylla* Willd., *Cocos nucifera* L, *Colubrina arborescens* (P. Mill) Sarg., *Pimenta dioica* L., *Cecropia obtusifolia* Bertol, and Roystonea dunlapiana P. H. Allen.
Muñoz et al. (2006)	México	*P. saman, G. sepium*, Ficus spp. *Spondias mombin* L., *D. robinioides, Manilkara zapota* (L.) P. Royen, *M. indica, Bursera simaruba* (L.) Sarg., *E. americana, C. obtusifolia, Selenicereus* sp, *T. amazonia, Paullinia pinnata* L., and *Syngonium podophyllum* Schott
Muñoz and Beer, 2001	Costa Rica	E. poeppigiana
Muñoz et al. (2005)	México	*Pithecellobium saman* (Jacq.) Benth., *E. americana, G. sepium*, and *Inga jinicuile* Schltdl.
Ramírez-Meneses et al., 2013	México	*Samanea saman* (Jacq.) Merr., *Diphysa robinioides* Benth., *Guazuma ulmifolia* Lam., *Erythrina* sp., Colubrína arborescens (Mill.) Sarg., *G. sepium, Diphysa robinioides* Benth., *Musa* sp, *Guazuma ulmifolia* Lam., *C. odorata*, C arborescens, *G. sepium*, Musa sp., *G. ulmifolia* y, *Artocarpus altilis* (Parkinson) Fosberg, *G. sepium*, and *C. arborescens*
Reitsma et al., 2001	Costa Rica	*C. alliodora* and *Inga* spp.
Salvador-Morales et al., 2020	México	*Citrus* spp, *C. nucifera*, among others…
Sánchez Gutiérrez et al., 2016	México	*E. americana* Mill., *E. poeppigiana* (Walp.) O.F. Cook, *Gliricidia sepium* (Jacq.) Kunth ex Walp., *C. odorata* Gürke and *Colutea arborescens* L.
Santana and Cabala-Rosand (1982)	Brazil	*E. glauca, E. poeppigiana, Genipa americana* L., and *Spondias lutea* L.
Somarriba et al., 2013	Nicaragua, Honduras, Costa Rica, Panamá, Guatemala (Alta Verapaz y Costa Sur)	*C. alliodora, C. odorata, Terminalia oblonga* (Ruiz & Pav.) Steud., *Iriartea* spp., *Bactris gasipaes* Kunth, *C. nucifera, Nephelium lappaceum* L., M. indica, *P. sapota, P. americana, G. sepium, Inga* spp, and Leucaena leucocephala (Lam.) de Wil

For example, the association of cacao with laurel (*C. alliodora*) is common in countries such as Peru (Goñas et al., 2022), Colombia (del Marín et al., 2016), and Costa Rica (Reitsma et al., 2001). According to Somarriba et al. (2013), it is also found in Nicaragua, Honduras, Costa Rica, Panama, and Guatemala. Fruit trees such as avocado (*Persea* sp.) are common in AFS in Peru (Goñas et al., 2022), while oranges, lemons, mangoes, coconut palms, and other fruit trees are grown in Mexico (Salvador-Morales et al., 2020).

Additionally, published literature mentions that diverse agroecosystems, known as successional AFS, tend to have a higher resilience capacity compared to simple AFS and monocultures (Jacobi et al., 2015).

6.2.6 Pests and diseases

The scarce availability of labor, restrictions related to the presence of heavy metals in beans, and the presence of diseases that affect cacao plants and pods are some of the challenges currently facing the crop (Antolinez-Sandoval et al., 2020). Cacao production worldwide is threatened by pests and diseases, regardless of the farming system used (Armengot et al., 2020). Frosty pod rot (*Moniliophthora roreri*) and Witches' broom (*M. perniciosa*) are responsible for causing more than 50% of cacao yield losses on non-technified farms (Meinhardt et al., 2008; Sánchez-Mora & Garcés-Fiallos, 2012).

In Bolivia, cacao mirid (*Monalonion dissimulatum*), witches' broom (*M. perniciosa*), black pod rot (*Phytophthora* spp.), and frosty pod rot (*M. roreri*) are the main biotic factors (Andres et al., 2016). *Carmenta theobromae* and *C. foraseminis* are important in Colombia. The latter mainly limits cultivation in Panama, Venezuela, Brazil, and Colombia. Diseases caused by *M. roreri* and *Phytophthora* spp and Cacao Swollen Shoot Virus Disease (CSSVD) are important diseases in Colombia (Antolinez-Sandoval et al., 2020). In Ecuador, farmers consider shade to be important for soil moisture and fertility management, as well as the control of some weeds and diseases (Bentley et al., 2004).

One of the most important factors within the AFS is the shade provided by forest species, which would affect the plant–pathogen/pest relationship. Therefore, the transdisciplinary development of decision support tools should aim to determine optimal shade levels and appropriate tree species composition to minimize disease and pest damage (Andres et al., 2016). Despite all, it is difficult to determine the appropriate shade levels and tree species composition that will reduce the intensity and damage caused by diseases and pests, while providing favorable growing conditions for cacao trees (Andres et al., 2016). Cacao-based AFS with around 30% cover or sometimes more can have similar levels of bean production to monocultures and a lower pest and disease pressure (Blaser et al., 2018). Farmers should be aware that in a cacao-based AFS, they need to prune the canopy trees to achieve just the right amount of shade; otherwise, the incidence of diseases such as Phytophthora pod rot disease, and perhaps the intensity of other diseases increase (Bentley et al., 2004).

Factors that could be associated with shade in cacao-based AFS include high light, soil, and air moisture, and soil characteristics such as fertility. Of these, high light and soil moisture would be more associated with shade in AFS than soil fertility (Andres et al., 2018). In fact, cacao plants established in full sun are much more stressed than those not exposed to this environmental factor due to excessive leaf transpiration and increased soil evaporation, which significantly affects bean yield (Oro et al., 2012; Andres et al., 2018).

The incidence of diseases and pests in cacao-based AFS does not increase compared to monocultures when efficient cultural management practices are applied, that is regular tree pruning, frequent pod harvesting, regular removal of infested pods, weed management, thereby increasing the productivity of cacao plantations (Armengot et al., 2020). For example, removing diseased pods reduces disease incidence and increases cacao bean yield by 25 and 46%, respectively.

Another relevant aspect is the significant changes in soil fungal diversity in the different cacao production systems. The greater or lesser abundance and diversity of soil fungal communities in the AFS would be associated with the different changes in the physical and chemical properties of the soil under cacao. Even the presence of the pathogenic genera Phytophthora and Rhizoctonia tends to disappear in the soil of cacao-based AFS over time, probably due to the increase of biocontrollers such as Trichoderma (Arévalo-Gardini et al., 2020). Shading of cacao-based AFS not only improves fungal diversity but also bird diversity and community composition. Previous studies have shown that shade trees of cacao are important habitats for inter-American migratory birds (Bentley et al., 2004). In this sense, each shade tree species can host a different community of birds, which act as natural predators of pests (Haggar et al., 2019) helping to strengthen the resilience of the system in terms of health. Moreover, some crops can act as natural barriers, repelling harmful insects or attracting natural enemies that control pests. This highlights the need to conserve forested landscapes (land sparing) and invest in mixed agroforestry (land sharing), as each strategy benefits a diverse and distinct biological community (Bennett et al., 2022).

6.2.6.1 Carbon stocks

Trees and other woody crops promote the capture and storage of large amounts of carbon (Fatumah et al., 2023) in both aboveground biomass (AGB) and belowground organic matter. Trees, as they grow and develop, absorb carbon dioxide from the atmosphere during photosynthesis, converting it into organic matter in the form of leaves, branches, and trunks. This tree biomass acts as a long-term carbon reservoir, helping to reduce the concentration of greenhouse gases in the atmosphere (Pan et al., Bonan, 2008; 2011; Malhi & Grace, 2000; Brown et al., 1999).

The decomposition of leaves, roots, and other plant residues in the soil contributes to the formation of humus. This humus accumulates in the upper layers of the soil, increasing its organic carbon content (Bradford et al., 2016; Berg & McClaugherty, 2014). In addition, the presence of a dense and diverse tree cover promotes microbial activity and the decomposition of organic matter, accelerating the carbon cycle in the soil.

The ability of cacao-based AFS to sequester carbon not only benefits the environment but also farmers. Reducing carbon emissions and capturing carbon dioxide from the atmosphere can generate carbon credits, which can be sold on international carbon markets. This represents an additional economic opportunity for cacao farmers, while incentivizing sustainable agricultural practices (Garrity et al., 2010).

At this point, it is important to keep in mind that carbon sequestration in cacao-based AFS with cacao varies according to different factors, such as shade tree species composition (Salas et al., 2020; Andrade-Castañeda et al., 2017; Luyssaert et al., 2008), agricultural management practices (Lal, 2006), and local climatic conditions (Wall et al., 2023). To maximize carbon sequestration potential, appropriate management strategies, such as the promotion of fast-growing tree species and the adoption of sustainable soil management practices, must be implemented. Table 6.4 shows some published data on carbon stocks in cacao-based AFS in Latin America.

This carbon sequestration potential makes cacao-based AFS valuable allies in the fight against climate change as they not only help reduce greenhouse gas emissions but also improve climate resilience by reducing the vulnerability of cacao plants to extreme weather events.

6.2.7 Socioeconomics

Agroforestry techniques tend to align with indigenous agricultural practices while respecting local knowledge and traditions (Jacobi et al., 2017; Altieri et al., 2004; Thapa et al., 1995). These systems foster a sense of identity and pride in agricultural communities by preserving the cultural heritage of cacao production. The shade trees of the AFS provide gathering spaces and refuge for farmers and their families, promoting social cohesion and well-being (Meli et al., 2023).

According to Schneider et al. (2017), the yield disparity between monoculture and agroforest systems has led some people to believe that AFS are incapable of producing sufficient cacao in a reasonable amount of time. Numerous studies have been conducted to compare the yield of cacao grown in monoculture and AFS; for example, in Bolivia, Jacobi et al., (2015) found no statistical differences between monoculture cacao cultivation systems (350–124.2 kg dry beans ha^{-1}), AFS (423kg ha^{-1}), and successional AFS (510–55.2 kg ha^{-1}).

Thus, cacao monocultures have been established in open fields (full sun) to maximize productivity and profitability in the short term, but have resulted in soil erosion and degradation, loss of biodiversity, and susceptibility to climate

Table 6.4 Some published data on carbon stocks in cacao-based agroforestry systems in Latin America

Carbon stocks (Mg C ha-1)	*Carbon pool*	*Reference*	*Country*	*Observation*
28.8 ± 4.1c	AGB	Núñez et al., 2021	Colombia	>3,5 years
28.8 ± 4.1c	SOC	Núñez et al., 2021	Colombia	>3,5 years
49.2 – 88.5	AGB	Ballesteros-Possú et al., 2022	Colombia	10 years
13.64 ± 3.69	AGB	Goñas et al., 2022	Perú	8 – 15 years
20.50 ± 5.63	AGB	Goñas et al., 2022	Perú	16 – 29 years
24.86 ± 19.91	AGB	Goñas et al., 2022	Perú	30 – 40 years
119.96 ± 24.07	SOC	Goñas et al., 2022	Perú	8 – 15 years
128.10 ± 42.39	SOC	Goñas et al., 2022	Perú	16 – 29 years
131.96 ± 30.50	SOC	Goñas et al., 2022	Perú	30 – 40 years
5.30	AGB	Mena-Mosquera & Andrade, 2021	Colombia	4 years
50.60	SOC	Carvajal-Agudelo & Andrade, 2021	Colombia	
75	AGB	Salvador-Morales et al., 2020	México	70 years
36	AGB	Salvador-Morales et al., 2020	México	10 years
95.30	AGB	Middendorp et al., 2018	Ecuador	50 years simulation
28.80	AGB	Andrade et al., 2013	Colombia	18 years
33.60	AGB	Andrade et al., 2013	Colombia	35 years
61	AGB	del Marín et al., 2016	Colombia	
65.61	AGB	Pocomucha et al., 2016	Perú	
65.57	SOC	Pocomucha et al., 2016	Perú	
Approx. 200 avg.	SOC	Monroe et al., 2016	Brazil	4 – 40 years
46 – 87	AGB	Schroth et al., 2015	Brazil	
52.70 ± 7.80	AGB	Jadán et al., 2015	Ecuador	
15.30 ± 2	BGB	Jadán et al., 2015	Ecuador	
69.20 ± 5	SOC	Jadán et al., 2015	Ecuador	
4.10 ± 0.,40	Necromass	Jadán et al., 2015	Ecuador	
141.40 ± 11	Total carbon	Jadán et al., 2015	Ecuador	
128.40 – 143,70	Total carbon	Jacobi et al., 2014	Bolivia	
93 ± 30	Total carbon	Somarriba et al., 2013	Nicaragua	
96 ± 37	Total carbon	Somarriba et al., 2013	Honduras	
106 ± 25	Total carbon	Somarriba et al., 2013	Guatemala (Alta Verapaz)	
122 ± 24	Total carbon	Somarriba et al., 2013	Costa Rica	
132 ± 60	Total carbon	Somarriba et al., 2013	Panamá	
155 ± 58	Total carbon	Somarriba et al., 2013	Guatemala (Costa Sur)	
12.18	Total carbon	Orjuela-Chaves et al., 2014	Colombia	1 – 7 years
53.24	Total carbon	Orjuela-Chaves et al., 2014	Colombia	8 – 12 years
104.66	Total carbon	Orjuela-Chaves et al., 2014	Colombia	>20 years

change, diseases, and pests (Andres et al., 2016; Armengot et al., 2016). In contrast, cacao-based AFS allow a differentiated productive use of land and the preservation of agroclimatic conditions such as the physical, chemical, and biological properties of soils in the tropical cacao-growing regions of the world (Niether et al., 2019; Arévalo-Gardini et al., 2020). In this sense, Armengot et al. (2016) noted that while monocultures may boost early cacao yields, the positive impact of by-crops on labor returns may decline, but timber tree sales could balance this effect.

The value added to the product obtained through AFS may increase the farm's profitability. According to Jacobi et al. (2015), local groups that facilitate organic certification promote agroecological variety and, as a result, contribute to strengthening agroecosystem resistance to climate change impacts. More sustainable production techniques, such as agroforestry and organically managed systems, are predicted to yield less cacao, although by-crops and premium pricing, respectively, may compensate for the reduced yields (Armengot et al., 2016). In addition, chocolate manufacturers, consumers, and environmental activists can encourage farmers to maintain AFS by paying a premium for this traditional practice (Bennett et al., 2022).

For the yield of cacao in monoculture versus cacao associated with timber or fruit species, comprehensive comparisons must be made of several components. Vaast and Somarriba (2014) stated that worldwide, cacao is mostly produced by smallholders with little capital; therefore, low investment capacity for technical innovation and the use of superior genotypes are essential for increasing cacao yields, limiting the incidence of pests and diseases, and producing high-quality chocolate. According to Andres et al. (2016), cacao genotyping programs are needed in the context of agroforestry management, as well as appropriate tree spacing and pruning practices at critical times in the production cycle. Suitable combinations of different trees with complementary leaf phenology and local species must be elaborated to improve functional biodiversity. In this sense, Notaro et al. (2021) found that cacao tree productivity increased in the presence of Fabaceae, even at short distances (<3 m), and to a lesser extent in the presence of timber trees.

Finally, in terms of economics, the inclusion of complementary and diverse crops not only reduces agriculturists' reliance on cacao as a sole source of income (Cialdella et al., 2023) but also provides greater economic stability in the face of potential fluctuations in yields as a result of climate variability, thereby opening up new market opportunities, which in turn strengthens the agriculturists' overall resilience (Vaast & Somarriba, 2014).

6.3 Conclusions

Agroforestry research related to cacao is a relevant and active field in Latin America. Countries, such as Brazil, Colombia, Mexico, Costa Rica, and Peru, have made significant contributions in terms of quantity and quality of research in this field. In addition, there is significant collaboration between countries in the fields of agroforestry and cacao. This suggests common interests and the importance of sharing knowledge and experiences in the development of sustainable practices in cacao production.

Cacao-based AFS provide a range of ecosystem services that benefit the environment and local communities. Recognizing and supporting the adoption of these systems is crucial for policymakers, farmers, and agroforestry experts to ensure the long-term viability of cacao production. To further strengthen the resilience of cacao-based AFS in Latin America, it is essential to promote adaptation strategies, best practices, and research that contribute to closing the knowledge gaps around these agroecosystems. This implies promoting research and development of cacao varieties resistant to different threats from climate variability, as well as the use of sustainable and environmentally friendly farming techniques.

Training farmers in the implementation of these practices and disseminating knowledge through extension programs and collaboration between researchers and farmers are also key elements for improving the resilience of the system. In addition, knowledge sharing among different actors is crucial to promoting the resilience of cacao-based AFS. Collaboration between research institutions, agricultural extension services, and farmers' organizations allows for the transfer of information, best practices, and lessons learned.

Agricultural extension services, on the other hand, are responsible for bringing scientific knowledge to farmers practically and understandably. These services provide direct technical assistance, working closely with farmers to implement innovative practices and overcome specific challenges on their farms.

By actively involving farmers, local communities, and stakeholders in decision-making processes, the creation of more robust and sustainable systems is encouraged. Organizations, such as agricultural cooperatives or cacao producer associations, work closely with farmers to encourage the adoption of sustainable agroforestry practices and the implementation of climate change adaptation measures. They also provide technical support, access to markets, and financial resources, thus strengthening farmers' capacity to face climate challenges. However, it is still necessary for cacao farmers to know some aspects, such as the relationship between shade and disease in the crop, because there are still doubts and lack of knowledge about the important aspects of AFS.

References

Alcívar, T. A., García, V. G., Cadena, P. D., & Sánchez, V. V. (2019). Evaluación y planificación de sistemas agroforestales sustentables de cacao (*Theobroma cacao* L.) y bambú (*Guadua angustifolia* K.), Montalvo, Ecuador. *Journal of Science and Research*, 4(4), 10–21.

Alpízar, L., Fassbender, H. W., Heuveldop, J., Fölster, H., & Enríquez, G. (1986). Modelling agroforestry systems of cacao (*Theobroma cacao*) with laurel (*Cordia alliodora*) and poro (*Erythrina poeppigiana*) in Costa Rica – I. Inventory of organic matter and nutrients. *Agroforestry Systems*, 4(3) , 175–189. https://doi.org/10.1007/BF02028353

Andres, C., Comoé, H., Beerli, A., Schneider, M., Rist, S., & Jacobi, J. (2016). Cocoa in monoculture and dynamic agroforestry. In: Lichtfouse, E. (eds) *Sustainable Agriculture Reviews*, vol 19, 121–153. Springer, Cham. https://doi.org/10.1007/978-3-319-26777-7_3

Andres, C., Blaser, W. J., Dzahini-Obiatey, H. K., Ameyaw, G. A., Domfeh, O., Awiagah, M. A., Gattinger, A., Schneider, M., Offei, S. K., & Six, J. (2018). Agroforestry systems can mitigate the severity of cacao swollen shoot virus disease. *Agriculture, Ecosystems & Environment*, 252, 83–92. http://dx.doi.org/10.1016/j.agee.2017.09.031

Antolinez-Sandoval, E., Almanza-Merchán, P., Barona-Rodriguez, A., Polanco-Díaz, E., & Serrano-Cely, P. (2020). Estado actual de la cacaocultura: una revisión de sus principales limitantes. *Revista Ciencia y Agricultura*, 17(2), 1–11. https://doi.org/10.19053/01228420.v17.n2.2020.10729

Arévalo-Gardini, E., Canto, M., Alegre, J. C., Arévalo-Hernández, C. O., Loli, O., Julca, A., & Baligar, V. C. (2020). Cacao agroforestry management systems effects on soil fungi diversity in the Peruvian Amazon. *Ecological Indicators*, 115, 106404. https://doi.org/10.1016/j.ecolind.2020.106404

Armengot, L., Barbieri, P., Andres, C., Milz, J., & Schneider, M. (2016). Cacao agroforestry systems have higher return on labor compared to full-sun monocultures. *Agronomy for Sustainable Development*, 36, 70. https://doi.org/10.1007/s13593-016-0406-6

Armengot, L., Ferrari, L., Milz, J., Velásquez, F., Hohmann, P., & Schneider, M. (2020). Cacao agroforestry systems do not increase pest and disease incidence compared with monocultures under good cultural management practices. *Crop Protection*, 130, 105047. https://doi.org/10.1016/j.cropro.2019.105047

Ahmad Dar, M., Singh, R., Ahmad, M., Kaur, S., Singh, H. P., & Batish, D. R. (2023). Socioeconomic and ecological sustainability of agroforestry in mountain regions. In *Understanding Soils of Mountainous Landscapes*, 375–394. https://doi.org/10.1016/b978-0-323-95925-4.00009-1

Altieri, M. A. (2004). Linking ecologists and traditional farmers in the search for sustainable agriculture. *Frontiers in Ecology and the Environment*, 2(1), 35–42. https://doi.org/10.1890/1540-9295(2004)002[0035:LEATFI]2.0.CO;2

Andrade, H. J., del Figueroa, J. M. P., & Silva, D. P. (2013). Almacenamiento de carbono en cacaotales (*Theobroma cacao*) en Armero-Guayabal (Tolima, Colombia). *Scientia Agroalimentaria*, 1, 2339–4684.

Andrade-Castañeda, H. J., Segura-Madrigal, M. A., Canal-Daza, D. S., Huertas-Gonzales, A., & Mosos-Torres, C. A. (2017). Composición florística y reservas de carbono en bosques ribereños en paisajes agropecuarios de la zona seca del Tolima, Colombia. *Revista de Biologia Tropical*, 65(4), 1245–1260. https://doi.org/10.15517/rbt.v65i4.27007

Apshara, S. E., Rajesh, M. K., & Balasimha, D. (2013). Assessment of morphological, physiological and molecular characteristics of cacao accessions from Central and South America in relation to drought tolerance. *Journal of Plantation Crops*, 41(3), 389–397.

Asante, W. A., Acheampong, E., Kyereh, E., & Kyereh, B. (2017). Farmers' perspectives on climate change manifestations in smallholder cacao farms and shifts in cropping systems in the forest-savannah transitional zone of Ghana. *Land Use Policy*, 66, 374–381. https://doi.org/10.1016/j.landusepol.2017.05.010

Avadí, A. (2023). Environmental assessment of the Ecuadorian cacao value chain with statistics-based LCA. *International Journal of Life Cycle Assessment*, 28, 1495–1515. https://doi.org/10.1007/s11367-023-02142-4

Baffour-Ata, F., Antwi-Agyei, P., Boakye, L., Tettey, L. S. N. A., Forson, M. N. E. F., Abiwu, A. E., Gyenin, E., & Larbi, R. N. M. (2023). Assessing the adaptive capacity of smallholder cacao farmers to climate variability in the Adansi South District of the Ashanti Region, Ghana. *Heliyon*, 9(3), e13994. https://doi.org/10.1016/j.heliyon.2023.e13994

Bai, H. S., Trueman, S. J., Nevenimo, T., Hannet, G., Bapiwai, P., Poienou, M., & Wallace, H. M. (2017). Effects of shade-tree species and spacing on soil and leaf nutrient concentrations in cacao plantations at 8 years after establishment. *Agriculture, Ecosystems & Environment*, 246, 134–143. https://doi.org/10.1016/j.agee.2017.06.003

Ballesteros-Possú, W., Valencia, J. C., & Navia Estrada, J. F. (2022). Assessment of a cocoa-based agroforestry system in the southwest of Colombia. *Sustainability* (Switzerland), 14(15), 9447. https://doi.org/10.3390/su14159447

Beer, J., Bonnemann, A., Chavez, W., Fassbender, H. W., Imbach, A. C., & Martel, I. (1990). Modelling agroforestry systems of cacao (*Theobroma cacao*) with laurel (*Cordia alliodora*) or poro (*Erythrina poeppigiana*) in Costa Rica. *Agroforestry Systems*, 12(3), 229–249. https://doi.org/10.1007/bf00137286

Beer, J., Harvey, C., Muhammad, I., Harmand, J. M., Somarriba, E., & Jiménez, F. (2003). Environmental services of agroforestry systems. / Servicios ambientales de los sistemas agroforestales. *Agroforestería En Las Américas*, 10.

Bennett, R. E., Sillett, T. S., Rice, R. A., & Marra, P. P. (2022). Impact of cacao agricultural intensification on bird diversity and community composition. *Conservation Biology*, 36(1), e13779. https://doi.org/10.1111/cobi.13779

Bentley, J. W., Boa, E. & Stonehouse, J. (2004). Neighbor trees: Shade, intercropping, and cacao in Ecuador. *Human Ecology*, 32, 241–270. https://doi.org/10.1023/B:HUEC.0000019759.46526.4d

Berg, B., & McClaugherty, C. (2014). Plant litter: Decomposition, humus formation, carbon sequestration. In *Plant Litter: Decomposition, Humus Formation, Carbon Sequestration*. https://doi.org/10.1007/978-3-642-38821-7_9

Biswas, B., Chakraborty, D., Timsina, J., Bhowmick, U. R., Dhara, P. K., Ghosh (Lkn), D. K., Sarkar, A., Mondal, M., Adhikary, S., Kanthal, S., Patra, K., Sarkar, S., Parsad, R., & Ray, B. R. (2022). Agroforestry offers multiple ecosystem services in degraded lateritic soils. *Journal of Cleaner Production*, 365, 132768. https://doi.org/10.1016/j.jclepro.2022.132768

Blaser-Hart, W. J., Hart, S. P., Oppong, J., Kyereh, D., Yeboah, E., & Six, J. (2021). The effectiveness of cacao agroforests depends on shade-tree canopy height. *Agriculture, Ecosystems and Environment*, 322. https://doi.org/10.1016/j.agee.2021.107676

Boinot, S., Poulmarc'h, J., Mézière, D., Lauri, P. É., & Sarthou, J. P. (2019). Distribution of overwintering invertebrates in temperate agroforestry systems: Implications for biodiversity conservation and biological control of crop pests. *Agriculture, Ecosystems and Environment*, 285. https://doi.org/10.1016/j.agee.2019.106630

Bonan, G. B. (2008). Forests and climate change: Forcings, feedbacks, and the climate benefits of forests. *Science*, 320(5882), 1444–1449. https://doi.org/10.1126/science.1155121

Bradford, M. A., Berg, B., Maynard, D. S., Wieder, W. R., & Wood, S. A. (2016). Understanding the dominant controls on litter decomposition. *Journal of Ecology*, 104(1), 229–238. https://doi.org/10.1111/1365-2745.12507

Brito-Vega, H., Salaya-Domínguez, J. M., Gómez-Méndez, E., Gómez-Vázquez, A., & Antele-Gómez, J. B. (2018). Physico-chemical properties of soil and pods (*Theobroma cacao* L.) in cacao agroforestry systems. *Journal of Agronomy*, 17(1), 48–55. https://doi.org/10.3923/ja.2018.48.55

Brown, S. L., Schroeder, P., & Kern, J. S. (1999). Spatial distribution of biomass in forests of the eastern USA. *Forest Ecology and Management*, 123(1), 81–90. https://doi.org/10.1016/S0378-1127(99)00017-1

Burgess, S. S. O., Adams, M. A., Turner, N. C., & Ong, C. K. (1998). The redistribution of soil water by tree root systems. *Oecologia*, 115(3), 306–311. https://doi.org/10.1007/s004420050521

Carvajal-Agudelo, B. N., & Andrade, H. J. (2021). Organic carbon in soils with different systems of use in Tacarimena Yopal, Colombia. *Revista U.D.C.A Actualidad and Divulgacion Cientifica*, 24(1), e1921. https://doi.org/10.31910/rudca.v24.n1.2021.1921

Cassamo, C. T., Draper, D., Romeiras, M. M., Marques, I., Chiulele, R., Rodrigues, M., Stalmans, M., Partelli, F. L., Ribeiro-Barros, A., & Ramalho, J. C. (2023). Impact of climate changes in the suitable areas for *Coffea arabica* L. production in Mozambique: Agroforestry as an alternative management system to strengthen crop sustainability. *Agriculture, Ecosystems and Environment*, 346 108341. https://doi.org/10.1016/j.agee.2022.108341

Celi-Delgado, L., & Aguirre-Mendoza, Z. (2022). Caracterización de los sistemas agroforestales tradicionales de la parroquia Zumba, cantón Chinchipe, Ecuador. *Ciencia Latina Revista Científica Multidisciplinar*, 6(4), 814–837. https://doi.org/10.37811/cl_rcm.v6i4.2626

Centeno-Alvarado, D., Lopes, A. V., & Arnan, X. (2023). Fostering pollination through agroforestry: A global review. In *Agriculture, Ecosystems and Environment*, 351, 108478. https://doi.org/10.1016/j.agee.2023.108478

Cialdella, N., Jacobson, M., & Penot, E. (2023). Economics of agroforestry: Links between nature and society. *Agroforestry Systems*, 97, 273–277. https://doi.org/10.1007/s10457-023-00829-z

Concha, J. Y., Alegre, J. C., & Pocomucha, V. (2007). Determinación de las reservas de carbono en la biomasa aérea de sistemas agroforestales de *Theobroma cacao* L. en el Departamento de San Martìn, Peru. *Ecología Aplicada*, 6(1–2), 75–82. https://doi.org/10.21704/rea.v6i1-2.343

da Cruz Silva, A., Teixeira, H. M., Victer, M. C. A., Bahia, L. B., Torres, C. M. M. E., Villa, P. M., & Cardoso, I. M. (2023). Diversity and function of tree species in human-modified Atlantic Forest landscapes. *Agroforestry Systems*, 97, 799–812. https://doi.org/10.1007/s10457-023-00826-2

Danso-Abbeam, G., Amin, K. M., & Ogundeji, A. A. (2023). Enhancing household welfare through perennial crop production in Northern Ghana. *Sustainability (Switzerland)*, 15(1), 451. https://doi.org/10.3390/su15010451

Daymond, A. J., & Hadley, P. (2008). Differential effects of temperature on fruit development and bean quality of contrasting genotypes of cacao (*Theobroma cacao*). *Annals of Applied Biology*, 153(2), 175–185. https://doi.org/10.1111/j.1744-7348.2008.00246.x

de Almeida, J., Tezara, W., & Herrera, A. (2016). Physiological responses to drought and experimental water deficit and waterlogging of four clones of cacao (*Theobroma cacao* L.) selected for cultivation in Venezuela. *Agricultural Water Management*, 171, 80–88. https://doi.org/10.1016/j.agwat.2016.03.012

de la Hoz-M, J., Fernández-Gómez, M. J., & Mendes, S. (2021). LDAShiny: An R package for exploratory review of scientific literature based on a Bayesian probabilistic model and machine learning tools. *Mathematics*, 9(14), 1671.

de Lima, R. F., Viegas, L., Solé, N., Soares, E., Dallimer, M., Atkinson, P. W., & Barlow, J. (2014). Can management improve the value of shade plantations for the endemic species of são tomé island? *Biotropica*, 46(2). 238–247. https://doi.org/10.1111/btp.12092

Delabie, J. H. C., Jahyny, B., do Nascimento, I. C., Mariano, C. S. F., Lacau, S., Campiolo, S., Philpott, S. M., & Leponce, M. (2007). Contribution of cacao plantations to the conservation of native ants (Insecta: Hymenoptera: Formicidae) with a special emphasis on the Atlantic Forest fauna of southern Bahia, Brazil. *Biodiversity and Conservation*, 16(8), 2359–2384. https://doi.org/10.1007/s10531-007-9190-6

de Sousa, D. C., Medeiros, J. C., Rosa, J. D., de Lacerda, J. J. J., Mafra, Á. L., & de Mendes, W. S. (2017). Chemical attributes of agricultural soil after the cultivation of cover crops. *Australian Journal of Crop Science*, 11(11). https://doi.org/10.21475/ajcs.17.11.11.pne799

Deheuvels, O., Avelino, J., Somarriba, E., & Malezieux, E. (2012). Vegetation structure and productivity in cacao-based agroforestry systems in Talamanca, Costa Rica. *Agriculture, Ecosystems and Environment*, 149, 181–188. https://doi.org/10.1016/j.agee.2011.03.003

Delgado-Ospina, J., Molina-Hernández, J. B., Chaves-López, C., Romanazzi, G., & Paparella, A. (2021). The role of fungi in the cacao production chain and the challenge of climate change. *Journal of Fungi*, 7(3), 202. https://doi.org/10.3390/jof7030202

Díaz-José, O., Aguilar-Ávila, J., Rendón-Medel, R., & Santoyo-Cortés, V. H. (2013). Current state of and perspectives on cacao production in Mexico. *Ciencia e Investigación Agraria*, 40(2), 279–289. https://doi.org/10.4067/s0718-16202013000200004

Domínguez-Núñez, J. A. (2022). Leguminous trees for sustainable tropical agroforestry. In *Advances in Legumes for Sustainable Intensification*, 483–504. https://doi.org/10.1016/B978-0-323-85797-0.00006-9

Donato, J. W., Lucio-Quintana, A.,Valverde, H., Quijije, R., Bayas, F., & Oderay Merino, O. (2016). Evaluación de la fauna inséctil en tres sistemas agroforestales utilizando como base el cultivo de cacao (*Theobroma cacao* L) en el cantón Caluma (Ecuador). *Revista Enlace Universitario*, 15(1), 42–46.

dos Reis, S. M., Talamini, E., da Silva Neto, P. J., Geraldo Augusto, S., de Melo, A. C. G., & Dewes, H. (2019). Growth and yield of mahogany wood in cacao-based agroforestry systems of two soil types in the Brazilian Amazon. *Agroforestry Systems*, 93(6), 2163–2172. https://doi.org/10.1007/s10457-018-0327-2

dos Santos, M. L. S., França, S., Gomes, F. P., do Nascimento, J. L., dos Anjos Silva, L., & Mielke, M. S. (2012). Low light availability affects leaf gas exchange, growth and survival of Euterpe edulis seedlings transplanted into the understory of an anthropic tropical rainforest. *Southern Forests*, 74(3), 167–174. https://doi.org/10.2989/20702620.2012.722833

End, M. J., Hadley, P., & Pettipher, G. L. (1988). Long and shortterm studies of the growth of cacao (*Theobroma cacao* L.) pods in relation to photo-thermal environment. IN *Proceedings of the 10th International Cocoa Research Conference*, Santo Domingo, Dominican Republic, pp. 219–223.

Fatumah, N., Mohammed, S., Ashraf, N., Abasi, K., & Shadia, N. (2023). Adoption of novel climate-smart farming systems for enhanced carbon stock and carbon dioxide equivalent emission reduction in cattle corridor areas of Uganda. *Heliyon*, 9(3), e14114. https://doi.org/10.1016/j.heliyon.2023.e14114

Fremier, A. K., DeClerck, F. A. J., Bosque-Pérez, N. A., Carmona, N. E., Hill, R., Joyal, T., Keesecker, L., Klos, P. Z., Martínez-Salinas, A., Niemeyer, R., Sanfiorenzo, A., Welsh, K., & Wulfhorst, J. D. (2013). Understanding spatiotemporal lags in ecosystem services to improve incentives. *BioScience*, 63(6), 472–482. https://doi.org/10.1525/bio.2013.63.6.9

García-Cruzatty, L., Arteaga-Alcívar, F., Vera-Pinargote, L., & Pérez-Almeida, I. (2023). Water deficit influence upon pollen grain production in cacao genotypes (*Theobroma cacao*). *Bioagro*, 35(2), 167–174. https://doi.org/10.51372/bioagro352.9

Garrity, D. P., Akinnifesi, F. K., Ajayi, O. C., Weldesemayat, S. G., Mowo, J. G., Kalinganire, A., Larwanou, M., & Bayala, J. (2010). Evergreen agriculture: A robust approach to sustainable food security in Africa. *Food Security*, 2(3), 197–214. https://doi.org/10.1007/s12571-010-0070-7

Gateau-Rey, L., Tanner, E. V. J., Rapidel, B., Marelli, J. P., & Royaert, S. (2018). Climate change could threaten cacao production: Effects of 2015-16 El Niño-related drought on cacao agroforests in Bahia, Brazil. *PLoS ONE*, 13(7), e0200454. https://doi.org/10.1371/journal.pone.0200454

Goñas, M., Rojas-Briceño, N. B., Culqui-Gaslac, C., Arce-Inga, M., Marlo, G., Pariente-Mondragón, E., & Oliva-Cruz, M. (2022). Carbon sequestration in fine aroma cocoa agroforestry systems in Amazonas, Peru. *Sustainability (Switzerland)*, 14(15), 1–12. https://doi.org/10.3390/su14159739

Habib-ur-Rahman, M., Ahmad, A., Raza, A., Hasnain, M. U., Alharby, H. F., Alzahrani, Y. M., Bamagoos, A. A., Hakeem, K. R., Ahmad, S., Nasim, W., Ali, S., Mansour, F., & el Sabagh, A. (2022). Impact of climate change on agricultural production; Issues, challenges, and opportunities in Asia. *Frontiers in Plant Science*, 13. https://doi.org/10.3389/fpls.2022.925548

Haggar, J., Pons, D., Saenz, L., & Vides, M. (2019). Contribution of agroforestry systems to sustaining biodiversity in fragmented forest landscapes. *Agriculture, Ecosystems and Environment*, 283, 106567. https://doi.org/10.1016/j.agee.2019.06.006

Harvey, C. A., & González Villalobos, J. A. (2007). Agroforestry systems conserve species-rich but modified assemblages of tropical birds and bats. *Biodiversity and Conservation*, 16(8), 2257–2292. https://doi.org/10.1007/s10531-007-9194-2

Hashmiu, I., Agbenyega, O., & Dawoe, E. (2022). Cash crops and food security: Evidence from smallholder cacao and cashew farmers in Ghana. *Agriculture and Food Security*, 11(1), 12. https://doi.org/10.1186/s40066-022-00355-8

Jacobi, J., Andres, C., Schneider, M., Pillco, M., Calizaya, P., & Rist, S. (2014). Carbon stocks, tree diversity, and the role of organic certification in different cacao production systems in Alto Beni, Bolivia. *Agroforestry Systems*, 88(6), 1117–1132. https://doi.org/10.1007/s10457-013-9643-8

Jacobi, J., Mathez-Stiefel, S. L., Gambon, H., Rist, S., & Altieri, M. (2017). Whose knowledge, whose development? Use and role of local and external knowledge in agroforestry projects in Bolivia. *Environmental Management*, 59(3), 464–476. https://doi.org/10.1007/s00267-016-0805-0

Jacobi, J., Schneider, M., Bottazzi, P., Pillco, M., Calizaya, P., & Rist, S. (2015). Agroecosystem resilience and farmers' perceptions of climate change impacts on cacao farms in Alto Beni, Bolivia. *Renewable Agriculture and Food Systems*, 30(2), 170–183. https://doi.org/10.1017/S174217051300029X

Jadán, O., Cifuentes, M., Torres, B., Selesi, D., Veintimilla, D., & Günter, S. (2015). Influence of tree cover on diversity, carbon sequestration and productivity of cacao systems in the Ecuadorian Amazon. *Bois et Forets Des Tropiques*, 325, 35–47. https://doi.org/10.19182/bft2015.325.a31271

Jadán, O., Günter, S., Torres, B., & Selesi, D. (2014). Riqueza y potencial maderable en sistemas agroforestales tradicionales como alternativa al uso del bosque nativo, Amazonia del Ecuador. *Revista Forestal Mesoamericana Kurú*, 12(28), 13–22. https://doi.org/10.18845/rfmk.v12i28.2096

Jaimez, R., Loor, R., Arteaga, F., Márquez, V., & Tezara, W. (2021). Differential response of photosynthetic activity, leaf nutrient content and yield to long-term drought in cacao clones. *Acta Agronomica*, 70(3), 274–284. https://doi.org/10.15446/acag.v70n3.92252

Jones, T. (2019). textmineR: Functions for text mining and topic modeling. R package version 3.0. 4. https://doi.org/10.32614/CRAN.package.textmineR

Kassa, G., Bekele, T., Demissew, S., & Abebe, T. (2023). Plant species diversity, plant use, and classification of agroforestry homegardens in southern and southwestern Ethiopia. *Heliyon* 9(6), e16341. https://doi.org/10.1016/j.heliyon.2023.e16341

Kheiri, M., Kambouzia, J., Sayahnia, R., Soufizadeh, S., Mahdavi Damghani, A., & Azadi, H. (2023). Environmental and socioeconomic assessment of agroforestry implementation in Iran. *Journal for Nature Conservation*, 72, 126358. https://doi.org/10.1016/j.jnc.2023.126358

Kummerow, J., Kummerow, M., & Souza da Silva, W. (1982). Fine-root growth dynamics in cacao (*Theobroma cacao*). *Plant and Soil*, 65(2), 193–201. https://doi.org/10.1007/BF02374650

Lagneaux, E., Andreotti, F., & Neher, C. M. (2021). Cacao, copoazu and macambo: Exploring Theobroma diversity in smallholder agroforestry systems of the Peruvian Amazon. *Agroforestry Systems*, 95(7), 1359–1368. https://doi.org/10.1007/s10457-021-00610-0

Lal, R. (2006). Enhancing crop yields in the developing countries through restoration of the soil organic carbon pool in agricultural lands. *Land Degradation and Development*, 17(2), 197–209. https://doi.org/10.1002/ldr.696

Lalremsang, P., Gopichand, B., Upadhaya, K., Singson, L., & Sahoo, U. K. (2022). Leaf litter decomposition and nutrient release dynamics of *Flemingia semialata* Roxb. - A potential woody perennial species for mountain agroforestry. *Journal of Sustainable Forestry*, 42(5), 214–472. https://doi.org/10.1080/10549811.2022.2045504

Litschel, J., Berendt, F., Wagner, H., Heidenreich, S., Bauer, D., Welp, M., & Cremer, T. (2023). Key Actors' Perspectives on Agroforestry's Potential in North Eastern Germany. *Land*, 12(2), 458. https://doi.org/10.3390/land12020458

Lori, M., Armengot, L., Schneider, M., Schneidewind, U., Bodenhausen, N., Mäder, P., & Krause, H. M. (2022). Organic management enhances soil quality and drives microbial community diversity in cacao production systems. *Science of the Total Environment*, 834, 155223. https://doi.org/10.1016/j.scitotenv.2022.155223

Luyssaert, S., Schulze, E. D., Börner, A., Knohl, A., Hessenmöller, D., Law, B. E., Ciais, P., & Grace, J. (2008). Old-growth forests as global carbon sinks. *Nature*, 455(7210), 213–215. https://doi.org/10.1038/nature07276

Malek, Ž., Loeffen, M., Feurer, M., & Verburg, P. H. (2022). Regional disparities in impacts of climate extremes require targeted adaptation of Fairtrade supply chains. *One Earth*, 5(8), 917–931. https://doi.org/10.1016/j.oneear.2022.07.008

Malhi, Y., & Grace, J. (2000). Tropical forests and atmospheric carbon dioxide. *Trends in Ecology and Evolution*, 15(8), 332–337. https://doi.org/10.1016/S0169-5347(00)01906-6

del Marín, Q.M. P., J. H. Andrade, and A.P. Sandoval, (2016). Fijación de carbono atmosférico en la biomasa total de sistemas de producción de cacao en el departamento del Tolima, Colombia. *Revista U.D.C.A Actualidad & Divulgación Científica*, 19(2), 351–360. https://doi.org/10.31910/rudca.v19.n2.2016.89

McQueen, J. P., & Treonis, A. M. (2020). Cacao agroforestry in Belize: Effects on soil nematode community structure. *Agroforestry Systems*, 94(3), 1123–1132. https://doi.org/10.1007/s10457-019-00477-2

Meinhardt, L. W., Rincones, J., Bailey, B. A., Aime, M. C., Griffith, G. W., Zhang, D. & Pereira, G. A. G. (2008). *Moniliophthora perniciosa*, the causal agent of witches' broom disease of cacao: What's new from this old foe? *Molecular Plant Pathology*, 9(5), 577–588. https://doi.org/10.1111/j.1364-3703.2008.00496.x

Meli, P., Vieli, L., Spirito, F., Reyes-Riveros, R., Gonzalez-Suhr, C., & Altamirano, A. (2023). The importance of considering human well-being to understand social preferences of ecosystem services. *Journal for Nature Conservation*, 72, 126344. https://doi.org/10.1016/j.jnc.2023.126344

Mena-Mosquera, V. E., & Andrade, H. J. C. (2021). Potential for reducing emissions and carbon sequestration in forests and cacao agroforestry systems in the Colombian Pacific. *Revista de Biologia Tropical*, 69(4), 1252–1263. https://doi.org/10.15517/rbt.v69i4.45927

Middendorp, R. S., Vanacker, V., & Lambin, E. F. (2018). Impacts of shaded agroforestry management on carbon sequestration, biodiversity and farmers income in cacao production landscapes. *Landscape Ecology*, 33(11), 1953–1974. https://doi.org/10.1007/s10980-018-0714-0

Molua, E. L., Sonwa, D., Bele, Y., Foahom, B., Mate Mweru, J. P., Wa Bassa, S. M., Gapia, M., Ngana, F., Joe, A. E., & Masumbuko, E. M. (2023). Climate-smart conservation agriculture, farm values and tenure security: Implications for climate

change adaptation and mitigation in the Congo Basin. *Tropical Conservation Science*, 16, 19400829231169980. https://doi.org/10.1177/19400829231169980
Monroe, P. H. M., Gama-Rodrigues, E. F., Gama-Rodrigues, A. C., & Marques, J. R. B. (2016). Soil carbon stocks and origin under different cacao agroforestry systems in Southern Bahia, Brazil. Agriculture, *Ecosystems and Environment*, 221, 99–108. https://doi.org/10.1016/j.agee.2016.01.022
Morán-Villa, V. L., Monterroso-Rivas, A. I., Gómez-Díaz, J. D., Márquez-Berber, S. R., & Valdes-Velarde, E. (2022). Composición florística y arreglo de sistemas agroforestales de cacao en tabasco, México. *Tropical and Subtropical Agroecosystems*, 25(2). https://doi.org/10.56369/tsaes.3840
Mortimer, R., Saj, S., & David, C. (2018). Supporting and regulating ecosystem services in cacao agroforestry systems. *Agroforestry Systems*, 92(6), 1639–1657. https://doi.org/10.1007/s10457-017-0113-6
Muñoz, D., Estrada, A, & Naranjo, E. (2005). Monos aulladores (*Alouatta palliata*) en una plantación de cacao (*Theobroma cacao*) en Tabasco, México: Aspectos de la ecología alimentaria. *Universidad y Ciencia*, II, 35–44. https://www.redalyc.org/articulo.oa?id=15421204
Muñoz, D., Estrada, A., Naranjo, E., & Ochoa, S. (2006). Foraging ecology of howler monkeys in a cacao (*Theobroma cacao*) plantation in Comalcalco, Mexico. *American Journal of Primatology*, 68(2), 127–142. https://doi.org/10.1002/ajp.20211
Muñoz, F., & Beer, J. (2001). Fine root dynamics of shaded cacao plantations in Costa Rica. *Agroforestry Systems*, 51(2), 19–130. https://doi.org/10.1023/A:1010651203815
Ndoumbè-Nkeng, M., Efombagn, M. I. B., Nyassé, S., Nyemb, E., Sache, I., & Cilas, C. (2009). Relationships between cacao Phytophthora pod rot disease and climatic variables in Cameroon. *Canadian Journal of Plant Pathology*, 31(3), 309–320. https://doi.org/10.1080/07060660909507605
Niether, W., Glawe, A., Pfohl, K., Adamtey, N., Schneider, M., Karlovsky, P., & Pawelzik, E. (2020). The effect of short-term vs. long-term soil moisture stress on the physiological response of three cacao (*Theobroma cacao* L.) cultivars. *Plant Growth Regulation*, 92(2), 295–306. https://doi.org/10.1007/s10725-020-00638-9
Niether, W., Jacobi, J., Blaser, W. J., Andres, C., & Armengot, L. (2020). Cocoa agroforestry systems versus monocultures: A multi-dimensional meta-analysis. *Environmental Research Letters*, 15(10), 104085. https://doi.org/10.1088/1748-9326/abb053
Niether, W., Schneidewind, U., Fuchs, M., Schneider, M., & Armengot, L. (2019). Below- and aboveground production in cacao monocultures and agroforestry systems. *Science of the Total Environment*, 657, 558–567. https://doi.org/10.1016/j.scitotenv.2018.12.050
Notaro, M., Collado, C., Depas, J. K., Dumovil, D., Denis, A. J., Deheuvels, O., Tixier, P., & Gary, C. (2021). The spatial distribution and height of associated crops influence cacao tree productivity in complex agroforestry systems. *Agronomy for Sustainable Development*, 41, 60. https://doi.org/10.1007/s13593-021-00716-w
Núñez, A. P. B., Gutiérrez-Montes, I., Hernández-Núñez, H. E., Suárez, D. R. G., García, G. A. G., Suárez, J. C., Casanoves, F., Flora, C., & Sibelet, N. (2023). Diverse farmer livelihoods increase resilience to climate variability in southern Colombia. *Land Use Policy*, 131, 106731. https://doi.org/10.1016/j.landusepol.2023.106731
Núñez, H. E. H., Andrade, H. J., Salazar, J. C. S., Sánchez, A. J. R., Gutiérrez, S. D. R., García, G. A. G., Trujillo, E. T., & Casanoves, F. (2021). Carbon storage in agroforestry systems in Colombia's eastern plains. *Revista de Biologia Tropical*, 69(1), 352–368. https://doi.org/10.15517/RBT.V69I1.42959
Nurhayati, Siregar, C., Akbar, A., & Sinaga, D. (2022). Determine the effects of drought stress on the cacao seedlings (*Theobroma cacao* L.) with rice straw compost. *Asian Journal of Plant Sciences*, 21(2), 215–220. https://doi.org/10.3923/ajps.2022.215.220

Oh, C., Tomczak, J., Gavves, E., & Welling, M. (2019). Combinatorial bayesian optimization using the graph cartesian product. *Advances in Neural Information Processing Systems*, 32. arXiv:1902.00448v2.

Orjuela-Chaves, J. A., Andrade C. H. J., & Vargas-Valenzuela, Y. (2014). Potential of carbon storage of rubber (Hevea brasiliensis Müll. Arg.) plantations in monoculture and agroforestry systems in the Colombian Amazon. *Tropical and Subtropical Agroecosystems*, 17(2), 231–240.

Oro, Z. F., Bonnot, F., Ngo Bieng, M. A., Delaitre, E., Dufour, B. P., Ametefe, K. E., Mississo, E., Wegbe, K., Muller, E., & Cilas, C. (2012). Spatiotemporal pattern analysis of cacao swollen shoot virus in experimental plots in Togo. *Plant Pathology*, 61, 1043–1051. https://doi.org/10.1111/j.1365-3059.2011.02589.x

Ortiz-Rodríguez, O. O., Villamizar-Gallardo, R. A., Naranjo-Merino, C. A., García-Caceres, R. G., & Castañeda-Galvís, M. T. (2016). Carbon footprint of the Colombian cacao production. *Engenharia Agricola*, 36(2), 260–270. https://doi.org/10.1590/1809-4430-Eng.Agric.v36n2p260-270/2016

Pan, Y., Birdsey, R. A., Fang, J., Houghton, R., Kauppi, P. E., Kurz, W. A., Phillips, O. L., Shvidenko, A., Lewis, S. L., Canadell, J. G., Ciais, P., Jackson, R. B., Pacala, S. W., McGuire, A. D., Piao, S., Rautiainen, A., Sitch, S., & Hayes, D. (2011). A large and persistent carbon sink in the world's forests. *Science*, 333(6045), 988–993. https://doi.org/10.1126/science.1201609

Papalexandratou, Z., Vrancken, G., de Bruyne, K., Vandamme, P., & de Vuyst, L. (2011). Spontaneous organic cacao bean box fermentations in Brazil are characterized by a restricted species diversity of lactic acid bacteria and acetic acid bacteria. *Food Microbiology*, 28(7), 1326–1338. https://doi.org/10.1016/j.fm.2011.06.003

Pérez, G. A., & Freile, J. A. (2017). Adaptabilidad de clones promisorios de cacao nacional (*Theobroma cacao* L.), en el cantón Arosemena Tola de Ecuador. *Centro Agrícola*, 44(2), 44–51.

Pérez-Neira, D., Schneider, M., Esche, L., & Armengot, L. (2023). Sustainability of food security in different cacao production systems: A land, labour, energy and food quality nexus approach. *Resources, Conservation and Recycling*, 190, 106874. https://doi.org/10.1016/j.resconrec.2023.106874

Perin, A., Santos, R. H. S., Urquiaga, S., Guerra, J. G. M., & Cecon, P. R. (2004). Produção de fitomassa, acúmulo de nutrientes e fixação biológica de nitrogênio por adubos verdes em cultivo isolado e consorciado. *Pesquisa Agropecuária Brasileira*, 39(1), 35–40. https://doi.org/10.1590/s0100-204x2004000100005

Pocomucha, V. S., Alegre, J., & Abregú, L. (2016). Análisis socio económico y carbono almacenado en sistemas agroforestales de cacao (*Theobroma cacao* L.) en Huánuco. *Ecología Aplicada*, 15(2), 107–114. https://doi.org/10.21704/rea.v15i2.750

Prieto-Benavides, O. O., Blezaca-Pinargote, C. E., Mora-Silva, W. F., Garcés-Fiallos, F. R., Sabando-Ávila, F. A., & Cedeño-Loja, P. E. (2012). Identificación de hongos micorrízicos arbusculares en sistemas agroforestales con cacao en el trópico húmedo ecuatoriano. *Agronomía Mesoamericana*, 23(2), 233–239. https://doi.org/10.15517/am.v23i2.6482

Quandt, A., Neufeldt, H., & Gorman, K. (2023). Climate change adaptation through agroforestry: opportunities and gaps. *Current Opinion in Environmental Sustainability*, 60, 101244. https://doi.org/10.1016/j.cosust.2022.101244

Quiroz, J. (2010). *Sistema de sombra de cacao con maderables*. Instituto Nacional de Investigaciones Agropecuarias del Ecuador, INIAP, Boletín Técnico No. 151. 2010. 11p.

Ramírez-Meneses, A., García-López, E., Obrador-Olán, J. J., Ruiz-Rosado, O., & Camacho-Chiu, W. (2013). Diversidad florística en plantaciones agroforestales de cacao en Cárdenas, Tabasco, México. *Universidad y Ciencia*, 29(3), 215–230.

Ramírez, O. A., Somarriba, E., Ludewigs, T., & Ferreira, P. (2001). Financial returns, stability and risk of cacao-plantain-timber agroforestry systems in Central America. *Agroforestry Systems*, 51(2), 141–154. https://doi.org/10.1023/A:1010655304724

Reitsma, R., Parrish, J. D., & Mclarney, W. (2001). The role of cacao plantations in maintaining forest avian diversity in southeastern Costa Rica. *Agroforestry Systems*, 53(2), 185–193. https://doi.org/10.1023/A:1013328621106

Rousseau, G. X., Deheuvels, O., Rodriguez Arias, I., & Somarriba, E. (2012). Indicating soil quality in cacao-based agroforestry systems and old-growth forests: The potential of soil macrofauna assemblage. *Ecological Indicators*, 23, 535–543. https://doi.org/10.1016/j.ecolind.2012.05.008

Salas, C., Montes, K., Sanchez, G., Alcívar, W., Murillo, A., Vera, F., Bolcato, D., & Iglesias, S. (2020). Influencia del gradiente altitudinal sobre la estimación del carbono almacenado en biomasa aérea viva y en suelos del "Bosque y vegetación protector El Artesan - EcuadorianHands". Joa, Jipijapa. *Ecosistemas*, 29(2), 1973.

Salvador-Morales, P., Martínez-Sánchez, J. L., Cabrales, L. C., & Ramos, C. Z. (2020). Structure and specific carbon in a chronosequence of *Teobroma cacao* L. Agroforestry systems in Tabasco, México. *Madera y Bosques*, 26(3), e2632131. https://doi.org/10.21829/myb.2020.2632131

Sánchez Gutiérrez, F., Pérez-Flores, J., Obrador Olan, J. J., Sol Sánchez, Á., & Ruiz-Rosado, O. (2016). Estructura arbórea del sistema agroforestal cacao en Cárdenas, Tabasco, México TT - Tree structure of cacao agroforestry system in Cárdenas, Tabasco, Mexico. *Revista Mexicana de Ciencias Agrícolas*, 7(spe14), 2695-2709.

Santana, M. B. M., & Cabala-Rosand, P. (1982). Dynamics of nitrogen in a shaded cacao plantation. *Plant and Soil*, 67(1–3), 271–281. https://doi.org/10.1007/BF02182774

Schneider, M., Andres, C., Trujillo, G., Alcon, F., Amurrio, P., Perez, E., Weibel, F., & Milz, J. (2017). Cocoa and total system yields of organic and conventional agroforestry vs. monoculture systems in a long-term field trial in Bolivia. *Experimental Agriculture*, 53(3), 351–374. https://doi.org/10.1017/S0014479716000417

Schnell, R. J., Brown, J. S., Kuhn, D. N., Cervantes-Martinez, C., Borrone, J. W., Olano, C. T., Motamayor, J. C., Lopes, U., Phillips, W., Johnson, E., Monteverde-Penso, E. J., & Amores, F. (2007). Current challenges of tropical tree crop improvement: Integrating genomics into an applied cacao breeding program. *Acta Horticulturae*, 738. https://doi.org/10.17660/ActaHortic.2007.738.11

Schroth, G., Bede, L. C., Paiva, A. O., Cassano, C. R., Amorim, A. M., Faria, D., Mariano-Neto, E., Martini, A. M. Z., Sambuichi, R. H. R., & Lôbo, R. N. (2015). Contribution of agroforests to landscape carbon storage. *Mitigation and Adaptation Strategies for Global Change*, 20(7), 1175–1190. https://doi.org/10.1007/s11027-013-9530-7

Schroth, G., Garcia, E., Griscom, B. W., Teixeira, W. G., & Barros, L. P. (2016). Commodity production as restoration driver in the Brazilian Amazon? Pasture re-agro-forestation with cacao (*Theobroma cacao*) in southern Pará. *Sustainability Science*, 11(2), 277–293. https://doi.org/10.1007/s11625-015-0330-8

Schroth, G., Jeusset, A., da Gomes, A. S., Florence, C. T., Coelho, N. A. P., Faria, D., & Läderach, P. (2016). Climate friendliness of cacao agroforests is compatible with productivity increase. *Mitigation and Adaptation Strategies for Global Change*, 21(1), 67–80. https://doi.org/10.1007/s11027-014-9570-7

Somarriba, E., Cerda, R., Orozco, L., Cifuentes, M., Dávila, H., Espin, T., Mavisoy, H., Ávila, G., Alvarado, E., Poveda, V., Astorga, C., Say, E., & Deheuvels, O. (2013). Carbon stocks and cacao yields in agroforestry systems of Central America. *Agriculture, Ecosystems and Environment*, 173, 46–57. https://doi.org/10.1016/j.agee.2013.04.013

Staton, T., Walters, R. J., Breeze, T. D., Smith, J., & Girling, R. D. (2022). Niche complementarity drives increases in pollinator functional diversity in diversified agroforestry systems. *Agriculture, Ecosystems and Environment*, 336, 108035. https://doi.org/10.1016/j.agee.2022.108035

Suazo, Y., Davidov-Pardo, G., & Arozarena, I. (2014). Effect of Fermentation and Roasting on the Phenolic Concentration and Antioxidant Activity of Cocoa from Nicaragua. *Journal of Food Quality*, 37(1), 50–57. https://doi.org/10.1111/jfq.12070

Subía, C., Caicedo, C., Calderón, D., Tinoco, L., Pico, J., Vargas, Y., Fernández, Y., Paredes, N., Vera, A., Díaz, A., Bastidas, S., Sotomayor, D., Lima, L., Intriago, J., Congo, C., & Sánchez, M. (2017). Establecimiento de un ensayo en sistemas agroforestales de cacao con diferentes niveles de manejo en la Amazonía ecuatoriana. In *2017 International Symposium on Cocoa Research (ISCR)*. Lima, Peru.

Suh, N. N., & Molua, E. L. (2022). Cocoa production under climate variability and farm management challenges: Some farmers' perspective. *Journal of Agriculture and Food Research*, 8, 100282. https://doi.org/10.1016/j.jafr.2022.100282

Talipov, K., & Mukimov, T. (2022). Creation of an agricultural security system: An effective model of sustainable land use. *American Journal of Plant Sciences*, 13(05), 613–622. https://doi.org/10.4236/ajps.2022.135041

Tapia-Vera, C. A., Sánchez-Mora, F. D., Vásconez-Montúfar, G. H., Barrera-Alvarez, A. E., Mora-Yela, R. V., Díaz-Coronel, G. T., & Garcés-Fiallos, F. R. (2021). Evaluación temporal de sistemas agroforestales de cacao en el trópico húmedo ecuatoriano. *Revista Bionatura*, 6(4), 2295–2300. https://doi.org/10.21931/RB/2021.06.04.27

Tezara, W., De Almeida, J., Valencia, E., Cortes, J., & Bolaños, M. (2015). Actividad fotoquímica de clones élites de cacao (*Theobroma Cacao* L.) ecuatoriano en el norte de la Provincia de Esmeraldas. *Revista Científica Interdisciplinaria Investigación Y Saberes*, 4(3), 37–51. http://revistasdigitales.utelvt.edu.ec/revista/index.php/investigacion_y_saberes/article/view/90

Tezara, W., Pereyra, G., Ávila-Lovera, E., & Herrera, A. (2020). Variability in physiological responses of Venezuelan cacao to drought. *Experimental Agriculture*, 56(3), 407–421. https://doi.org/10.1017/S0014479720000058

Thapa, B., Sinclair, F. L., & Walker, D. H. (1995). Incorporation of indigenous knowledge and perspectives in agroforestry development. *Agroforestry Systems*, 30(1–2), 235–248. https://doi.org/10.1007/bf00708924

Toledo-Hernández, M., Wanger, T. C., & Tscharntke, T. (2017). Neglected pollinators: Can enhanced pollination services improve cacao yields? A review. *Agriculture, Ecosystems and Environment*, 247, 137–148. https://doi.org/10.1016/j.agee.2017.05.021

Torres de la Cruz, M., Ortiz García, C. F., Téliz Ortiz, D., Mora Aguilera, A., & Nava Díaz, C. (2011). Temporal progress and integrated management of frosty pod rot (*Moniliophthora roreri*) of cacao in Tabasco, Mexico. *Journal of Plant Pathology*, 93(1), 31–36.

Torres, E., Torres, A., & Sánchez, A. (2018). Agro-ecosistemas tradicionales con cacao: Análisis de casos de pequeños productores en Los Ríos, Ecuador. *Revista Amazónica Ciencia y Tecnología*, 7(2), 38–95.

Tschora, H., & Cherubini, F. (2020). Co-benefits and trade-offs of agroforestry for climate change mitigation and other sustainability goals in West Africa. *Global Ecology and Conservation*, 22, e00919. https://doi.org/10.1016/j.gecco.2020.e00919

Vaast, P., & Somarriba, E. (2014). Trade-offs between crop intensification and ecosystem services: the role of agroforestry in cacao cultivation. *Agroforestry Systems*, 88(6), 947–956. https://doi.org/10.1007/s10457-014-9762-x

Vargas Tierras, Y. B., Prado-Beltrán, J. K., Nicolalde-Cruz, J. R., Casanoves, F., de Virginio-Filho, E. M., & Viera Arroyo, W. F. (2018). Caracterización y rol de los frutales amazónicos en fincas familiares en las provincias de Sucumbíos y Orellana (Ecuador). *Ciencia y Tecnología Agropecuaria*, 19(3), 485–499. https://doi.org/10.21930/rcta.vol19_num3_art:812

Vásquez, E. C., Hipólito-Romero, E., Ricaño-Rodríguez, J., & Ramos-Prado, J. M. (2022). Ecophysiological plasticity of *Theobroma cacao* L. clones in response to the structure and microclimate of agroforestry systems in Mexico. *Botanical Sciences*, 100(4). https://doi.org/10.17129/botsci.2925

Vaupel, A., Bednar, Z., Herwig, N., Hommel, B., Moran-Rodas, V. E., & Beule, L. (2023). Tree-distance and tree-species effects on soil biota in a temperate agroforestry system. *Plant and Soil*. 487, 355–372. https://doi.org/10.1007/s11104-023-05932-9

Vera, M., Rosales, H., & Urena, N. (2000). Caracterización físico-química de algunos suelos cacaoteros de la zona sur del lago de Maracaibo, Venezuela. *Revista Geografica Venezolana*, 41(2), 257–270.

Verdezoto Vargas, V. H., Valdiviezo Freire, E. W., Durán Mera, C. A., & Illicachi Gusñay, R. C. (2023). Especies maderables en sistemas agroforestales con cacao (*Theobroma cacao* L.), cantón yaguachi, Provincia del Guayas. *LATAM Revista Latinoamericana de Ciencias Sociales y Humanidades*, 4(2), 1080–1093. https://doi.org/10.56712/latam.v4i2.670

Wall, A. M., Goodrich, J. P., Campbell, D. I., Morcom, C. P., & Schipper, L. A. (2023). The carbon balance of a temperate grazed pasture following periodic maize silage cropping depends on climate and management. *Agriculture, Ecosystems & Environment*, 352, 108523. https://doi.org/10.1016/j.agee.2023.108523

Wartenberg, A. C., Blaser, W. J., Janudianto, K. N., Roshetko, J. M., van Noordwijk, M., & Six, J. (2018). Farmer perceptions of plant–soil interactions can affect adoption of sustainable management practices in cacao agroforests: A case study from Southeast Sulawesi. *Ecology and Society*, 23(1). https://doi.org/10.5751/ES-09921-230118

Wong, E., Pajuelo, G., Sotelo, F., & Raymundo, C. (2020). Implementation of a quality management system (QMS) based on TQM principles for Cocoa MSEs: A case study in Cusco, Perú. *Advances in Intelligent Systems and Computing*, 1018, 910–916. https://doi.org/10.1007/978-3-030-25629-6_142

Wongnaa, C. A., & Babu, S. (2020). Building resilience to shocks of climate change in Ghana's cacao production and its effect on productivity and incomes. *Technology in Society*, 62, 101288. https://doi.org/10.1016/j.techsoc.2020.101288

Xiong, H., Cheng, Y., Zhao, W., & Liu, J. (2019). Analyzing scientific research topics in manufacturing field using a topic model. *Computers & Industrial Engineering*, 135, 333–347.

Zhang, R., Tian, D., Wang, J., & Niu, S. (2023). Critical role of multidimensional biodiversity in contributing to ecosystem sustainability under global change. *Geography and Sustainability*, 4(3), 232–243. https://doi.org/10.1016/j.geosus.2023.05.002

Zhu, X., Yuan, X., Lu, E., Yang, B., Wang, H., Du, Y., Singh, A. K., & Liu, W. (2023). Soil splash erosion: An overlooked issue for sustainable rubber plantation in the tropical region of China. *International Soil and Water Conservation Research*, 11(1), 30–42. https://doi.org/10.1016/j.iswcr.2022.05.005

7 Vulnerability of small cacao producers in Manabí, Ecuador, to climate change from the exposure dimension

Jose Ricardo Macias Barberan

7.1 Introduction

Since ancient times, nature has undergone a series of changes that have forced human beings to adapt, allowing their subsistence. Over the centuries, man has developed a series of actions that have influenced these changes, causing phenomena that have threatened his own existence and caused thousands of deaths.

It was in the mid-50s of the last century when a group of scientists alerted the international community about a set of changes that had been accumulating. The main protagonists are CO_2 emissions, erosion, and damage to the ozone layer, generating a phenomenon that increases the temperature of the atmosphere due to the accumulation of gasses called the 'greenhouse effect'.

Consequently, the concern of large organizations of global scope such as the United Nations Organization arises, which on different occasions brings all its members together with the purpose of taking actions to reduce CO_2 emissions, contributing to the care of the environment and therefore to the preservation of animal and plant species in the short term and even to the human species in the long term.

At present, there is a great concern regarding climate change, causing a series of meteorological alterations that bring with them innumerable phenomena such as drought, floods, melting glaciers, rise in sea level, and heat waves among others, altering the stability of ecosystems and even destroying areas that a few years ago were considered safe.

Consequently, the effects of climate change are increasingly worrisome. The projections of scientific organizations are not at all encouraging for humanity, which is why an increase in global GHGs of between 25% and 90% is expected between the years 2000 and 2030, assuming that fossil fuels maintain their dominant position in the global mix of energy sources until at least the 2030 Intergovernmental Panel on Climate Change (IPCC, 2021).

Similarly, it is estimated that by the year 2050, the frequency of extreme warm values, heat waves, and intense precipitation will increase, as well as the intensity of tropical cyclones, displacements towards the poles of the trajectories of extra-tropical storms, accompanied by changes in wind, precipitation,

DOI: 10.4324/9781003381761-9

and temperature patterns, and precipitation will most likely increase at high latitudes, as well as decrease over most subtropical land regions (IPCC, 2021).

One of the sectors most prone to the effects of climate change is agriculture, since the natural factor determines the success or productive debacle, with water, soil, and sunlight being the essential elements to develop a crop. On the contrary, if the variations in temperature and rainfall were of great proportions, it would threaten the food sovereignty of the people.

In the Ecuadorian context, climatic variations have forced farmers to start taking measures related to the supply of water to plantations through irrigation systems, the most vulnerable being those who own small extensions and who do not have money to purchase such service. In the province of Manabí, in recent years, there has been evidence of changes in the amount of rainfall and traditional rainy seasons, generating losses on a small scale, especially in small producers.

The abovementioned implies significant variations in short-term meteorological conditions, making communities increasingly vulnerable to the presence of natural phenomena related to climate change. Therefore, in this chapter, a series of elements that determine this vulnerability are analyzed; in this case, those of cacao producers in the province of Manabí belonging to the 'Fortaleza del Valle' Association.

7.2 Materials and methods

7.2.1 Location

The geographical site survey for vulnerability determination was carried out in 269 production units (PUs) located in five cantons of the province of Manabí—Bolívar, Chone, Tosagua, Junín, and Portoviejo—located at latitude: 0° 25' 00" N up to 1° 57' S and longitude: 79° 24' W up to 80° 55' W, with an average annual rainfall of 1,000 to 2,000 mm, with temperatures ranging between 25 and 36 °C, with predominantly sandstone soils, on steep hills, are well developed, with typical horizons of alteration, shallow and in the process of erosion, with pH fluctuation between 5.9 and 8.2, and with a silty texture vegetation with scrub, cacti, and thorny bushes.

7.2.2 Selection of sampling sites

As previously mentioned, a single random probabilistic sampling was conducted to ensure the inclusion of members from all the cantons that are part of the 'Fortaleza del Valle' Corporation using the criteria.

In the study area, soil sampling sites were selected considering the spatial variability of the terrain, especially color, and slope, for respective physical and chemical analyses. In each selected site, a test pit was made to describe the profile and sampling for the characterization of the properties was used in the vulnerability assessment.

7.2.3 Processes

For this chapter, the methodology was guided by the proposal of Monterroso et al. (2012), where all the indicators had to be standardized to eliminate the different scales and make them comparable among themselves as presented below **(Eq 7.1)**:

$$\mathbf{zv} = (X_i X) / DS \tag{7.1}$$

In this regard, Z represents the standardized value of the variable v, X_i is the observed value of v, X is the mean value of the set of values i, and DS is the standard deviation of the set of values i. With the standardized data, each of the subindices of exposure, sensitivity, and adaptive capacity were calculated. The process followed was to algebraically add the standardized values of the indicators of each subindex to later estimate vulnerability. Once the subindices were calculated, their distances were normalized to values ranging from **0 to 100%** according to the following formula **(Eq 7.2)**:

$$\mathbf{Ip} = (\text{Iobs} - \text{minimumI}) / (\text{maximumI} - \text{minimumI}) \tag{7.2}$$

where

I = subscript of the variable P,
Iobs = value of the subscript for the given region,
Minimum = minimum observed value of the set of IP values, and
Maximum = maximum observed value of the same set of IP values.

In a later step, the final score in percentage was obtained by multiplying the normalized values by 100. The vulnerability to climate change was determined using climatic, edaphic, and social variables, in response to the proposals of Olivares (2015), Monterroso et al. (2012), Jimenez (2012), and Zevallos (2004), considering 52 indicators, distributed across exposure (18), sensitivity (16), and adaptive capacity (18), algebraically combined according to the following equation **(Eq 7.3)**:

$$\mathbf{Vulnerability} = (\text{exposure} + \text{sensitivity}) - \text{adaptive capacity} \tag{7.3}$$

$$\mathbf{Exposure} = \sum \text{indicators}(\%) / \text{n}$$

$$\mathbf{Sensibility} = \sum \text{indicators}(\%) / \text{n}$$

$$\mathbf{Adaptative\ Capacity} = \sum \text{indicators}(\%) / \text{n}$$

where n is the number of indicators for each case expressed in %

For the algebraic treatment of the information, a database was generated with the indicators registered and processed in the conventional units of each variable and then all the indicators were normalized and standardized between 0 and 1, by the maximum value method, to eliminate the different scales and make them comparable to each other. According to Monterroso et al. (2012), the indicators can be converted into %, obtaining values between **0 and 100**, and are reclassified into five vulnerability categories as shown in **Table 7.1**.

7.2.4 Physical and chemical properties

In each selected site, soil samples were collected according to the proposal of the Colombian Agricultural Institute (CAI, 1992). Each sample was dried and sieved using a 2 mm sieve at the Technical University of Manabí (TUM), and then taken to the National Institute of Agricultural Research (NIAR) for its chemical characterization and to the National University of Palmira for its physical characterization. The properties that were determined are presented in Tables 7.2 and 7.3:

7.2.5 Laboratory methods

7.2.5.1 pH

It was determined using the potentiometry method with a glass electrode in a volumetric soil–water ratio of 1:1; that is, 10 cm^3 of distilled water was added to 10 g of soil. The mixture was allowed to stand for one hour, stirred, and then the pH was read using the potentiometer (AOAC, 1999).

Table 7.1 Categorization of vulnerability levels and percentage by vulnerability level

Vulnerability Levels	*Percentage by Vulnerability Level*
Very high	0–20
High	21–40
Half	41–60
Low	61–80
Very low	81–100

Table 7.2 Chemical soil properties and methods of chemical analysis

Chemical Soil Properties	*Medium or Extraction*	*Detection Technique*
pH	1:1 Water:Soil Ratio	Potentiometry
Organic carbon	Concentrated H_2SO_4	Colorimetry
Phosphorus	Sodium bicarbonate solution	Colorimetry
Calcium, magnesium, and potassium	Soil-Solution 1:5	Ammonium Acetate

Table 7.3 Physical soil properties and techniques or methods of determination

Physical soil properties	*Technique or Method of Determination*
Texture	Pipette method for the pits and fertility trunks
Apparent density	Cylinder method
Actual density	Pycnometer method
Moisture retention (Field Capacity and Wilting point	Strain Pots and Plates Method
Saturated hydraulic conductivity	Load drop. Localized source of water
Water stored in the soil	Available water sheet

7.2.5.2 *Organic material*

It was calculated indirectly by determining organic carbon. Organic carbon was oxidized using chromic acid in the presence of an excess of concentrated H_2SO_4 to later determine the organic material by colorimetric determination (Mackean, 1993).

7.2.5.3 *Phosphorus*

The Olsen Method (1982) was used, based on extraction with sodium bicarbonate solution, under the principle that in calcareous or alkaline soils bicarbonate ions cause calcium precipitation and therefore calcium activity in the solution decreases. This facilitates the extraction of the more soluble calcium phosphates. In more acidic soils, bicarbonate ions replace aluminum and iron phosphates, thus increasing the pH of the solution and facilitating the extraction of phosphate from surfaces that have a pH-dependent charge (Carrero et al., 2015).

7.2.5.4 *Sodium, potassium, calcium, and magnesium exchangeable cations*

Ammonium acetate 1N of pH=7 was used with a soil–solution ratio of 1:5. The extraction of cations from soils is done with a neutral salt solution. Ammonium acetate in a normal solution and its pH was adjusted to 7, which is the most widely used in the world. When extraction is done with a soil–solution ratio of 1:5 and shaken for 5 min, the extracted cations are expressed as 'exchangeable' cations, which are expressed internationally as such and the values obtained are more representative of soil evaluation (Mackean, 1993).

7.2.5.5 *Mechanical physical analysis*

The physical properties were analyzed under the parameters described by the USDA Soil Health and Quality Evaluation Guide (AOAC, 1999), using the following techniques and methods: apparent density ($g.cm^3$) by the known volume method using the cylinder and saturated hydraulic conductivity using a constant head permeameter (Forsythe, 1980); gravimetric humidity and volumetric humidity in percentage (Montenegro and Malagón, 1990); real density using the pycnometer method, and water stored in the soil and moisture retention by the strain pot and plates method (Klute, 1986). Texture analysis was carried out by the pipette or hydrometer method (Israelsen and Hansen, 1979).

7.2.6 Vulnerability variables

In the present investigation, a set of variables and subvariables through indicators are presented below:

7.2.6.1 Variable 'exposure' (18)

The indicators to determine exposure are distributed in four dimensions:

7.2.6.1.1 FREQUENCY OF EXTREME EVENTS (5)

This subvariable refers to a certain alteration of any previous condition that exceeds the response capacity of a community. Consequently, in this investigation the following indicators are considered extreme events:

- **Water available to plants**: To calculate this indicator, precipitation and temperature records from three meteorological stations that cover all the PUs under study were used. Water balance is calculated through real and potential evapotranspiration to determine the number of days with water deficit and the level of vulnerability. It is necessary to highlight that the meteorological readings of the 269 PUs considered as the study sample were located near the stations: Chone, Portoviejo, and Rocafuerte.
- **Flood report**: A cartographic analysis of the database of the National Secretariat of Risk Management (SNGR, 2017) was performed to geographically locate the PUs prone to flooding. They were weighed in terms of floodable and non-floodable percentages and presented. Subsequently, a flood index with dimensionless values between 0 and 1 was built using the normalization method based on the maximum value and reclassifying vulnerability levels.
- **Landslide**: For landslide determination, the cartography available in the SNGR was consulted (SNGR, 2017) and in this way, those PUs prone to landslides were located through percentages. Once the results were obtained, a landslide index was developed with dimensionless values between 0 and 1, using the normalization method based on the maximum value and reclassifying vulnerability levels.
- **Heavy rains**: To determine its vulnerability, rainfall data available at the National Institute of Meteorology and Hydrology of Ecuador were extracted through the yearbooks of the last 20 years available (1993–2013) and thus know the anomaly in that period. It is important to point out that

the studies were carried out through the observations of the PUs located in the stations Chone, Portoviejo, and Rocafuerte. Thus, the scale to classify vulnerability due to rainfall anomalies is known, and the results are shown by the weather stations.

- **Other types of natural phenomena**: Reported in different databases of official organizations
  - **Fires**: Regarding the fires produced in the study area, it was possible to know through records from the Ministry of the Environment (MAGAP, 2017) that in the Province of Manabí, 1,173 hectares were burned, a figure that was contrasted by the number of hectares consumed at the national level (16,908.14). Subsequently, the national–provincial relationship was made and thus determined the degree of vulnerability, starting from values between 0 and 1, which was 0.93.
  - **Percentage affected by pests and diseases**: To know this indicator, the producers under study were consulted if they had suffered attacks from pests and diseases through the surveys. To define the assessment, ranges between 0 and 1 were used, where 0 indicates those PUs with attacks and 1 indicates those that do not report such a situation.
  - **Biodiversity loss**: To know the percentage, the 'Red Book of Species' was investigated through the registry of endangered species, yielding the results of the province of Manabí.
  - **Percentage of PU with losses due to lack of soil fertility**: This indicator was obtained by consulting the producers about the loss of production due to the lack of any of the chemical elements necessary for the development of the crops. The answers obtained were through a simple option, which was then weighted according to the scale between 0 and 1, taking the one with the highest production as a reference to measure its vulnerability.

7.2.6.1.2 Environmental problems (2)

- **Deforestation**: It was calculated using the information from the Ministry of the Environment (MAGAP, 2017) and the Normalized Difference Vegetation Index (NDVI), allowing the identification of forested and deforested areas, evaluating the state of the vegetation and its degree of stress, discriminating different types of plant masses, and delimiting areas occupied by water.

It is important to highlight that for the calculation of the NDVI the five cantons with incidence in the investigation were taken as reference: Chone, Portoviejo, Tosagua, Junín, and Bolívar. Therefore, the vulnerability levels of each one of the areas do not reflect the reality of the PUs, since their total area is equal to that of cultivation. In this sense, the NDVI was used through dimensionless values that vary between -1 and 1. Its calculation was made by means of the infrared and red bands, in addition to an image, using the following equation **(Eq 7.7)**:

$$\mathbf{NDVI} = (\text{Inf} - \text{R}) / (\text{Inf} + \text{R}) \tag{7.7}$$

where

info = Infrared band of the image
R = Red Band of the image.

- **Percentage of Production Units that report losses due to weather issues**: It was obtained through the survey, where the producers were asked if the climatic conditions affected their production. Likewise, the level of productivity was taken as a reference. Subsequently, the answers obtained were weighted according to the scale between 0 and 1, taking the one with the highest production as a reference to measure their vulnerability.

7.2.6.1.3 MODELED CHANGES IN CLIMATE (3)

- **Variations in average temperature**: To know the variations in the average temperature, we used the historical records of the meteorological stations of the National Institute of Meteorology and Hydrology (2017). In this sense, it is necessary to point out that the measurements were made at three points, Chone, Portoviejo, and Rocafuerte, and then the PUs were located taking as reference the proximity in relation to said unit.
- **Variations in average precipitation**: For the calculation of this indicator, the historical records of the National Institute of Meteorology and Hydrology (2017) from the Chone, Portoviejo, and Rocafuerte stations were used, and the 269 PUs under study were located by geographical criteria. Subsequently, the vulnerability value of each of the stations was determined, taking rainfall variations as a reference.
- **Standardized Pluviometric Drought Index**: This indicator was calculated by analyzing historical rainfall data in the study area, defining spatial variations, as well as temporal trends with GIS and mathematical models. After obtaining the data, the proposal of Thielen (2015) was used to estimate the temporal variability of rainfall and know the index for each of the PUs **(Eq 7.8)**.

$$\text{IESP}_i = \text{APA}_{ci} \text{-} \text{APA}_c / \sigma \text{APA}_c \tag{7.8}$$

where

IESPi = Standardized Pluviometric Drought Index for month i

APAc = Mean value of the accumulated rainfall anomalies of all the months of the series.
APAci = Anomaly accumulated rainfall for month i
σAPAc = Standard deviation of the accumulated rainfall anomalies for all the months of the series.

After applying the formula, the Standardized Pluviometric Drought Index was determined for each of the PUs based on the meteorological readings of the corresponding unit.

7.2.6.1.4 SOIL PHYSICOCHEMICAL PROPERTIES (8)

- **Amount of Oxygen available to the roots**: This indicator is related directly to the percentage of existing macropores and the types of drainage, being valued by each PU. To calculate its valuation, the PU with the highest percentage of macropores (in this case 31.55%) was taken as the norm and divided by each of the values of the others.
- **Total Porous Space for Soil**: It is related to the macro, meso, and microspaces of the soil according to its texture. For its calculation, the results at the laboratory level were extracted and classified according to the proposal of Flores and Alcalá (2010).
- **Soil Hydraulic Conductivity**: It is related to the texture of the soil and its values in the field and given laboratory, referring to the ease or not with which the soil allows water to pass through units that are linear or turbulent. Likewise, it has to do with a derived quantity such as speed that concerns a length and time. Its unit of measurement is $cm.s^1$ (Whitlow, 1994). For its vulnerability, the results were known at the laboratory level, which is directly related to the type of soil.
- **Usable Water Sheet (UWS)**: It is related to field capacity, root depth, wilting point, apparent density, and usable water. It was calculated for each PU included in the study through the following mathematical expression **(Eq 7.9)**.

$$\mathbf{UWS} = (\mathrm{Fc} - \mathrm{Pm})\,\mathrm{x}\,\mathrm{Ad}\,\mathrm{x}\,\mathrm{Rd} \tag{7.9}$$

where

UWS = Usable Water Sheet
Fc = field capacity
Ad = apparent density
Rd = required depth

- **Texture and ease of rooting for Cacao**: It depends on the texture in relation to the height of the cacao plant, soil moisture, and rainy season (García,

2005; FAO, 1999). In this regard, the vulnerability assessment in terms of texture and ease of rooting for Cacao by PUs is presented.

- **Availability of Organic Matter for Cacao cultivation**: For the calculation of this indicator, the organic matter content and its relationship with soil respiration, the general indices for respiration, and the state of the soil in optimal conditions in relation to temperature and humidity, primarily for agricultural use, were considered (Woods End Research, 1997). In this sense, at the laboratory level, soil respiration was assessed through the amount of kilograms of carbon dioxide (CO_2) and thus the vulnerability was determined.
- **Soil slope for Cacao cultivation**: To develop this indicator, the slope level of each of the PUs was located using satellite methods and later the types of soils suitable for cacao cultivation in the area were related to its topography as stated by FAO (1999). Soil Type versus Topography of study area.
- **Fertility levels for Cacao**: The presentation of this indicator was carried out considering elements such as acidity (pH), organic matter (OM), phosphorus oxide (P_2O_5), calcium (Ca), magnesium (Mg), and potassium (K), according to the proposal of the SIG-CORPOICA and the 'Agroecosystems' Program (García, 2005).

7.3 Results

To obtain each vulnerability indicator, detailed procedures were developed that generated information in specific units for each case. Then, each indicator was normalized by the maximum value method, dividing the data of each indicator by the maximum record and obtaining dimensionless magnitudes, between 0 and 1, to compare them with each other and categorize them into vulnerability levels according to the class limits indicated in each case.

7.3.1 Frequency of extreme events

- **Water available to plants**: For every reason, three results are presented in relation to the table of vulnerability mentioned above. 87.36% (Chone), equivalent to 235 PUs, having 161.2 days of water deficit present high vulnerability, 10.03% (Rocafuerte), equivalent to 27 PUs, having 241.40 days of water deficit present very high vulnerability, and 2.60% (Portoviejo), equivalent to seven PUs, having 217.72 days of water deficit present very high vulnerability.

According to Araque et al. (2014), a cacao plant of productive age (>5 years) needs 4–6 mm of water/day; therefore, the soil must receive at least 1800 mm of water per year. The timing of drought or water scarcity during the annual crop cycle has key effects on expected production; if drought occurs during the fruit-filling phase, the size of the almonds will be smaller, and if it occurs during the vegetative growth phase, the end of the harvest will be severely affected, thus reducing its photosynthetic capacity (Orozco and López, 2018).

- **Flood report**: According to the results obtained, and in relation to the previous table, the vulnerability of the PUs in relation to flooding is as follows: 74.38%, equivalent to 201 PUs, is floodable with a vulnerability value of 0 and 25.62%, equivalent to 68 Pus, is classified as 'non-floodable', presenting a vulnerability of 1.

The PUs with high risk correspond to plots located in the low areas of the river basin that tend to be at risk due to flooding in the rainy season, which would cause damage from pests and soil diseases as well as lack of oxygen. In this sense, Suárez et al. (2015) point out that excess water in the soil causes aeration deficit, asphyxiation, and death of roots. The crop reduces its capacity to absorb nutrients which leads to delayed growth and abortion of small fruits called 'gherkins'.

- **Landslide**: In relation to the results obtained, the vulnerability of the PUs with respect to landslides is as follows: 11.90%, equivalent to 31 PUs, have a 'high' vulnerability range of 0.33 and 88.10%, equivalent to 238 PUs, have a 'very low' vulnerability range of 1.

In the rainy season, there is a tendency to lose the structural stability of the soil with greater sediment carryover and a considerable reduction in organic matter deposits (Suarez et al., 2015).

- **Heavy rains**: Based on what is indicated in the previous scale, the vulnerability of the PUs in relation to the Rocafuerte meteorological station is 1, considered 'very low'. The rainfall anomalies are high in the seven PUs located in the Portoviejo weather station. It is important to highlight that from the agricultural point of view, rainfall anomalies are considered to be of 'very low' vulnerability, since the millimeters of water added are not enough to cover the soil storage.

Under this guideline, upon calculating the vulnerability between said ranges, for the Chone weather station, a total of 235 PUs was geographically referenced. It is essential to consider two important factors: one, the rainfall anomalies due to excess rainfall are modest such that they do not affect vulnerability on a large scale, and two, the slopes of the soils are an essential factor for the level of flooding. In this sense, the PU gave an index of 0.75, which means low vulnerability.

The PU located in areas with high rainfall may be more predisposed to the presence of phenomena such as 'El Niño' caused by climate change. In this sense, Moreno, et al. (2013) state that one of the climatic factors that most affects the intensity of crop growth is the amount and distribution of rainfall since rainfall variations could affect vegetative growth, flowering, and fruit development.

- **Other types of natural phenomena**
  - Fires: By contrasting the result obtained with the level categorization table, it was possible to know that there is a very low vulnerability to fire consequences.

In this situation, most PU are prone to great risks due to forest fires, which will cause increases in soil temperature, loss of plant cover, and reduction in moisture content, which alter some of their physical and biological characteristics (Moreno et al., 2013).

  - Percentage affected by pests and diseases: In this case, the number of PUs that reported attacks was 169, obtaining 'very high' vulnerability. As for the PUs that did not report attacks, there were 100, with 'very low' vulnerability.
  - Biodiversity loss: A loss of 6.88% in the biodiversity was recorded, according to the map which was extrapolated to the relationship 0 and 1, resulting in 0.07. It can be deduced that there exists very high vulnerability.
  - Percentage of PU with losses due to lack of soil fertility: Once the vulnerability indices of each PU were obtained, they were determined for each category. In this sense, 230 PUs presented 'very high' vulnerability to the phenomenon. Similarly, it was determined that 36 PUs are 'Highly' vulnerable in terms of loss due to lack of fertility in the soil. Likewise, the vulnerability index of another two PUs was calculated in relation to the degree of susceptibility, categorizing vulnerability as 'half'. Accordingly, one PU turned out to have a 'very low' vulnerability to loss due to the lack of soil fertility.
  - Integration of the sub indicators inherent to the indicator 'Other types of natural phenomena': Once the values are related to each of the subindicators that make up this indicator, we proceeded to integrate the vulnerability index, guided by the nature of this type of phenomenon. In this regard, it was known that 163 PUs are 'very highly' vulnerable to the natural phenomena specified above. On the other hand, five PUs were found to be 'Highly vulnerable'. Likewise, it was known that 96 PUs reported having 'half' vulnerability to other types of phenomena. In the same way, five PUs were classified as having 'low' vulnerability.

7.3.2 Environmental problems

- **Deforestation**: Once the equation was applied, the following values were known. 11.89% of the PUs had a deforestation level between 0.3 and ≥0.5. Therefore, it presents very high vulnerability, and 88.11% of PUs showed deforestation values less than 0.15 (considered acceptable). In summary, some plots have a tendency to high risk of impact from human activities or natural causes, which could lead to the loss of soil vegetation cover, reduction of the water table, and exposure to wind and solar radiation due to lack of natural shade for plantations since cacao is a shade-loving crop (Almeida and Valle, 2008).

- **Percentage of PUs that report losses due to weather issues**: The vulnerability value of each PU according to its level of production is presented. 39.77%, equivalent to 107 PUs, have a very low vulnerability, 28.99%, equivalent to 78 PUs, presented low vulnerability, 23.79%, equivalent to 64 Pus, presented half vulnerability, and 0.74%, equivalent to 20 PUs, have high vulnerability.

7.3.3 Modeled changes in climate

7.3.3.1 Variations in average temperature

- **PUs studied at the Rocafuerte, Portoviejo, and Chone Meteorological Stations.**

By geographic range, 27 PUs are located at Rocafuerte weather station. The temperature variation in the period between 1993 and 2013 shows a moderately worrisome anomaly in the years 1998 and 1999, on average varying between 1.2 and 1.5°C. However, the said variation is modest in the other years. Subsequently, the vulnerability level was calculated by comparing the historical data concerning the average, locating it between 0 and 1, and determining it for the PUs related to the weather station, which was 0.7. Thus, it was determined that the vulnerability is low.

A total of seven PUs were located near the Portoviejo weather station in the years ranging from 1993 to 2013, which show temperature variations between -4 and 3.8°C. It is decisive from the agricultural point of view that it directly affects the productivity levels. Therefore, in comparison to the annual averages of the values considered normal for the area, it was determined that there is a high vulnerability with a value of 0.

For the Chone weather station, in 235 PUs the temperature ranges during the period from 1993 to 2013 showed deviations between 2 and -1.4° C, with the years of greatest magnitude being 1994 at 1.9, 1999 at 1.4, and 2000 at 1.5° C. The other years presented ranges considered to be of low impact. In comparison with the history and the anomalies, it was determined that the vulnerability is 0.7, which according to the scale is low.

The problem could be aggravated by the presence of the 'La Niña' phenomenon as a result of climate change, since according to Saldaña (2021) the World Meteorological Organization published a report where the probability of occurrence for the year 2022 is 90%, which would cause a lot of evaporation, few clouds, direct radiation from the Sun, and temperature anomalies of up to -1.1 °C in the Equatorial Pacific. Consequently, a small variation in temperature would disturb some physiological processes sensitive to these fluctuations, such as flowering, fruit setting, and fruit formation (Snoeck et al., 2016).

7.3.3.2 Variations in average precipitation

- **PUs studied at the Rocafuerte, Portoviejo, and Chone Meteorological Stations**

By geographic range, 27 PUs are under the Rocafuerte weather station. For the years valued in the historical data, a considerable variation is observed in 1998 with around 1000 mm of excess, as well as in 2002 and 2012. Although to a lesser extent, for the years 1995, 1996, 2000, 2003, 2004, 2007, 2009, and 2011, the rainfall was negative. In this regard, the assessment of the PUs belonging to this meteorological station is 0.31, with 'high vulnerability'.

A total of seven PUs were located at the Portoviejo weather station from the years 1993 to 2013. An important variation was observed in the amount of rain, between ranges 262 and -290, except for the years 1997 and 1998, when the rainfall varied between 738 and 1082 mm, which is considered an alarming figure. The negative values oscillated between -290 and -127. Considering the data obtained, it was calculated that there is 'very high' vulnerability with a value of 0 concerning this indicator.

The pluviometric variations measured in the Chone weather station reflect anomalies within the historical average. The years with the greatest amount of precipitation are 1997, 1998, and 2001, with a maximum of 1129 mm. On the other hand, the years with the greatest drought were 1996, 2004, 2009, and 2011 with ranges between -584 and -447 mm. In this regard, the vulnerability of the PUs observed from this unit is 0.96, interpreted as 'very low'.

Cacao has a low tolerance to lack of water and in months with less than 100 mm of rain, water deficit occurs (MAGAP, 2017). The rest of the farms recorded values less than 120 days without water.

7.3.3.3 Standardized pluviometry drought index

Once the scale to locate the levels of vulnerability in relation to the Standardized Rainfall Drought Index is known, the following results are shown according to the meteorological stations. All have moderate drought and a vulnerability level of 'half'.

According to Orozco and López (2018), for cacao, with prolonged dry periods, flowering is low and therefore fruit production is low. In addition, vegetative growth suffers a reduction in the leaf sprouting process, reducing the photosynthetic area, which has an impact on fruit filling.

7.3.4 Physicochemical properties of the soil

- **Amount of oxygen available to the roots**: It was known that 205 PUs have waterlogged soil from 8 days to 2 months a year and presented 'half' vulnerability to the amount of oxygen available. On the other hand, it was found that 48 PUs have well-drained soil, hydromorphism below 60 cm, and registered low vulnerability. Likewise, it was known that 16 PUs have well-drained soils, hydromorphism below 150 cm, and reported having 'very low' vulnerability.

It is considered that a minimum of 10% of the volume of pores filled with air must exist in the soil so that the roots do not suffer anoxia problems (Barrezueta, 2020). In this situation of lack of oxygen, the production of abscisic acid increases, which is involved in the abscission of organs, due to its ability to induce the synthesis of ethylene (Taíz and Zeiger, 2012).

- **Total Porous Space for Soil**: For this indicator, the vulnerability values for the different PUs are presented. It was found that 260 PUs had values between 40% and 50% and low vulnerability to TPS. Likewise, it was shown that seven PUs had values between 51% and 53% and 'half' vulnerability to this condition. On the other hand, it was found that two were between 38% and 39% and had very low vulnerability to TPS.

In some PUs, there is a tendency for soil compaction, which reduces the volume of the pores and modifies the structure, reducing the water storage capacity and adequate development of the plant's root system (Ruíz et al., 2016). Undoubtedly, in these plots, the apparent density will increase and the hydraulic conductivity will decrease.

- **Soil Hydraulic Conductivity**: It was found that 131 PUs with frank soil presented 0.00069 cm.s^{1} and half vulnerability to the hydraulic conductivity of the soil. Likewise, it was known that 54 PUs with clay loam soil registered 0.00025 cm.s^{1} and reported having 'half' vulnerability to this condition. On the other hand, it was found that 19 PUs with sandy loam soil presented 0.00347 cm.s^{1} and had low vulnerability. Finally, one PU with sandy soil registered 0.01736 cm.s^{1} and presented very high vulnerability. The rest registered half vulnerability.

Generally, soil with hydraulic conductivity less than 2.5 cm/H usually has some problems such as low pore space, low organic matter content, and the presence of expansive clays such as montmorillonites (Barrezueta, 2020); therefore, its ability for moisture retention is reduced, disturbing the growth of cacao plants (Ruíz et al., 2016).

- **Usable Water Sheet (UWS)**: The result indicated that 42.75% of the PUs evaluated have water depths of 10 cm or less; that is, they tend to have little water availability for plants, considering that the optimum is 30 cm (INIAP, 2016). The rest of the PU (57.25%) register UWS between 20 and 25 cm. Possibly, the low total pore space found in almost half of the plots shows that their capacity to store water is reduced. According to Pire and Pereira (2018), soils with low contents of fine particles have a deficit for micropores, which can affect moisture retention.

- **Texture and ease of rooting for Cacao**: It was found that 131 PUs with soil type frank have very low vulnerability to soil texture. Likewise, it was known that 54 PUs with clay loam soil reported having 'half' vulnerability to this parameter. On the other hand, it was found that 19 PUs with sandy loam soil have low vulnerability. Finally. nine PUs with sandy loam or clayey soil presented very high or high vulnerability, respectively. The rest registered low vulnerability.
- **Availability of Organic Matter for Cacao cultivation**: The result indicated that the organic matter values were recorded between 3.0% and 2.5% or less, that is, they are below the minimum range for cacao which is ≥4% (Pico et al., 2012). Only 2.6% of the PU presented organic matter values above the optimal level. The rest were between 3% and 4%. Therefore, it presents high vulnerability. These results would affect the structural stability of the soil and pore space, limiting the exchange complex and hindering the penetration of water and its retention (Pire and Pereira, 2018).
- **Soil slope for Cacao cultivation**: In this case, 198 PUs with <2% slope in topography recorded very low vulnerability, 40 PUs showed between 2–5% slope and presented low vulnerability, and 31 PUs registered between 5–10%, showed half vulnerability.

In this situation, the risks of erosion can increase, as landslides and avalanches could determine their tendency to generate loss of soil fertility. However, with proper management, it can be solved (Mirabal, 2016).

- **Fertility levels for Cacao**: In this case, the statistical results showed that 2.23% of PUs registered a moderately alkaline pH, 3.35% indicated a slightly alkaline pH, 81.78% presented a neutral pH, 6.32% registered a slightly acidic pH, 4.46% had a moderately acidic pH, and 1.49% had a strongly acidic pH. The pH values were located within the optimal range for the development of cacao between 6.5 and 7.3 (INIAP, 2016). Therefore, its vulnerability is very low.

The 83.27% of PU recorded P values between 33.5 and ≥40 $mg.kg^1$, that is, they were above the optimal level for cacao, which is 33.5 $mg.kg^1$ (Pico et al., 2012) with low vulnerability. The rest of the PUs presented P values below that parameter, with high vulnerability.

The Ca^{++} content determined was above the optimal level, which is >6 $Cmol.\ kg^1$ (Pico et al., 2012) causing an imbalance of other elements such as K^+ and Mg^{++} (González et al., 2018). Therefore, it presents a high vulnerability.

The 94.05% of the PUs recorded Mg^{++} values between 3.0 and 5.99 $mg.kg^1$, that is, they were above the optimal level for cacao, which is >2.5 $mg.kg^1$ (Pico et al., 2012), and their vulnerability is low. The rest of the PUs presented Mg^{++} valuesbelow that parameter and their vulnerability is high.

The statistical results showed maximum values for K^+ of 4.4 $Cmol.kg^1$ and a minimum of 0.54 $Cmol.kg^1$. In this sense, 100% of the PU presented K^+ levels greater than the optimum, which is 0.41 $Cmol.kg^1$ (Pico et al., 2012). Therefore, it presents a very low vulnerability.

The high variability observed between the mean and standard deviation of Ca^{++} in the soils of the area indicates that this element is in a nutritional imbalance with K^+ and Mg^{++} in some of the PUs evaluated and a balance between them is required, it could be one of the causes of low performance (González et al., 2018). Knowing the cation exchange capacity (CEC) of a soil is essential, since this value indicates the potential of the soil to retain and exchange nutrients, particularly cations (Barrezueta, 2020).

However, it can be seen from its standard deviation that some soils presented CEC values below the optimum, which may be related to the presence of 1:1 type clays and the low-OM content determined in these soils, characteristic of tropical regions rainy (Osorio, 2012).

7.4 Conclusions

The production of cacao cultivation in the Fortaleza del Valle Association is in a situation of medium vulnerability, being affected by indicators related to environmental problems, and modeled changes in climate and some physico-chemical properties of the soil.

The vulnerability index between the range of 0 and 100% allowed the area to be classified as a medium vulnerability class with soils suitable for cacao cultivation, which with appropriate agricultural practices represent minor risks to soil conservation.

The progressive change in land use as a result of deforestation and the replacement of natural ecosystems, as well as changes in the capacity to adapt to anthropic phenomena, could generate a future risk for the change in the vulnerability of cacao producers in the region in the face of climate change.

References

Association of Official Analytical Chemists (AOAC) (1999). *Official Methods of Analysis*. USDA, Washington, DC.

Araque, O., Carr, L., & Lokwood, M. (2014). Comparative photosynthesis, water relations, growth and survival rates in juvenile Criollo cacao cultivars during the dry and wet seasons. *Experimental Agriculture*, 48(4): 513–522.

Baker, J. L., & Laflen, J. M. (1983). Water quality consequences of conservation tillage. *Journal of Soil and Water Conservation*, 38.

Barrezueta, U.S. (2020). Propiedades de algunos suelos cultivados con cacao en la Provincia El Oro, Ecuador. *Revista Ciencia UAT*, 14(1): 25–34.

Bourdieu, P. (1980). The social capital. Provisional notes. *Actes de la Recherche en Sciences Sociales*, 31: 23.

Carrero, A., Zambrano, A., Hernández, E., Contreras, F., Machado, D., Bianchi, G., & Varela R. (2015). Comparación de dos métodos de extracción de fósforo disponible en un suelo ácido. *Avances de Química*, 10 (especial): 29–33.

Colombian Agricultural Institute (CAI) (1992). Fertilization in various crops. Fifth approximation, Bogotá: ICA Technical Assistance Manual No. 25, pp. 1–26.

Flores Delgadillo, L., & Alcalá Martínez, J. R. (2010). Manual de procedimientos analíticos, laboratorio de Física de Suelos. Universidad Nacional Autónoma de México. Instituto de Geología, México, D. F.

Food and Agriculture Organization of the United Nations (FAO) (1999). Guide for the Description of Soils. Available at http://www.fao.org/3/a-a0541s.pdf

Forsythe, W. (1980). *Física de suelos: manual de laboratorio*. 2ª reimp. IICA, San José, Costa Rica, 212 pp.

García, J., Romero, M., & Ortiz, L. (2005). Edaphoclimatic evaluation of the lands of the Colombian low tropics for the cultivation of cacao. Project: "Genetic improvement and characterization of the conditions of competitiveness of cacao cultivation in Colombia." COMPONENT: Characterization and zoning of potential areas for cacao cultivation in Colombia. CORPOICA-MINAGRICULTURE AGREEMENT N° 034/2003.

González, R., Alvares, E., & Castañeda, D. (2018). Evaluación de la calidad química del suelo en agroecosistemas cacaoteros de la subregión del Nordeste y Urabá Antioqueño. *Revista Colombiana de Investigaciones Agroindustriales*, 5(1): 41–52.

Intergovernmental Panel on Climate Change (IPCC) (2021). Synthesis Report. Contribution of Working Groups I, II and III to the Third Assessment Report of the Intergovernmental Panel on Climate Change. Summary for Policymakers. Robert T. Watson (ed.). Available at https://www.ipcc.ch/pdf/special-reports/srex/IPCC_SREX_ES_web.pdf

International Institute for Food Policy Research (2015). Global Nutrition Report 2015: Actions and Responsibilities for Advancing Nutrition and Sustainable Development. Report.

Israelsen, O. W., & Hansen, V. E. (1979). Principios y aplicaciones del riego. Barcelona, ES, Reverté, 336 pp.

Manabí Province Reforestation Commission (2016). Available at: http://www.manabi.gob.ec/ecuador/ven-a-manabi/page/63

Mackean, S. (1993). *Manual de Análisis de Suelos y Plantas*. Centro Internacional de Agricultura Tropical, Cali, Colombia.

Ministry for the Ecological Transition of Spain (2017). Types of Drought. Available at https://www.miteco.gob.es/ca/agua/temas/observatorio-nacional-de-la-sequia/que-es-la-sequia/Observatorio_Nacional_Sequia_1_1_tipos_sequia.aspx

Ministry of Agriculture and Livestock (MAGAP) (2017). Agricultural Public Information System. Available at http://sipa.agricultura.gob.ec/index.php/cifras-agroproductivas

Mirabal, O. (2016). Relación entre la variabilidad de las propiedades físicas de un suelo y el rendimiento del cultivo de cacao (*Theobroma cacao* L.). Availabe at http://saber.ucv.ve/handle/10872/20847/2016.

Montenegro, H., & Malagón, D. (1990). Propiedades físicas de los suelos. IGAC. Bogotá, 811 pp.

Monterroso, A., Conde, C., Gay, C., Gómez, J., & López, J. (2012). Vulnerability indicators and climate change in agriculture in Mexico. National Autonomous University of Mexico.

Moreno, M., Pirela, M., Medina, N., & Molina, V. (2013). Evaluación de las propiedades físicas, químicas y biológicas de los suelos cultivados con cacao (*Theobroma cacao* L.) en la sierra de Perijá. Revista de la Universidad del Zulia. Ciencias del Agro, Ingeniería y Tecnología. Año 4 N° 8: 8–28.

Nieto, M. (2015). Cycle of dissolved organic matter in a coastal upwelling system. Doctoral Thesis. University of Vigo. Available at http://digital.csic.es/bitstream/10261/109813/1/Nieto%20Cid_Tesis%20Doctoral_2005.pdf

Olivares (2015). Characterization of meteorological drought in the network of agro-meteorological stations of the National Institute of Agricultural Research (INIA) Venezuela. Master's Thesis. International University of Andalusia.

Olsen, S. R., & Sommers, L. E. (1982). Phosphorus. In: Page, A. L., et al. (eds), *Method of Soil Analysis*, Part 2, 2nd ed., Agron. Monogr. 9, ASA and ASSA, Madison, WI, pp. 403–430.

Organization for Economic Cooperation and Development (OECD) (2016). *Environmental Performance Assessments*. Chili.

Orozco, L., & López, A. (2018). Balance de agua y requerimientos de riego en cacao. Proyecto Cultivo/Manejo de Cacao. Informe N°20. CATIE-CEI-VECOMA.

Osorio, N. W. (2012). pH del suelo y disponibilidad de nutrientes. *Revista Manejo Integral del Suelo y Nutrición Mineral*, 1(4).

Pico, J., Calderón, D., Fernández, F., & Díaz, A. (2012). Guía del manejo integrado de enfermedades del cultivo de cacao (*Theobroma cacao* L.) en la Amazonia. INIAP. Guayaquil, Ecuador. 12–15 pp.

Pire, R., & Pereira, A. (2018). Tamaño de los poros del suelo y crecimiento de raíz y vástago del chile jalapeño (*Capsicum annum* L.). *Agrociencia*, 52(5).

Ruíz, H., Sarli, O., Goncalvez, C., Fliquero, R., & Silva, F. (2016). La superficie específica de oxisoles y su relación con la retención hídrica. *Revista de la Facultad de Ciencias Agrarias de La Universidad del Cuyo*, 48(2): 95–105.

Saldaña, S. (2021). El fenómeno de la niña enciende focos rojos en México. Available at https://www.xataca.com.mx

Snoeck, D., Koko, L., Joffre, J., Bastide, P., & Jagoret, P. (2016). Cacao nutrition and fertilization. In: Lichtfouse, E. (ed.), *Sustainable Agriculture Review*. Springer International Publishing, Montpellier, pp. 1–34.

Taíz, L., & Zeiger, E. (2012). Abscisic acid: A seed maturation and antistress signal. 539–558.

Thielen, D. (2015). Dynamics of extreme climatic events in the Portoviejo river basin, Manabí, Ecuador. Technical magazine no. 14.

Walls, M. (2003). Cacao Cultivation Manual. Available at http://www.infocafes.com/descargas/biblioteca/215.pdf

Whitlow, M., Filpula, D., Rollence, M., Feng, S., & Woods, J. F. (1994). *Protein Engng*, 7, 1017–1026.

World Health Organization (WHO) (2017). Climate Change and Health. Available at http://www.who.int/es/news-room/fact-sheets/detail/cambio-clim%C3%A1tico-y-salud

Part III

Pests and disease management

8 Cacao diseases caused by fungi and fungi-like in Latin American agrosystems

Felipe R. Garcés-Fiallos, Anthony A. Moreira-Morrillo, Danilo I. Vera-Coello, Zoila K. Solís-Hidalgo, Hayron F. Canchignia-Martínez, and Ángel V. Cedeño-Moreira

8.1 Introduction

Cacao (*Theobroma cacao* L.) is a native crop of Latin America, currently being cultivated in most intertropical regions of the world (Cilas and Bastide, 2020). During the last decade, an exponential increase in new cacao crops has been reported. Cacao beans used as a raw material for the food industry are valued at more than 47,000 million dollars (Vásquez et al., 2019). Despite this, climate change and variations in water regimes seem to have generated alterations in the physiological behavior of the cacao crop, which has favored the occurrence of pests and diseases (Cilas and Bastide, 2020). Diseases considerably affect cacao production (Sánchez-Mora et al., 2015) and can even kill the plant (Moreira-Morrillo et al., 2021; Paladines-Rezabala et al., 2022).

In this chapter, we approach the main diseases present in Latin American cacao agrosystems caused by fungi and fungi-like (Figure 8.1), that is witches' broom [*Moniliophthora perniciosa* (Stahel) Aime & Phillips-Mora] (Figure 8.1 A–C), frosty pod [synonymous: Monilia pod rot and watery pod rot; *Moniliophthora roreri* H.C. Evans et al.] (Figure 8.1 D–F), root and crown rot and pod rot [*Phytophthora palmivora* (Butler), *P. megakarya* Brasier & M.J. Griffin, and *P. capsici* Leonian] (Figure 8.1. G–I), dieback and pod rot [*Lasiodiplodia theobromae* (Pat.) Griffon & Maul] (Figure 8.1. J–L), and lethal wilt [*Ceratocystis cacaofunesta* Engelbrecht & Harrington] (Figure 8.1. M–O). Considering the importance of the cacao crop worldwide and the diseases that affect it, restricting its production and even killing the plants, and based on the experience of the authors, we present technical information on the etiology, symptomatology, life cycle, and crop management based on phytosanitary measures used in both monocultures and production systems.

8.2 Witches' broom

Witches' broom was first described in 1895 in Suriname (Figure 8.1 A–C). Later, in 1900 it spread to the rest of South America, where it became an endemic (CABI, 2022; Evans, 1978). This disease is widespread in Tropical

DOI: 10.4324/9781003381761-11

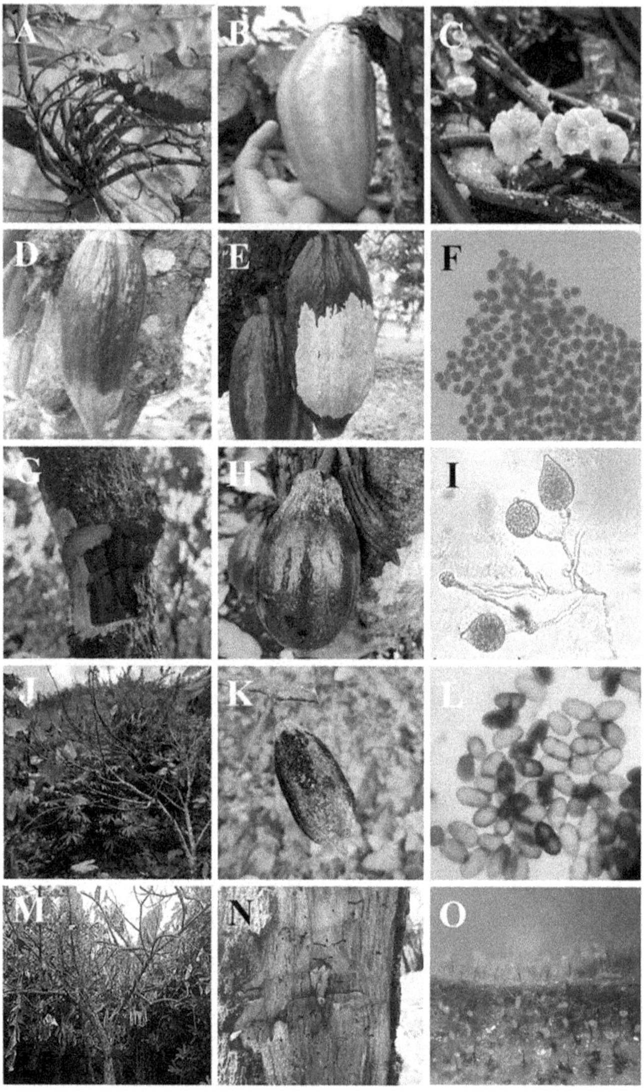

Figure 8.1 Symptoms and signals (vegetative or reproductive structures) produced by plant pathogens affecting cacao plants. (A–C) witches' broom caused by *Moniliophthora perniciosa*: necrotic sprouts named witches broom (A), malformed pods (B), and basidiocarps on infected plant residue (C). (D–F) Frosty pod rot produced by *Moniliophthora roreri*: pods showing brown spots (D), abundant sporulation (E), and microscopic view of unicellular conidia production (F). (G–I) Root and crown rot and pod rot caused by *Phytophthora* spp.: stem canker (G), black pod rot (H) symptoms, and microscopic view of sporangiophore and sporangia (I). (J–L) Dieback and pod rot produced by *Lasiodiplodia theobromae*: dieback (J) and rot (K) on plants and pods, respectively, and microscopic view of immature (hyaline and unicellular) and mature (pigmented and bicellular) conidia (L). (M–O) Lethal wilt caused by *Ceratocystis cacaofunesta*: dead plant (M), galleries built inside stems by bark beetles (N), and stereomicroscopic view of perithecia and ascospores on cacao tissues (O).

America from where it has apparently not progressed to other countries. In Ecuador, the disease was reported in 1918 and today is considered an endemic, affecting up to 80% of cacao production, pods, sprouts, and flowering bearing (Suárez, 1993). Up to the present date, the disease continues to be a problem for cacao farmers.

8.3 The life cycle of witches' broom caused by *Moniliophthora perniciosa*

Witches' broom is caused by the basidiomycete fungus *M. perniciosa* (Aime and Phillips-Mora, 2005). This pathogen is a hemibiotrophic microorganism with a biotrophic or parasitic phase that colonizes actively growing cacao tissue, such as vegetative shoots (Figure 8.1 A), floral bearings, or pods (Figure 8.1 B), as well as the saprophytic phase, which occurs on dead tissues (dry brooms or fruits) of its host (Suárez, 1993; Solis et al., 2021). *M. perniciosa* produces basidiocarp (Figure 8.1 C), a sexual structure that grows on dry broom, like purple-pink umbrellas, releasing the basidiospores (sexual spores) during the night with the help of wind and rain. Basidiospores germinate on the surface of the plant (growing tissues: floral cushions, fruits, leaves, and branches) and penetrate through the stomata, forming a mycelium that propagates intercellularly (Purdy and Schmidt, 1996) (Figure 8.2).

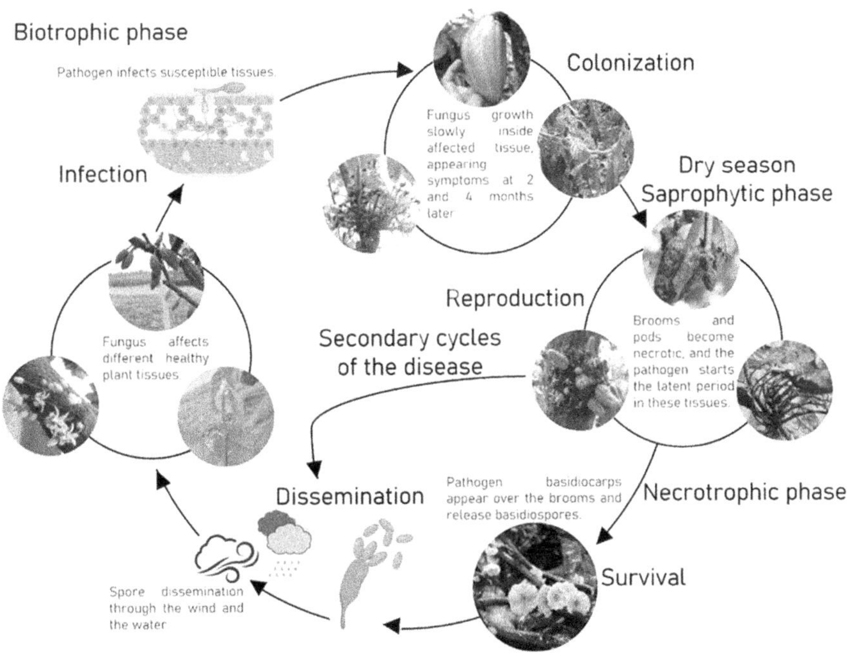

Figure 8.2 Witches' broom life cycle caused by *Moniliophthora perniciosa* on tissues of cacao plants.

The fungus *M. perniciosa* colonizes internally through its mycelium inside tissues between two and three weeks later, while the surface of the tissue appears healthy. Then, the fungus causes hypertrophy and hyperplasia in tissues from the third week to the 14th (about four months), being able to observe symptoms such as swellings, deformations, and ramifications (Figure 8.2). Due to their appearance, the disease is called witches' broom. However, these symptoms could vary according to the genotype susceptibility and pathogen aggressiveness, as well as favorable environmental conditions such as 1) high precipitation (hours or days or periods of rain and drought); 2) high temperature (24–28 °C); and 3) relative humidity >80% (Suárez, 1993).

Once the sprouts are infected, the proliferation of swelled branches with short internodes begins, and the bark becomes necrotic (dry) and split, causing the symptom known as canker (Figure 8.2). Vegetative tissue brooms are formed on the infected flower cushions; flowers and parthenocarpic pods could appear at their bases (Figure 8.2 C). Individual flowers show thick and long peduncles, remaining attached to the bearing that is called star flowers, and sometimes they produce strawberry-shaped parthenocarpy fruits called chirimoyas (Suárez, 1993; Silva et al., 2002). Cacao pods are susceptible to this pathogen at any stage of development, being critical in the first three months. Infected fruits less than a month old become elongated and swollen, named carrots. The infected pods could show variable symptoms such as deformations, premature yellowing, and necrotic spots (Figure 8.2). Sometimes pods could reach maturity without superficial symptoms, but internally they may be affected (Suárez, 1993; Silva et al., 2002).

Later, the infected tissues or brooms become necrotic and dry, beginning the saprophytic phase of the pathogen. The necrotic tissue can remain on the tree indefinitely. The fungus forms mycelium masses and basidiocarps on the brooms when dry (Figure 8.2), under conditions of high relative humidity (>80%), precipitation (1000 to 2000 mm), and temperatures between 22 and 28°C, coinciding with the beginning of the rainy season (Purdy and Schmidt, 1996). In this way, basidiocarps generate basidiospores that initiate new infections on fruits, shoots, and flower cushions during the all-life cycle of the plant, starting the point of the life cycle of the pathogen (Griffith et al., 2003). Witches' broom caused by *M. perniciosa* can complete more than one life cycle in the tissue of the same plant, thus turning it into a polycyclic disease.

8.4 Prevention measures for witches' broom

For the management of the witches' broom, the following should be considered (Suárez, 1993):

- Perform phytosanitary pruning at least once a year.
- Remove diseased tissues (fruits, branches, floral cushions, etc.), a task that is recommended to be carried out together with scheduled pruning.

- The removed fruits and diseased material should be chopped into the ground to speed up their decomposition.
- During the rainy season, apply two cycles of chlorothalonil-based fungicides at an interval of 15 days depending on the intensity of the rains to protect the sprouts of the plant during the period of greatest infection.
- Efficient control of witches' broom must integrate neighboring farms because the release of spores and their dissemination with the wind is rapid.

8.5 Frosty pod rot

Frosty pod rot is an endemic disease that specifically attacks cacao pods. It is caused by the hemibiotrophic Basidiomycota *M. roreri* (Figure 8.1 D–F). There is circumstantial evidence that flowers and flower cushions may harbor this pathogen (Bailey et al., 2018). It was observed for the first time in 1817 in Colombia (Evans, 2016), and was later reported in Ecuador in 1916, causing estimated damages between 50 and 80% of the total annual production (Suárez, 1993). The disease is restricted to Central and Tropical America, affecting cacao crops in producing countries except for the Dominican Republic of the Caribbean (Evans, 2016) and Brazil. Thus, frosty pod rot was first reported in Cruzeiro do Sul, Northern Brazil in 2021 (Estrela Junior et al., 2022). In cacao-producing countries like Ecuador, the disease is very important due to the pod production damage (Sánchez Mora and Garcés Fiallos, 2012).

8.6 The life cycle of Frosty pod rot caused by *Moniliophthora roreri*

The fungus is very specialized with slow development affecting cacao pods at any growing stage (Figure 8.1 D). This pathogen produces asexual spores (conidia) on the infected fruit, which become the main inoculum source (Figure 8.1 E–F). The life cycle begins with the sporulated pods (Figure 8.3). Conidia are easily detached with the help of the wind and dispersed to healthy pods. Once the spores were deposited on the surface of the pod tissues, they germinate in the presence of a water film and then penetrate the pod causing infection (Figure 8.3) (Suárez, 1993; Evans, 2016; Solis et al., 2021).

During the first stage of infection, *M. roreri* colonized intracellularly the epidermis and cortex of pods, taking between six to ten weeks. The first symptoms showed in pods are wilting, deformation, excessive presence of water in the cells or tissues (hyperhidrosis), irregular maturity, necrosis, and oily spots (Figure 8.3). Later, this symptom turns into a chocolate-colored spot (Figure 8.2 D), on which, after two to three days, a white mycelium (stroma) grows, which will give rise to conidia, a powdery mass of cream or gray-brown color (Figure 8.3) (Evans, 2016). *M. roreri* survives up to nine months on mummified pods that remain attached to the tree. Whether pods are cut or dropped to the ground, the viability of the spores will be reduced after a couple of months by the action of antagonistic microorganisms that inhabit the soil (Figure 8.3) (Suárez, 1993).

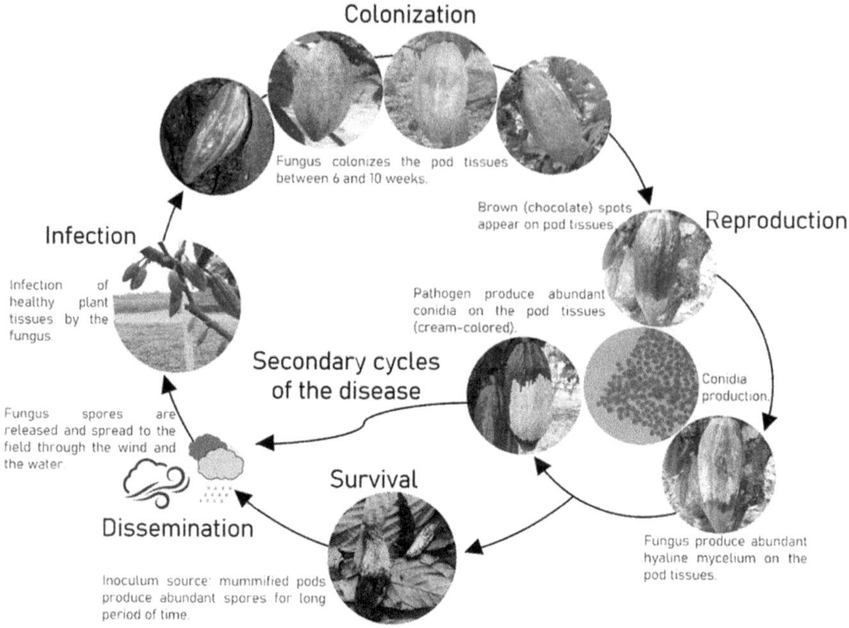

Figure 8.3 Frosty pod rot life cycle caused by *Moniliophthora roreri* on pod tissues of cacao plants.

8.7 Prevention measures for frosty pod rot

For the management of frosty pod rot, the following are recommended (Suárez, 1993):

- Maintaining shade of 30–40% (depending on the place or planting area) that allows the entrance of light and air to reduce the percentage of humidity in the environment.
- Pruning the plantation to maintain adequate plant distribution and avoid microclimate for the development of diseases.
- Harvesting mature pods every two weeks to avoid infections at the final stages of development or every seven days depending on the disease pressure.
- Carrying out sanitation practices, which consist of the weekly removal of affected pods (hypertrophied, stained, deformed, or swollen), before sporulation occurs, thus reducing the risk of infection in healthy pods.
- Diseased pods could be left on the ground (if possible, cover with leaf litter) for decomposition. Do not move sporulated diseased pods within the plantation to avoid dispersion of the spores.
- During high production periods and greater frosty pod rot infection, especially during the rainy season, copper-based fungicides (copper oxide or copper hydroxide) could be applied according to the technical

recommendation (in frequency bases of 21 days) and azoxystrobin for protection from frosty pod rot and witches' broom diseases.
- Maintaining an operational drainage system to avoid flooding areas that could favor the development of diseases.

8.8 Management of witches' broom and frosty pod rot

The causal agents of witches' broom and frosty pod rot in cacao are hemibiotrophs recognized as sister basidiomycetes of the genus *Moniliophthora*, and represent a current threat to unaffected regions in the Americas, as well as to those in the Paleotropics (Marelli et al., 2019). The main strategies for managing cacao diseases include the use of tolerant varieties, adequate planting design, nutritional management, environmental conditions that reduce the presence of diseases, and phytosanitary measures (elimination of diseased tissue, fungicide application, and biological control). However, none of these measures alone completely control the pathogens (Ploetz, 2007). The use of tolerant genotypes is the most economical and promising long-term control alternative. Tests for selecting genotypes are key to the improvement of cultivars, but it is the pathogen that causes the disease that exerts pressure, due to its high genetic diversity (Suárez, 1993; Solis et al., 2021; Evans, 2016).

When diseases cause losses of more than 50% of potential cacao production, investment in a health prevention program is justified, and it is necessary to adopt preventive management since pods and/or sprouts are infected and there is no possibility of curative control. A sanitary prevention program must be integral; it means combining pruning (elimination of diseased tissues: floral cushions, pods, and sprouts with a broom), weekly removal of diseased pods, and protection during affordable environmental conditions for the development of the pathogen with copper-based fungicides, chlorothalonil or azoxystrobin, as well as an adequate fertilization plan (Amores et al., 2010).

The growth of cacao is constant. Every three to four months, cacao plants produce new tissue, depending on their food reserves. This indefinite growth must be managed to maintain a plant size no greater than three meters. Also, it is important to have a balanced cup to facilitate cropping management. In fact, management of the cacao tree top based on pruning must be the initial step to improve the health of the crop and increase production. The height of the plantation decisively affects the presence of diseases; the higher the size of the plant, the greater the difficulty of managing properly the inoculum sources. New tender tissues (leaves, shoots, flower buds, and fruits) are susceptible to infection by *M. perniciosa*, due to their higher capacity to maintain humidity and the fact that self-shading blocks the entry of sunlight, providing conducive conditions to the development of diseases (Sánchez Mora and Garcés Fiallos, 2012). In this sense, pruning is a specialized practice recommended to be carried out during the dry period to help reduce the risk of disease infections (Amores et al., 2010).

Research has been carried out toward the collection, characterization, evaluation, and use of microorganisms as biological control agents. These include several species of *Trichoderma*, which can also improve the mechanisms of plant defense, offering alternatives for the management of frosty pod rot and witches' broom as is demonstrated in several essays with the use of *T. stromaticum*, which is a mycoparasite of basidiocarps of *M. perniciosa* or *T. koningiopsis* and *T. ovalisporum* isolates, which are antagonists of *M. roreri* (Griffith et al., 2003; Samuels et al., 2006; Mejía et al., 2008; Evans, 2016; Ten Hoopen and Krauss, 2016; Vera et al., 2018; Terrero et al., 2018; Vera Loor et al., 2018; Peñaherrera et al., 2020). Also, it has been demonstrated under lab conditions the antagonistic activity of yeasts against *M. roreri* (Estrela Junior et al., 2022). The management of both diseases by Biological Control Agents (BCA) such as bacteria, fungi, and yeasts could be mainly related to their production of various production antifungal components.

8.9 Diseases caused by *Phytophthora*

Among the most destructive cacao pathogens are several species of *Phytophthora* (*P capsici, P citrophthora, P heveae, P. megakarya, P palmivora* and *P. theobromicola*) (Surujdeo-Maharaj et al., 2016; Decloquement et al., 2021), which could cause black pod rot and root and crown rot (Figure 8.1 G–I). This oomycete also could affect other plant tissues, such as branches and young shoots. *P. palmivora* was reported for the first time in 1727 on Trinidad Island, causing pod wilt and stem cancer on cacao plants during all their phenological stages (Tollenaar, 1958). The pathogenic species is a hemibiotrophic oomycete capable of infecting over 200 plant species, including cacao crops (Perrine-Walker, 2020). Although *P. megakarya* occurs mainly in Africa and *P. palmivora* is distributed worldwide (Deberdt et al., 2008; Puig et al., 2018), in Latin America, the second species causes root and crown rot and pod rot on cacao plants (Cedeño-Moreira et al., 2020). Either way, *P. megakarya* causes severe damage in the cacao-producing countries in West Africa, while *P. palmivora* causes significant losses globally (Evans, 2007).

8.10 The life cycle of pod rot and root and crown rot caused by *Phytophthora* spp.

In humid tropical regions like those in Latin America, *Phytophthora* survives in plant tissue (cacao or other crops). Soil from cacao plantations may be a massive and consistent inoculum source (Purwantara et al. 2015). The pathogen can infect tissues throughout the year, that is, roots, trunks, leaves, flowers, and pods (Figure 8.1 G–H), causing seasonal epidemic outbreaks. However, in dry periods, the pathogen survives (primary inoculum source) in asymptomatic tissues, mainly due to poor sanitary management (Figure 8.4) (Drenth & Guest 2013).

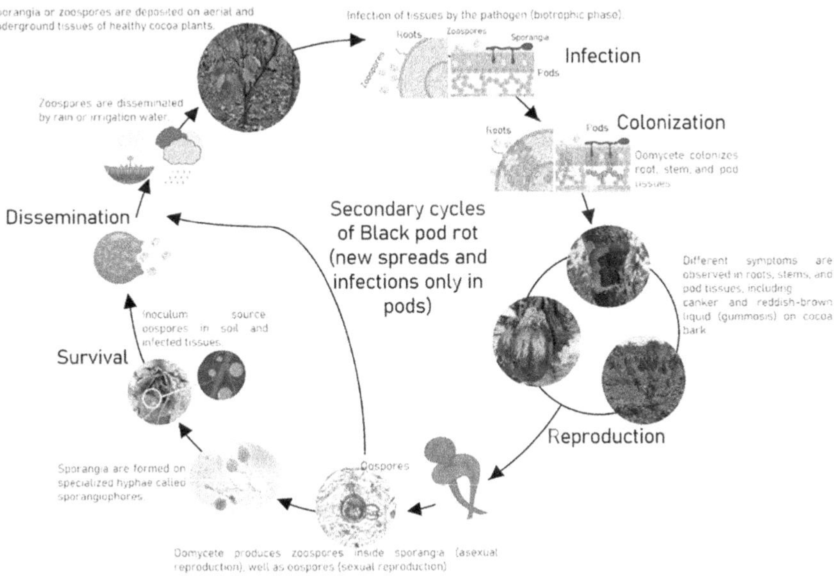

Figure 8.4 Root and crown rot and pod rot life cycle caused by *Phytophthora* spp. on tissues of cacao plants.

Water is a dissemination source high-up for the pathogen, spreading through branches and trunks and infected soil particles (Porras et al., 1990). Biflagellate zoospores facilitate the dispersal of the pathogen through rainwater driven by the wind or irrigation until it reaches susceptible tissues of roots (Bassani et al., 2020) or pods. *Phytophthora* can infect both the aboveground and belowground tissues of cacao plants. Although water is significant for the dissemination, formation, and dispersal of oomycete zoospores, the temperature is also transcendent. In fact, the temperature can affect *Phytophthora* species differently (Figure 8.4). For instance, the optimal temperature is lower for *P. megakarya* (25°C) than for *P. palmivora* (30°C) (Brasier and Griffin, 1979; Puig et al., 2018).

Although the production of zoospores by oomycetes makes it lethal, appressoria are other structures that could play an important role in cacao–*Phytophthora* interaction (Delgado-Ospina et al., 2021). Appressoria formation by *P. megakarya* makes it more aggressive due to its direct penetration of the epidermis, compared to *P. palmivora*, which penetrates through the stomata (Figure 8.4). During the infection and colonization phases, germ tube, appressorium, and haustorium formation enable the oomycete to penetrate the epidermis (stem and pods) and rhizodermis (root system) cells and contribute to cacao disease development. The infection and colonization process of *P. palmivora* in cacao tissues was summarized with modifications by Perrine-Walker (2020) (Figure 8.4): A) zoospores lose their motility and adhere to plant tissue surfaces; B) encysted zoospore germinates forming a germ tube and later an appressorium, and subsequently initiates the penetration through the epidermis or rhizodermis tissues; C) once inside, the pathogen hyphae grow intracellularly, forming a haustorium (biotrophic stage); D) pathogen secondary hyphae begin to kill host cells (necrotrophic stage); and E) oomycete develops chlamydospores and sporangia, providing a new inoculum source for future infections. Although *P. palmivora* can infect any place on pods or cherelles' (immature pods) surface, the initial infection is usually at the tip or stem end (Perrine-Walker, 2020). In stems, *P. palmivora* mycelia spread from infected pods to flower cushions and infect the woody tissues through wounds (Figure 8.4) (Appiah et al., 2004; Perrine-Walker, 2020).

Phytophthora species have homothallic (self-fertile) or heterothallic (self-sterile, presenting the mating types A1 and A2) mycelium, resulting in the production of oospores (Perrine-Walker, 2020). This sexual reproductive structure is produced by the union of antheridia (male gametes) with oogonia (female gametes), which due to their genetic recombination, generate high variability in aggressiveness and pathogenicity (Surujdeo-Maharaj et al., 2016). Although this oomycete requires different hyphae to form a new structure, their genetic differentiation can produce sporangia (Figure 8.1 I) and asexual spores inside as biflagellate zoospores (Figure 8.4) (Rojas-Rojas and Vega-Arreguín, 2021).

As a result of the tissue infection, colonization, and reproduction by *Phytophthora*, different symptoms might be observed in roots, stems, and pods. Also, *P. palmivora* can infect flower cushions, sprouts, leaves, seedlings, as well as roots, causing detached cauliflorous, reddish-white color, odorless, and self-fertile flower, suckers near the base of the trunk, cherelles (immature pods) at different stages of growth, and development arising from former flower cushions (Perrine-Walker, 2020). Another of the symptoms caused by *Phytophthora* is the wilting of plants or seedlings, mainly related to root and crown rot (Figure 8.4). Canker symptom is the formation of reddish water-soaked lesions with dark brown to black margins, and in some cases, reddish-brown liquid (gummosis) oozing from these lesions, usually through cracks in the bark (Appiah et al., 2004). Canker intensity in cacao stem by *P. palmivora* can increase during the wet season, especially in susceptible genotypes (Purwantara et al. 2015). In cacao pods, after *P. megakarya* penetrates its cuticle and epidermis under high humidity conditions (about 72 hours after infection), small necrotic spots that vary in color and appearance can be observed (Akrofi et al., 2015). These spots spread uniformly and rapidly in a few days until they completely cover the pod surface. *Phytophthora* also affects the pulp and seeds of the ear, which become finally necrotic and colonized by secondary fungi (Figure 8.4) (Purwantara et al. 2015).

8.11 Management of pod rot and root and crown rot

Phytophthora diseases on cacao crops could be managed using genotype selection and cultural and chemical techniques (Marelli et al., 2019). Regarding the genetic resistance of cacao to *Phytophthora* diseases, one of the genotypes with the lowest lesion expansion in plants is the clone IMC-67 (Okey et al., 1996). Although there is evidence of quantitative trait loci (QTLs) in cacao genotypes related to resistance against *P. capsici, P. megakarya*, and *P. palmivora* (Lanaud et al., 2009; Barreto et al., 2018), only when resistance sources for various *Phytophthora* isolates and species have been identified, cacao breeding for disease resistance can be carried out to meet the specific worldwide producer and consumer needs (Surujdeo-Maharaj et al., 2016). Nowadays, cacao breeding programs have had mixed successes due to the difficulty in resistance source identification against the three *Phytophthora* species.

In general, the management of *Phytophthora* diseases consists of reducing the pathogen population and preventing the disease's spread. Fungicide (oomyceticides) applications can generate a lower disease incidence and higher yields in cacao plants (Akrofi et al., 2003); for example, copper-based fungicides (Liu et al., 2020). One of the most used oomyceticides in cacao against *Phytophthora* diseases is Fosetyl-aluminum. However, fungicides can contribute to the appearance of oomycete populations less sensitive to molecules, and their effectiveness may decrease during high inoculum pressure, especially in rainy seasons and high relative humidity (Guest, 2007). On the other hand, the profitability of fungicide applications not only depends on the level of farm management but also on the nature of land tenure and labor arrangements for farm operations (Akrofi et al., 2003).

Plant extract application also could be used in the management of *Phytophthora* diseases because it is an ecological and friendly alternative to humans and the environment. For instance, Widmer and Laurent (2006) showed that lavender (*Lavandula angustifolia*) and rosemary (*Rosmarinus officinalis*) foliar extracts reduce the *P. capsici, P. megakarya*, and *P. palmivora* zoospore germination, and the latter extract negatively affects necrosis caused by *P. megakarya* zoospores in cacao leaf disks. Likewise, pods first treated with lavender extract showed no symptoms caused by *P. megakarya* infection. The caffeic acid found in both extracts might be involved in the observed negative effect against pathogens and their caused diseases. On the other hand, inhibition of *P. parasitica* and *P. capsici* by silver nanoparticles synthesized using an aqueous extract of absinthe (*Artemisia absinthium*) may also be feasible (Ali et al., 2015).

BCA use might also be an excellent alternative to the management of *Phytophthora* diseases. In fact, El-Sayed & Ali (2020) mentioned that the extract of *Aspergillus flavipes* inhibits mycelial growth inhibition, zoospore germination, and sporangial production of *P. parasitic*, possibly due to the active compound 3-hydroxy-2′,4,4′,6′-tetramethoxychalocone. Also, *T. asperellum* (strain PR11) application might negatively affect the pod rot rates, showing even a prolonged effect of biological agents up to 1 year after treatment. Biological controller propagules may remain in flower cushions of cacao trees than in soil or bark (Deberdt et al., 2008). The use of rhizobacteria also can be used in the biological fight against root and crown rot and pod rot, both caused by *P. palmivora*. In this regard, Cedeño-Moreira et al. (2020) found that *Acinetobacter calcoaceticus* (strain BMR2-12), *Serratia marcescens* (PM3-8), *Pseudomonas protegens* (CHA0), and *P. veronii* (R4) reduce the disease intensity in both pods and seedlings and promote growth in the latest.

Phytophthora has singular features that can be considered when planning its management. Although the microbial control against black pod disease is promising, this is not as effective as the chemical methods under high disease pressure and therefore needs to be integrated with other management methods to establish a sustainable management system for *Phytophthora* cacao diseases (Deberdt et al., 2008). Unfortunately, there is not a single solution or simple remedy to manage this oomycete on tropical tree crops (Surujdeo-Maharaj et al., 2016). A combination of cultural, chemical, and biological methods has been effective against *Phytophthora*. Meanwhile, cultural measures including shade management, pruning, basal sucker removal, and regular pod picking, could be sufficient for disease control (Asare-Nyako, 1974; Akrofi et al., 1997).

8.12 Dieback and pod rot caused by *Lasiodiplodia theobromae*

Among the *Lasiodiplodia* species reported to date, *L. theobromae* is the most studied in cacao and other crops (Figure 1 J–L) (Moreira-Morrillo et al., 2021). The fungus has a wide host range, behaving as an endophyte or phytopathogen (Salvatore et al., 2020). *L. theobromae* is a pleomorphic and plurivorous ascomycete fungus, present mainly in regions with tropical and subtropical climate (Slippers et al., 2013; Moreira-Morrillo et al., 2021). The pathogen usually affects plants that present some hydric nutritional stress, and their incidence may be influenced by temperatures above 30 °C (Khanzada et al., 2005). Although dieback (Figure 8.1J) is one of the characteristics symptoms caused by *L. theobromae* in cacao, death of young branches, leaf blight, cankers with gummosis on stems, and pod rot (Figure 8.1K) are also symptoms observed in cacao plants (Burgess et al., 2006; Kannan et al., 2010; Moreira-Morrillo et al., 2021; Huda-Shakirah et al., 2022).

8.13 The life cycle of dieback and pod rot caused by *Lasiodiplodia theobromae*

The fungus survives as pycnidia on infected plant debris (inoculum source) (Michereff et al., 2005) or by remaining as a saprophyte (Figure 8.5) (Moreira-Morrillo et al., 2021). The immature or mature conidia (Figure 1L) contained in the pycnidia are released into the environment under conditions of high rainfall and temperatures around 25 °C (Brown, 1970). Conidia are dispersed toward healthy cacao plant tissues by the action of wind (Chávez et al., 2016), mirid bugs (Hemiptera: Miridae) (Adu-Acheampong et al., 2012), bark beetles (Masood & Saeed, 2011), and possibly ants (Figure 8.5).

The fungus is considered a hemibiotrophic pathogen. The microorganism infects branches and stems through natural openings such as lenticels, stomata,

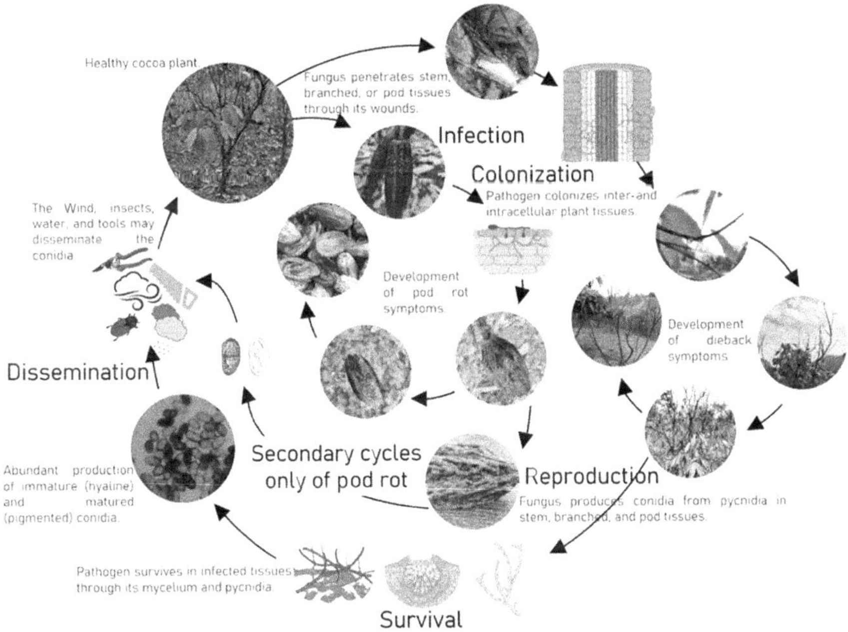

Figure 8.5 Dieback and pod rot life cycle caused by *Lasiodiplodia theobromae* on tissues of cacao plants.

or even wounds (Maftoonazad et al., 2007). In the case of pods, the pathogen penetrates the epidermis of these tissues directly (Puig et al., 2021). Once inside, the pathogen colonizes the cells of the host inter- and intracellularly until it reaches the xylem vessels, affecting the transport of water and nutrients (Muniz et al., 2011) from underground to aerial tissues (Figure 8.5). *L. theobromae* produces secondary metabolites that could be necessary during the infection and colonization processes in plant tissues (Salvatore et al., 2020).

In aerial tissues such as young branches and shoots, necrosis can be observed (Figure 8.5), while in leaves the main symptom is leaf blight (Mvondo et al., 2019; Huda-Shakirah et al., 2022). In pods, the pathogen rapidly infects these tissues until they become necrotic (Peter et al., 2014). Necrosis begins in the peduncle (Picos-Muñoz et al., 2015) and reaches the almonds, rendering them unviable (Moreira-Morrillo et al., 2021; Huda-Shakirah et al., 2022). Gummosis is another symptom observed in affected stems, due to the degradation of plant cell walls (Gao et al., 2016; Huda-Shakirah et al., 2022). In some cases, when the infection of plant tissues is intense and there is no plantation management, dead plants are observed, commonly referred to as dieback (Figure 8.5) (Alvindia & Gallema, 2017; Moreira-Morrillo et al., 2021). A significant aspect to consider regarding symptoms in plants would be related to drought-stressed and genotype response (Mullen, 1991).

Regarding the pathogen's reproduction, it produces pycnidia on infected tissues such as branches, shoots, petioles, leaves, and pods (Moreira-Morrillo et al., 2021; Huda-Shakirah et al., 2022). The high production of conidia inside pycnidia on pods could contribute to the dissemination of the pathogen to other tissues of the same plant or neighboring plants within the cacao plantation, creating countless cycles of the disease (Figure 8.5).

8.14 Management of dieback and pod rot

As in any crop, growers can use resistant or tolerant genotypes as one of the main tools for management, particularly for diseases like dieback (Moreira-Morrillo et al., 2021). Adu-Acheampong et al. (2012) recommend the use of resistant cacao cultivars such as CATIE 1000, T85/799, and MXC 67 to counteract the disease. These cultivars should be considered as a potential source for cacao genetic improvement programs against *L. theobromae*. Conversely, we know that the clone CCN-51, important for having a lower incidence of witches' broom and pod rot, and high productivity (Sánchez-Mora et al., 2015), would be highly susceptible to *L. theobromae*.

Cultural practices also should be considered in *L. theobromae* disease management, as they can interrupt the pathogen's biological cycle (Moreira-Morrillo et al., 2021). For example, pathogen-free plant material should be used and branches washed for propagation (Nurlaila et al., 2020). In the cacao farm, infected debris that act as a source of inoculum should be destroyed (Moreira-Morrillo et al., 2021), and infected tissues such as branches and pods should be removed (Akrofi et al., 2016). During these tasks, the tools used should be disinfected to avoid the spread of fungus from a diseased plant to a healthy one.

Other types of management, such as physical and alternative methods (botanical and biological), could also be employed. The farmer plays an important role in reducing the inoculum load of pathogens like *L. theobromae* in a substrate by solarizing the soil (Michereff et al., 2005), which is considered in the management of soil used in cacao seedling production (Moreira-Morrillo et al., 2021). On the other hand, the use of plants like aqueous extracts of *Azadirachta indica* seeds reduces the mycelial growth of *L. theobromae*, even matching the efficiency observed with the fungicide Carbendazim + chlorothalonil (Mvondo et al., 2019). BSA applications such as *T. harzianum* and *T. viride* could be a sustainable measure to consider, even for protecting wounds generated during the pruning (Rusin et al., 2020).

Fungicides, such as Mancozeb, Chlorothalonil + Carbendazim, copper hydroxide, Iprodione, methyl thiophanate, and copper oxychloride, among others, are used for the management of diseases such as pod rot (Al-Jabri et al., 2017; Mvondo et al., 2019). However, according to Moreira-Morrillo et al. (2021), the integrated management of this disease and dieback is the most appropriate control method, which includes implemented measures during seedling development, crop establishment, and adult cacao plant management, considering the use of fungicides as a last resource.

8.15 Lethal wilt caused by *Ceratocystis cacaofunesta*

Lethal wilt caused by *C. cacaofunesta* (formerly known as *C. fimbriata*) is one of the main phytosanitary problems restricting cacao production in tropical and subtropical countries (Johnson et al., 2005; Cabrera et al., 2016), especially in Latin America (Ploetz, 2007). This ascomycete fungus exclusively affects cacao plants (Figure 8.1 M–O) (Engelbrecht & Harrington, 2005). The disease was first reported affecting cacao in Ecuador at the beginning of the 20th century. It is believed that there is a strong relationship between lethal wilt and *Xyleborus* beetles, where the bark beetles transport the pathogen inside the cacao stems (Figure 8.1 M–N) (Saunders, 1965; Paladines-Rezabala et al., 2022). Once there, *C. cacaofunesta* colonizes the vascular bundles causing foliar chlorosis, vascular wilting, and then the death of the plant (End et al., 2017; Juzwik et al., 2011).

8.16 The life cycle of lethal wilt caused by *Ceratocystis cacaofunesta*

C. cacaofunesta propagules are found in tissues of infected roots, stems, and branches of diseased or dead plants (Hayasaka et al., 2015; Paladines-Rezabala et al., 2022), or even in the soil near the infected plant (Figure 8.6). Although the fungus could be spread by farm tools such as 'machete' (long-bladed knife), pruning of branches, harvesting pods (Hayasaka et al., 2015), or wounds in roots and stem bases (Santos et al., 2013), the pathogen is mainly disseminated by bark beetle insects of the genus *Xyleborus* (Coleoptera: Curculionidae: Scolytinae) (Paladines-Rezabala et al., 2022).

Possibly, adult virgin bark beetle females fly to healthy cacao plants in search of reproduction sites and build galleries, then they emit semiochemicals, pheromones, and volatile substances that attract male insects (Figure 8.6) (Kendra et al., 2011), possibly with spores attached to their exoskeletons as in the *Ophiostoma ulmi–Scolytus scolytus* interaction (Boddy, 2016). Ambrosia beetles perforate stems and branches (Saunders, 1965), and expel their excrement outside these tissues, accumulating mainly at the base of the cacao tree as small dusty masses (End et al., 2017). This event is considered the first sign of a possible infection caused by *C. cacaofunesta* and is observed before the plants present foliar wilting. Bark beetle complete their metamorphosis in the galleries inside the trunks of cacao trees (Figure 8.6).

Once the fungus reaches the cacao tissues, the spores germinate in around 25 hours (Santos et al., 2013), starting the infectious process (Figure 8.6). In this process, it is still unknown if the fungus forms specialized structures to penetrate the plant tissue (Cabrera et al., 2016). Either way, the aim of *C. cacaofunesta* is to reach the xylem cells (End et al., 2017). Although the colonization of susceptible tissues by the pathogen can take around seven days, this progress depends on its vertical rather than transversal colonization. Hyphae and spores invade considerably the parenchyma, cambium, and cortex tissues, observing lesions or cankers on stems or branches, especially in infected mature trees

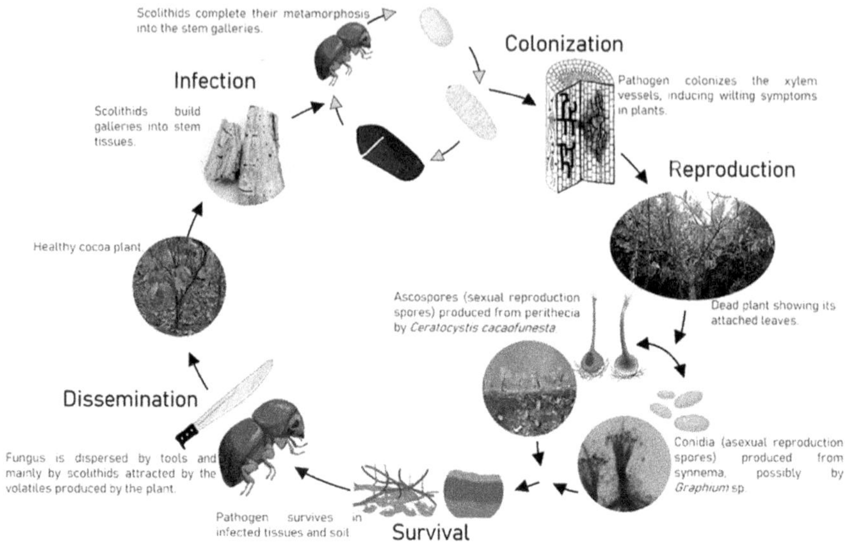

Figure 8.6 Lethal wilt life cycle caused by *Ceratocystis cacaofunesta* on cacao plants.

(MacDonald & Hindal, 1981; Santos et al., 2013). The fungus takes between 15 and 18 months to colonize cacao trees (Silva et al., 2012). It is important to consider that infection and colonization depend on the degree of resistance or the state of health of the plant (Figure 8.6) (End et al., 2017).

With the advance of colonization, the first symptoms appear, that is, wilting and vascular discoloration (Santos et al., 2013), which become more intense, observing the withered brown leaves on one or several branches of the tree, depending on the point of growth infection (branches or trunk) (Figure 8.6). After the first symptoms described above appear, the plant dies slowly, lasting between 10 and 30 days or even years (Juzwik et al., 2011; Cabrera et al., 2016). One way to recognize whether the plant is affected by *C. cacaofunesta* is that once the tree is dead, its leaves remain attached to the plant, and even upon shaking, do not fall (Figure 8.1M) (End et al., 2017).

The reproduction of *C. cacaofunesta* is like that of other species of the same genus, presenting a type of bipolar mating (Harrington & McNew, 1997). The fungus forms different structures; on the one hand, a set of hyphae probably produce conidiophores and conidia, and on the other, perithecia and ascospores (Figure 8.1O) (Santos et al., 2013). In species like *C. fimbriata*, the ascospores or aleuroconidia production is influenced by temperatures between 18 and 28 °C (Syazwan et al., 2021). *C. cacaofunesta* produces perithecia with an elongated neck, dark brown to black, and abundant ascospore mass on the ostiole (Engelbrecht & Harrington, 2005; Wilken et al., 2014; Paladines-Rezabala et al., 2022). Although the asexual state of *C. cacaofunesta* is unknown, we could infer the genus *Graphium* forming synnema and conidia. In fact, it is known that *Ophiostoma ulmi* (anamorph *Graphium ulmi*) is the causative agent of the Dutch elm disease (Figure 8.6) (Schwarz, 1922; de Beer et al., 2013).

8.17 Management of lethal wilt

Lethal wilt management focuses mainly on prevention, but if cacao plants are infected, exclusion and eradication measures are used (Cabrera et al., 2016). Among the methods are the genetic, cultural, and biological, which could be used to reduce the occurrence or interrupt the disease cycle and even the bark beetle *X. ferrugineus* management (Juzwik et al., 2011; Paladines-Rezabala et al., 2022). Before the cacao crop establishment, the correct choice of genotype is very important. There are different genotypes considered resistant to *C. cacaofunesta*, that is CEPEC-2008, IMC-67, JACA, and TSH-1188, among others used as rootstocks (Sanches et al., 2008; Magalhães et al., 2015; 2016; Santos et al., 2013; dos Santos et al., 2021). However, in large cacao areas in Latin America (especially in Ecuador), clone CCN-51 is widely used, and according to several researchers, it is susceptible to lethal wilt (Sanches et al., 2008; Santos et al., 2013; Magalhães et al., 2016). The resistance of the tissues in cacao plants against *C. cacaofunesta* would be associated with structural (Santos et al., 2013) and biochemical defense components (Neves dos Santos et al., 2021).

Cultural management within the farm is also very essential. For example, we must take quarantine measures during the obtaining process of grafted plants, releasing them only when we confirm that they do not present symptoms caused by *C. cacaofunesta* (Engelbrecht et al. 2007). The disinfestation of tools during activities such as pruning is also essential (Teviotdale and Harper, 1991). Beetle monitoring and diseased plant eradication would be an important management measure that we must consider (Paladines-Rezabala et al., 2022). In this regard, Mazón et al. (2013) recommend setting windowed bottle traps (placed 1 and 3 m above ground level) and funnel traps to collect bark and ambrosia beetles (Scolytinae).

BCA use is another interesting method that we could use for the lethal wilt management of cacao. For example, the antagonists *T. harzianum* (IMI-392432) and *T. pseudokoningii* (IMI-392431) reduced the mycelial growth of *C. paradoxa*, which causes the pineapple disease of sugarcane (Rahman et al., 2009). In cacao, isolates of *T. harzianum* (910) and *T. pseudokoningii* (1052) showed potential *in vitro* antagonistic activity against *M. perniciosa*, *P. palmivora*, and *C. cacaofunesta*, and can be recommended as possible biocontrollers under field conditions (de Oliveira et al., 2013). Either way, the positive results obtained under *in vitro* conditions can be contrary to those obtained in the field. In this regard, *T. virens* (T68), *T. harzianum* (2927), *T. loningiopsis* (Tc26), and *T. atroviride* (7CC) inhibit mycelial growth and perithecia formation of *C. cacaofunesta* by more than 90%, but the efficiency of these biocontrollers is not as expected when applied to cacao seedlings (Rodrigues et al., 2018).

To counteract lethal wilt disease on cacao plants, the ideal method would be to integrate the different management measures, that is making the area free of

C. cacaofunesta, use of resistant cultivars, disinfestation of tools, and use of BCA, among others. This set of practices should include monitoring the presence of bark and ambrosia beetles through traps used on cacao farms. Also, we could use semiochemicals because beetles are strongly attracted to volatile compounds released by trees (Kendra et al., 2011; Martini et al., 2015).

8.18 Conclusions

Diseases are one of the major biotic factors that significantly reduce bean production worldwide and cause millions of dollars in cacao crops worldwide. In Latin America, cacao agrosystems can be affected by fungi and fungi-like that is witches' broom (*M. perniciosa*), frosty pod (*M. roreri*), root and crown rot and pod rot (*P. palmivora*), dieback and pod rot (*L. theobromae*), and lethal wilt (*C. cacaofunesta*). Although witches' broom and frosty pod have been a problem for Latin American cacao farmers, they have been able to manage them through cultural practices. However, other emerging diseases caused by *P. palmivora* and *L. theobromae* that are negatively affecting cacao crops in countries such as Ecuador are becoming a stone in the shoe for the farmers. Both phytopathogens (especially the latter) can be spread within the crop by tools, intensifying their infection, colonization, and reproduction processes in cacao aerial and subterranean tissues. In order to reduce the intensity of all the diseases of the cacao crop presented in this chapter, various integrated management measures can be implemented, that is pruning the trees at least once a year, periodically removing diseased tissues (fruits, branches, flower cushions, etc.) or severely affected trees, triturate diseased pods on the ground to speed up their decomposition, reduce a shade between 30 and 40% that allows light and air to enter to reduce the percentage of humidity in the environment, maintain an operational drainage system to avoid water logging that favors the development of diseases, and harvest ripe pods every two weeks or every seven days, depending on disease pressure. As a last resort, especially in humid areas or during the rainy season, fungicides such as chlorothalonil, copper-based (copper oxide or hydroxide), azoxystrobin, fosetyl-aluminum, among others, could be used, depending on the etiological agent.

References

Adu-Acheampong, R., Archer, S. A., & Leather, S. R. (2012). Resistance to dieback disease caused by *Fusarium* and *Lasiodiplodia* species in cacao (*Theobroma cacao* L.) genotypes. *Experimental Agriculture*, *48*(1), 85–98. https://doi.org/10.1017/s0014479711000883

Aime, M. C., & Phillips-Mora, W. (2005). The causal agents of witches' broom and frosty pod rot of cacao (chocolate, *Theobroma cacao*) form a new lineage of Marasmiaceae. *Mycology*, *97*(5), 1012–1022. https://doi.org/10.3852/mycologia.97.5.1012

Akrofi, A. Y., Amoako-Atta, I., Assuah, M., & Asare, E. K. (2015). Black pod disease on cacao (*Theobroma cacao*, L) in Ghana: Spread of *Phytophthora megakarya* and role of economic plants in the disease epidemiology. *Crop Protection*, *72*, 66–75. https://doi.org/10.1016/j.cropro.2015.01.015

Akrofi, A. Y., Appiah, A. A., & Opoku, I. Y. (2003). Management of Phytophthora pod rot disease on cocoa farms in Ghana. *Crop Protection*, *22*, 469–477. https://doi.org/10.1016/j.cropro.2015.01.015

Akrofi, A. Y., Amoako-Atta, I., Acheampong, K., Assuah, M. K., & Melnick, R. L. (2016). Fruit and canopy pathogens of unknown potential risk. In Bailey, B. A., & Meinhardt, L. W. (eds), *Cacao Diseases: A History of Old Enemies and New Encounters* (pp. 361–382). Springer. https://doi.org/10.1007/978-3-319-24789-2_11

Al-Jabri, M. K., Al-Shaili, M., Al-Hashmi, M., Nasehi, A., AlMahmooli, I. H., & Al-Sadi, A. M. (2017). Characterization and evaluation of fungicide resistance among *Lasiodiplodia theobromae* isolates associated with mango dieback in Oman. *Journal of Plant Pathology*, *99*(3), 753–759. https://doi.org/10.4454/jpp.v99i3.3959

Ali, M., Kim, B., Belfield, K. D., Norman, D., Brennan, M., & Ali, G. S. (2015). Inhibition of *Phytophthora parasitica* and *P. capsici* by silver nanoparticles synthesized using aqueous extract of *Artemisia absinthium*. *Phytopathology*, *105*(9), 1183–1190. https://doi.org/10.1094/PHYTO-01-15-0006-R

Alvindia, D. G., & Gallema, F. L. M. (2017). *Lasiodiplodia theobromae* causes vascular streak dieback (VSD)-like symptoms of cacao in Davao region, Philippines. *Australasian Plant Disease Notes*, *12*(1), 54. https://doi.org/10.1007/s13314-017-0279-9

Amores, F., Suárez, C. C., & Garzón, C. I. (2010). Intensive production of national cacao with "top" flavor technology, budget and profitability. INIAP, EET-Pichilingue. Quevedo, Ecuador. Technical Manual No. 82, pp. 71–83.

Appiah, A., Opoku, I., & Akrofi, A. (2004). Natural occurrence and distribution of stem cankers caused by *Phytophthora megakarya* and *Phytophthora palmivora* on cacao. *European Journal of Plant Pathology*, *110*, 983–990. https://doi.org/10.1007/s10658-004-0811-6

Asare-Nyako, A. (1974). Towards an economic chemical control of the black pod disease of cacao in Ghana. *Ghana Journal of Agricultural Science*, *7*, 17–23.

Bailey, B. A., Evans, H. C., Phillips-Mora, W., Ali, S. S., & Meinhardt, L. W. (2018). *Moniliophthora roreri*, causal agent of cacao frosty pod rot. *Molecular Plant Pathology*, *19*(7), 1580–1594. https://doi.org/10.1111/mpp.12648

Barreto, M. A., Rosa, J. R. B. F., Holanda, I. S. A., Cardoso-Silva, C. B., Vildoso, C. I. A., Ahnert, D., Souza, M. M., Corrêa, R. X., Royaert, S., Marelli, J., Santos, E. S. L., Luz, E. D. M. N., Garcia, A. A. F., & Souza, A. P. (2018). QTL mapping and identification of corresponding genomic regions for black pod disease resistance to three *Phytophthora* species in *Theobroma cacao* L. *Euphytica*, *214*, 188. https://doi.org/10.1007/s10681-018-2273-5

Bassani, I., Larousse, M., Tran, Q. D., Attard, A., & Galiana, E. (2020). *Phytophthora* zoospores: From perception of environmental signals to inoculum formation on the host-root surface. *Computational and Structural Biotechnology Journal*, *18*, 3766–3773. https://doi.org/10.1016/j.csbj.2020.10.045

Boddy, L. (2016). Pathogens of autotrophs. In S. C. Watkiston, L. Boddy, & N. P. Money (eds), *The Fungi* (3rd ed., pp. 245–292). Academic Press.

Brasier, C., & Griffin, M. (1979). Taxonomy of *Phytophthora palmivora* on cacao. *Transactions of the British Mycological Society*, *72*(1), 111–143. https://doi.org/10.1016/S0007-1536(79)80015-7

Brown, G. (1970). Pycnidial released and survival of *Diplodia natalensis* spores. *Phytopathology*, *61*, 559–561.

Burgess, T. I., Barber, P. A., Mohali, S., Pegg, G., de Beer, W., & Wingfield, M. J. (2006). Three new *Lasiodiplodia* spp. from the tropics, recognized based on DNA sequence comparisons and morphology. *Mycologia*, *98*(3), 423–435. https://doi.org/10.3852/mycologia.98.3.423

CABI. (2022). Moniliophthora perniciosa (witches' broom disease of cacao). *CABI Compendium*. https://doi.org/10.1079/cabicompendium.16054

Cabrera, O. G., Molano, E. P. L., José, J., Álvarez, J. C., & Pereira, G. A. G. (2016). *Ceratocystis* wilt pathogens: History and biology—Highlighting *C. cacaofunesta*, the causal agent of wilt disease of cacao. In Bailey, B., & Meinhardt, L. (eds), *Cacao Diseases*. Springer. https://doi.org/10.1007/978-3-319-24789-2_12

Chávez, M. A., Espinosa, K. C. S., & Vázquez, L. D. (2016). *Lasiodiplodia theobromae* en la atmósfera de La Habana. *Revista Cubana de Ciencias Biológicas*, *5*(1), 130–134.

Cedeño-Moreira, Ángel V., Romero-Meza, R. F., Auhing-Arcos, J. A., Mendoza-León, A. F., Abasolo-Pacheco, F., & Canchignia-Martínez, H. F. (2020). Caracterización de *Phytophthora* spp. y aplicación de rizobacterias con potencial en biocontrol de la enfermedad de la mazorca negra en *Theobroma cacao* variedad CCN-51. *Scientia Agropecuaria*, *11*(4), 503–512. https://doi.org/10.17268/sci.agropecu.2020.04.05

Cilas, C., & Bastide, P. (2020). Challenges to cacao production in the face of climate change and the spread of pests and diseases. *Agronomy*, *10*(9), 1232. https://doi.org/10.3390/agronomy10091232

de Beer, Z. W., Seifert, K. A., & Wingfield, M. J. (2013). A nomenclator for ophiostomatoid genera and species in the Ophiostomatales and Microascales. *Biodiversity Series*, *12*, 245–322.

de Oliveira, B. F., Silva, S. D. V. M., & dos Santos, M. V. O. (2013). Antagonismo *in vitro* de *Trichoderma* spp. a patógenos do cacaueiro. *Agrotrópica 25*(2), 117–120. https://doi.org/10.21757/0103-3816.2013v25n2p117-120

Deberdt, P., Mfegue, C. V., Tondje, P. R., Bon, M. C., Ducamp, M., Hurard, C., Begoude, B. A. D., Ndoumbe-Nkeng, M., Hebba, P. K. & Cilas, C. (2008). Impact of environmental factors, chemical fungicide and biological control on cacao pod production dynamics and black pod disease (*Phytophthora megakarya*) in Cameroon. *Biological Control*, *44*(2), 149–159. https://doi.org/10.1016/j.biocontrol.2007.10.026

Decloquement, J., Ramos-Sobrinho, R., Elias, S. G., Britto, D. S., Puig, A. S., Reis, A., da Silva, R. A. F., Honorato-Júnior, J., Luz, E. D. M. N., Pinho, D. B., & Marelli, J. P. (2021). *Phytophthora theobromicola* sp. nov.: A new species causing black pod disease on cacao in Brazil. *Frontiers in Microbiology*, *12*, 537399. https://doi.org/10.3389/fmicb.2021.537399

Delgado-Ospina, J., Molina-Hernández, J. B., Chaves-López, C., Romanazzi, G., & Paparella, A. (2021). The role of fungi in the cacao production chain and the challenge of climate change. *Journal of Fungi*, *7*(3), 202. https://doi.org/10.3390/jof7030202

Drenth, A., & Guest, D. (2013). *Phytophthora*: la destructora de plantas. *Revista Palmas*, *34*, 49–56.

El-Sayed, A. S., & Ali, G. S. (2020). *Aspergillus flavipes* is a novel efficient biocontrol agent of *Phytophthora parasitica*. *Biological Control*, *140*, 104072. https://doi.org/10.1016/j.biocontrol.2019.104072

End, M. J., Daymond, A. J., & Hadley, P. (2017). Directrices técnicas para el movimiento seguro del germoplasma del cacao. Versión revisada de las Directrices Técnicas de FAO/IPGRI No. 20. Retrieved January 12, 2023, from https://www.bioversityinternational.org/fileadmin/user_upload/Directrices_tecnicas_2017_Spanish.pdf#page=53

Engelbrecht, C. J., & Harrington, T. C. (2005). Intersterility, morphology and taxonomy of *Ceratocystis fimbriata* on sweet potato, cacao, and sycamore. *Mycologia*, *97*(1), 57–69. https://doi.org/10.3852/mycologia.97.1.57

Engelbrecht, C. J., Harrington, T. C., & Alfenas, A. (2007). *Ceratocystis* wilt of cacao–A disease of increasing importance. *Phytopathology*, *97*(12), 1648–1649. https://doi.org/10.1094/PHYTO-97-12-1648

Estrela Junior, A. D. S., Solís, K., Pimenta Neto, A. A., Vera, D. I., Garzón, I., Peñaherrera, S., Diorato, V. S., Gramacho, K. P., & Laranjeira, D. (2022). Effect of antagonistic yeasts from cacao tissues on controlling growth and sporulation of *Moniliophthora roreri*. *Biological Control*, *172*, 104956. https://doi.org/10.1016/j.biocontrol.2022.104956

Evans, H. C. (1978). Witches' broom disease of cacao (*Crinipellis perniciosa*) in Ecuador. The fungus. *Annals of Applied Biology*, *89*(2), 185–192. https://doi.org/10.1111/j.1744-7348.1978.tb07689.x

Evans, H. C. 2007. Cacao Diseases—The Trilogy Revisited. *Phytopathology*, *97*(12), 1640–1643. https://doi.org/10.1094/PHYTO-97-12-1640

Evans, H. (2016). Frosty pod rot (*Moniliophthora roreri*). In Bailey, B., & Meinhardt, L. (eds), *Cacao Diseases*. (pp. 63–96). Springer. https://doi.org/10.1007/978-3-319-24789-2_3

Gao, L., Wang, Y., Li, Z., Zhang, H., Ye, J., & Li, G. (2016). Gene expression changes during the gummosis development of peach shoots in response to *Lasiodiplodia theobromae* infection using RNA-Seq. *Frontiers in Physiology*, *7*, 170. https://doi.org/10.3389/fphys.2016.00170

Griffith, G. W., Nicholson, J., Nenninger, A., Birch, R. N., & Hedger, J. N. (2003). Witches' brooms and frosty pods: Two major pathogens of cacao. *New Zealand Journal of Botany*, *41*, 423–435. https://doi.org/10.1080/0028825X.2003.9512860

Guest, D. (2007). Black pod: Diverse pathogens with a global impact on cacao yield. *Phytopathology*, *97*(12), 1650–1653. https://doi.org/10.1094/PHYTO-97-12-1650

Harrington, T. C., & McNew, D. (1997). Self-fertility and uni-directional mating-type switching in *Ceratocystis coerulescens*, a filamentous ascomycete. *Current Genetics*, *32*(1), 52–59. https://doi.org/10.1007/s002940050247

Hayasaka, D., Kuwayama, N., Takeo, A., Ishida, T., Mano, H., Inoue, M. N., Nagai, T., Sánchez-Bayo, F., Goka, K., & Sawahata, T. (2015). Different acute toxicity of fipronil baits on invasive *Linepithema humile* supercolonies and some non-target ground arthropods. *Ecotoxicology*, *24*(6), 1221–1228. https://doi.org/10.1007/s10646-015-1483-z

Huda-Shakirah, A. R., Mohamed Nor, N. M. I., Zakaria, L., Leong, Y. H., & Mohd, M. H. (2022). *Lasiodiplodia theobromae* as a causal pathogen of leaf blight, stem canker, and pod rot of *Theobroma cacao* in Malaysia. *Scientific Reports*, *12*(1), 8966. https://doi.org/10.1038/s41598-022-13057-9

Johnson, J. A., Harrington, T. C., & Engelbrecht, C. J. B. (2005). Phylogeny and taxonomy of the North American clade of the *Ceratocystis fimbriata* complex. *Mycologia*, *97*(5), 1067–1092. https://doi.org/10.3852/mycologia.97.5.1067

Juzwik, J., Appel, D. N., MacDonald, W. L., & Burks, S. (2011). Challenges and successes in managing oak wilt in the United States. *Plant Disease*, *95*(8), 888–900. https://doi.org/10.1094/PDIS-12-10-0944

Kannan, C., Karthik, M., & Priya, K. (2010). *Lasiodiplodia theobromae* causes a damaging dieback of cacao in India. *Plant Pathology*, *59*(2), 410–410. https://doi.org/10.1111/j.1365-3059.2009.02192.x

Khanzada, M. A., Lodhi, A. M., & Shahzad, S. (2005). Chemical control of *Lasiodiplodia theobromae*, the causal agent of mango declines in Sindh. *Pakistan Journal of Botany*, *37*, 1023–1030.

Kendra, P. E., Montgomery, W. S., Niogret, J., Peña, J. E., Capinera, J. L., Brar, G., Epsky, N. D., & Heath, R. R. (2011). Attraction of the redbay ambrosia beetle, *Xyleborus glabratus*, to avocado, lychee, and essential oil lures. *Journal of Chemical Ecology*, *37*(9), 932–942. https://doi.org/10.1007/s10886-011-9998-0

Lanaud, C., Fouet, O., Clément, D., Boccara, M., Risterucci, A. M., Surujdeo-Maharaj, S., Legavre, T., & Argout, X. (2009). A meta–QTL analysis of disease resistance traits of *Theobroma cacao* L. *Molecular Breeding*, *24*, 361–374. https://doi.org/10.1007/s11032-009-9297-4

Liu, H. F., Xue, X. J., Yu, Y., Xu, M. M., Lu, C. C., Meng, X. L., Zhang, B. G., Ding, X. H., & Chu, Z. H. (2020). Copper ions suppress abscisic acid biosynthesis to enhance defence against *Phytophthora infestans* in potato. *Molecular Plant Pathology*, *21*(5), 636–651. https://doi.org/10.1111/mpp.12919

MacDonald, W. L., & Hindal, D. F. (1981). Life cycle and epidemiology of *Ceratocystis*. In Mace, E., Bell, A., & Beckman, H. (eds), *Fungal Wilt Diseases of Plants* (pp. 113–144). Academic Press. https://doi.org/10.1016/B978-0-12-464450-2.50010-3

Maftoonazad, N., Ramaswamy, H., Moalemiyan, M., & Kushalappa, A. (2007). Effect of pectin-based edible emulsion coating on changes in quality of avocado exposed to *Lasiodiplodia theobromae* infection. *Carbohydrate Polymers*, *68*(2), 341–349. https://doi.org/10.1016/j.carbpol.2006.11.020

Magalhães, D. M. A., Luz, E. D. M. N., Lopes, U. V., Silva, S. D. V. M., Damaceno, V. O., & Niella, A. R. R. (2015). A new method for early detection of *Ceratocystis* spp. on various hosts. *Agrotrópica*, *27*, 209–214.

Magalhães, D. M. A., Luz, E. D. M. N., Lopes, U. V., Niella, A. R. R., & Damaceno, V. O. (2016). Leaf disc method for screening Ceratocystis wilt resistance in cacao. Trop. *Plant Pathology*, *41*, 155–161. https://doi.org/10.1007/s40858-016-0081-9

Marelli, J. P., Guest, D. I., Bailey, B. A., Evans, H. C., Brown, J. K., Junaid, M., Barreto, R. W., Lisboa, D. O., & Puig, A. S. (2019). Chocolate under threat from old and new cacao diseases. *Phytopathology*, *109*(8), 1331–1343. https://doi.org/10.1094/PHYTO-12-18-0477-RVW

Martini, X., Hughes, M. A., Smith, J. A., & Stelinski, L. L. (2015). Attraction of redbay ambrosia beetle, *Xyleborus glabratus*, to leaf volatiles of its host plants in North America. *Journal of Chemical Ecology*, *41*(7), 613–621. https://doi.org/10.1007/s10886-015-0595-5

Masood, A., & Saeed, S. (2011). Bark beetle, *Hypocryphalus mangiferae* Stebbing (Coleoptera: Curculionidae: Scolytinae) is a vector of mango sudden death disease in Pakistan. *Pakistan Journal of Botany*, *44*(2), 813–820. https://www.cabdirect.org/cabdirect/abstract/20123182876

Mazón, M., Díaz, F., & Gaviria, J. C. (2013). Effectiveness of different trap types for control of bark and ambrosia beetles (Scolytinae) in Criollo cacao farms of Mérida, Venezuela. *International Journal of Pest Management*, *59*(3), 189–196. https://doi.org/10.1080/09670874.2013.810794

Mejía, L. C., Rojas, E. I., Maynard, Z., Van Bael, S., Arnold, A. E., Hebbar, P., Samuels, G. J., Robbins, N., & Herre, E. A. (2008). Endophytic fungi as biocontrol agents of *Theobroma cacao* pathogens. *Biological Control*, *46*(1), 4–14. https://doi.org/10.1016/j.biocontrol.2008.01.012

Michereff, J., Peruch, A. M., & ANdrade, E. G. T. (2005). Manejo Integrado de Doenças Radiculares. In Michereff, J., Andrade, E. G. T., & Menezes, M. (eds), *Ecologia e Manejo de Patógenos Radiculares em Solos Tropicais* (pp. 367–388).

Moreira-Morrillo, A. A., Cedeño-Moreira, N. V., Canchignia-Martínez, F., & Garcés-Fiallos, F. R. (2021). *Lasiodiplodia theobromae* (Pat.) Griffon & Maubl [(syn.) *Botryodiplodia theobromae* Pat] in the cacao crop: Symptoms, biological cycle, and strategies management. *Scientia Agropecuaria*, *12*(4), 653–662. https://doi.org/10.17268/sci.agropecu.2021.068

Mullen, J. M. (1991). Canker of dogwood caused by *Lasiodiplodia theobromae*: A disease influenced by drought stress or cultivar selection. *Plant Disease*, *75*(9), 886–889. https://doi.org/10.1094/PD-75-0886

Muniz, C. R., Freire, F. C. O., & Viana, F. M. P. (2011). Colonization of cashew plant by *Lasiodiplodia theobromae*: Microscopical features. *Micron*, *42*, 419–428. https://doi.org/10.1016/j.micron.2010.12.003

Mvondo, N. D., Manga, E. F., Kone, N. A. N., Ndogho, P. A., & Ambang, Z. (2019). Pathogenicity and *in vitro* control of *Lasiodiplodia theobromae* and *Fusarium* sp., pathogens associated with cacao dieback in Cameroon. *International Journal of Current Research in Biosciences and Plant Biology*, *6*(7), 1–13. https://doi.org/10.20546/ijcrbp.2019.607.001

Neves dos Santos, F., Magalhães, D. M. A., Luz, E. D. M. N., Eberlin, M. N. & Simionato, A. V. C. (2021). Metabolite mass spectrometry profiling of cacao genotypes reveals contrasting resistances to *Ceratocystis cacaofunesta* phytopathogen. *Electrophoresis*, *42*(23), 2519–2527. https://doi.org/10.1002/elps.202100097

Nurlaila, N., Rosmana, A., & Dewi, V. S. (2020). The capability of *Trichoderma asperellum* in suppressing vascular streak diseases on five different cacao clones. *IOP Conference Series: Earth and Environmental Science*, *486*(1), 012158. https://doi.org/10.1088/1755-1315/486/1/012158

Okey, E. N., Duncan, E. J., Sirju-Charran, G., & Sreenivasan, T. N. (1996), Factors affecting the susceptibility of six cacao clones to *Phytophthora palmivora* (Butl.) Butler bark canker in Trinidad. *Plant Pathology*, *45*(1), 84–91. https://doi.org/10.1046/j.1365-3059.1996.d01-94.x

Paladines-Rezabala, A., Moreira-Morrillo, A. A., Mieles, A. E., & Garcés-Fiallos, F. R. (2022). Advances in understanding of the interaction between *Ceratocystis cacaofunesta* and *Xyleborus ferrugineus* (Coleoptera: Curculionidae: Scolytinae) on cocoa trees. *Scientia Agropecuaria*, *13*(1), 43–52. http://dx.doi.org/10.17268/sci.agropecu.2022.004

Peñaherrera, V. S., Cedeño, G. G., Solórzano, A. F., Cedeño-García, G., & Terrero, Y. P. (2020). Eficacia de mezclas de *Trichoderma* spp. y aceite de palma en el manejo de *Moniliophthora roreri* Cif & Par en cacao. *Centro Agrícola*, *47*(2), 5–15.

Perrine-Walker, F. (2020). *Phytophthora palmivora*–cacao interaction. *Journal of Fungi*, *6*(3), 167. https://doi.org/10.3390/jof6030167

Peter, T., Godfried, O. M., & Emmanuel, M. (2014). The rot fungus *Botryodiplodia theobromae* strains cross infect cacao, mango, banana and yam with significant tissue damage and economic losses. *African Journal of Agricultural Research*, *9*(6), 613–619. https://doi.org/10.5897/ajar2013.7528

Picos-Muñoz, P. A., García-Estrada, R. S., León-Félix, J., Sañudo-Barajas, A., & Allende-Molar, R. (2015). *Lasiodiplodia theobromae* in agricultural crops in México: Taxonomy, host, diversity and control. *Revista Mexicana de Fitopatología*, *33*(1), 54–74.

Ploetz, R. C. (2007). Cacao diseases: Important threats to chocolate production worldwide. *Phytopathology*, *97*(12), 1634–1639. https://doi.org/10.1094/PHYTO-97-12-1634

Porras, V. H., Cruz Chang, C. A., & Galindo, J. J. (1990). Manejo integrado de la mazorca negra y la moniliasis del cacao en el trópico húmedo bajo de Costa Rica. *Turrialba*, *40*(2), 238–245.

Puig, A. S., Ali, S., Strem, M., Sicher, R., Gutierrez, O. A., & Bailey, B. A. (2018). The differential influence of temperature on *Phytophthora megakarya* and *Phytophthora palmivora* pod lesion expansion, mycelia growth, gene expression, and metabolite profiles. *Physiological and Molecular Plant Pathology*, *102*, 95–112. https://doi.org/10.1016/j.pmpp.2017.12.002

Puig, A. S., Keith, L. M., Matsumoto, T. K., Gutierrez, O. A., & Marelli, J. P. (2021). Virulence tests of *Neofusicoccum parvum, Lasiodiplodia theobromae*, and *Phytophthora palmivora* on *Theobroma cacao*. *European Journal of Plant Pathology*, *159*(4), 851–862. https://doi.org/10.1007/s10658-021-02210-1

Purdy, L. H., & Schmidt, R. A. (1996). Status of cacao Witches' broom: Biology, epidemiology, and management. *Annual Review of Phytopathology*, *34*, 573–594. https://doi.org/10.1146/annurev.phyto.34.1.573

Purwantara, A., McMahon, P., Susilo, A. W., Sukamto, S., Mulia, S., Nurlaila, Saftar, A., Purung, H. B., Lambert, S., Keane, P., & Guest, D. (2015). Testing local cacao selections in Sulawesi: (ii) resistance to stem canker and pod rot (black pod) caused by *Phytophthora palmivora*. *Crop Protection*, *77*, 18–26. https://doi.org/10.1016/j.cropro.2015.07.005

Rahman, M. A., Begum, M. F., & Alam, M. F. (2009). Screening of *Trichoderma* isolates as a biological control agent against *Ceratocystis paradoxa* causing pineapple disease of sugarcane. *Mycobiology*, *37*(4), 277. https://doi.org/10.4489/myco.2009.37.4.277

Rodrigues, G. D. S., Magalhães, D. M. A., Costa, A. M. D., & Luz, E. D. M. N. (2018). Antagonismo de *Trichoderma* spp. ao agente etiológico da Murcha de *Ceratocystis* em cacaueiro. *Summa Phytopathologica*, *44*(1), 72–78. https://doi.org/10.1590/0100-5405/172774

Rojas-Rojas, F. U., & Vega-Arreguín, J. C. (2021). Epigenetic insight into regulatory role of chromatin covalent modifications in lifecycle and virulence of *Phytophthora*. *Environmental Microbiology Reports*, *13*(4), 445–457. https://doi.org/10.1111/1758-2229.12954

Rusin, C., Cavalcanti, F. R., Lima, P. C. G. D., Faria, C. M. D. R., Almança, M. A. K., & Botelho, R. V. (2020). Control of the fungi *Lasiodiplodia theobromae*, the causal agent of dieback, in cv. Syrah grapevines. *Acta Scientiarum. Agronomy*, *43*, e44785. https://doi.org/10.4025/actasciagron.v43i1.44785

Salvatore, M. M., Alves, A., & Andolfi, A. (2020). Secondary metabolites of *Lasiodiplodia theobromae*: Distribution, chemical diversity, bioactivity, and implications of their occurrence. *Toxins*, *12*(7), 457. https://doi.org/10.3390/toxins12070457

Samuels, G. J., Suarez, C., Solis, K., Holmes, K. A., Thomas, S. E., Ismaiel, A., & Evans, H. C. (2006). *Trichoderma theobromicola* and *T. paucisporum*: Two new species isolated from cacao in South America. *Mycological Research*, *110*(4), 381–392. https://doi.org/10.1016/j.mycres.2006.01.009

Sanches, C. L. G., Pinto, L. R. M., Pomella, A. W. V., Silva, S. D. V. M., & Loguercio, L. L. (2008). Assessment of resistance to *Ceratocystis cacaofunesta* in cacao genotypes. *European Journal of Plant Pathology*, *122*(4), 517–528. https://doi.org/10.1007/s10658-008-9319-9

Sánchez Mora, F. D., & Garcés Fiallos, F. R. (2012). *Moniliophthora roreri* (Cif y Par) Evans *et al.* in the crop of cacao. *Scientia Agropecuaria*, *3*, 249–258. https://doi.org/10.17268/sci.agropecu.2012.03.06

Sánchez-Mora, F. D., Medina-Jara, S. M., Díaz-Coronel, G. T., Ramos-Remache, R. A., Vera-Chang, J. F., Vásquez-Morán, V. F., Troya-Mera, F. A., Garcés-Fiallos, F. R., & Onofre-Nodari, R. (2015). Potencial sanitario y productivo de 12 clones de cacao en Ecuador. *Revista Fitotecnia Mexicana*, *38*(3), 265. https://doi.org/10.35196/rfm.2015.3.265

Santos, R. M. F., Silva, S. D. V. M., Sena, K., Micheli, F., & Gramacho, K. P. (2013). Kinetics and histopathology of the Cacao-*Ceratocystis cacaofunesta* interaction. *Tropical Plant Biology*, *6*(1), 37–45. https://doi.org/10.1007/s12042-012-9115-8

Schwarz, M. B. (1922). Das Zweigsterben der Ulmen, Trauerweiden und Pfirsichbäume, eine vergleichend-pathologische Studie. Doctor's dissertation, University Utrecht. Utrecht, The Netherlands: A. Oosthoek.

dos Santos, E., Magalhães, D. M. A., Lopes, U. V., & Luz, E. D. M. N. (2021). Selection of cacao trees resistant to Ceratocystis wilt by inoculation in leaf discs and field. *Tropical Plant Pathology*, *46*(5), 536–544. https://doi.org/10.1007/s40858-021-00452-2

Saunders, J. L. (1965). The Xyleborus-Ceratocystis complex of cacao. *Cacao*, *10*, 7–13.

Silva, S. D. V. M., Luz, E. D. M. N., Almeida, O. C., Gramacho, K. P., & Becerra, J. L. (2002). Redescrição da sintomatologia causada por *Crinipellis perniciosa* em cacaueiro. *Agrotropica*, *14*(1), 1–24.

Silva, S. D. V. M., Pinto, L. R. M., Oliveira, B. F. D., Damaceno, V. O., Pires, J. L., & Dias, C. T. D. S. (2012). Resistência de progênies de cacaueiro à murcha-de-Ceratocystis. *Tropical Plant Pathology*, *37*(3), 191–195. https://doi.org/10.1590/s1982-56762012000300005

Slippers, B., Boissin, E., Phillips, A., Groenewald, J., Lombard, L., Wingfield, M., Postma, A., Burgess, T., & Crous, P. (2013). Phylogenetic lineages in the Botryosphaeriales: A systematic and evolutionary framework. *Studies in Mycology*, *76*, 31–49. https://doi.org/10.3114/sim0020

Solís-Hidalgo, Z. K., Peñaherrera-Villafuerte, S. L., Vera-Coello, D. I. (2021). Las enfermedades del cacao y las buenas prácticas agronómicas para su manejo. Guía No.178, Mocache, EC: INIAP, Estación Experimental Tropical Pichilingue. 20 p.

Suárez, C. C. (1993). Diseases and their control. In Suarez, C. (ed.), *Manual of Cacao Cultivation* (pp. 90–106). 2nd ed. Quevedo, Ecuador. Tropical Experimental Station Pichilingue INIAP. Technical manual No. 25.

Surujdeo-Maharaj, S., Sreenivasan, T. N., Motilal, L. A., & Umaharan, P. (2016). Black pod and other *Phytophthora* induced diseases of cacao: History, biology, and control. In Bailey, B., & Meinhardt, L. (eds), *Cacao Diseases* (pp. 213–266). Springer. https://doi.org/10.1007/978-3-319-24789-2_7

Syazwan, S. A., Mohd-Farid, A., Wan-Muhd-Azrul, W. A., Syahmi, H. M., Zaki, A. M., Ong, S. P., & Mohamed, R. (2021). Survey, identification, and pathogenicity of *Ceratocystis fimbriata* complex associated with wilt disease on *Acacia mangium* in Malaysia. *Forests*, *12*(12), 1782. https://doi.org/10.3390/f12121782

Ten Hoopen, G. M., & Krauss, U. (2016). Biological control of cacao diseases. In Bailey, B., & Meinhardt, L. (eds), *Cacao Diseases* (pp. 511–566). Springer. https://doi.org/10.1007/978-3-319-24789-2_17

Teviotdale, B. L. (1991). Infection of pruning and small bark wounds in almond by *Ceratocystis fimbriata*. *Plant Disease*, *75*(10), 1026. https://doi.org/10.1094/pd-75-1026

Tollenaar, D. (1958). *Phytophthora palmivora* of cacao and its control. *Netherlands Journal of Agricultural Science*, *6*(1), 24–38. https://doi.org/10.18174/njas.v6i1.17721

Vásquez, Z. S., de Carvalho Neto, D. P., Pereira, G. V. M., Vandenberghe, L. P. S., de Oliveira, P. Z., Tiburcio, P. B., Rogez, H. L. G., Góes Neto, A., & Soccol, C. R. (2019). Biotechnological approaches for cocoa waste management: A review. *Waste Management*, *90*, 72–83. https://doi.org/10.1016/j.wasman.2019.04.030

Vera Loor, M. D. C., Cabrera, A. B., Mora, M. L., Loor, A. E. A. V., Coello, D. I. V., Villafuerte, S. L. P., Hidalgo, K. S., Yépez, P. I. T., & Guerrero, V. (2018). Microorganismos endófitos asociados a *Theobroma cacao* como agentes de control biológico de *Moniliophthora roreri*. *Centro Agrícola*, *45*(3), 81–87. http://scielo.sld.cu/pdf/cag/v45n3/0253-5785-cag-45-03-81.pd

Widmer, T. L., & Laurent, N. (2006). Plant extracts containing caffeic acid and rosmarinic acid inhibit zoospore germination of *Phytophthora* spp. pathogenic to *Theobroma cacao*. *European Journal of Plant Pathology*, *115*(4), 377–388. https://doi.org/10.1007/s10658-006-9024-5

Wilken, P. M., Steenkamp, E. T., Wingfield, M. J., de Beer, Z. W., & Wingfield, B. D. (2014). DNA Loss at the *Ceratocystis fimbriata* mating locus results in self-sterility. *PLoS ONE*, *9*(3), e92180. https://doi.org/10.1371/journal.pone.0092180

9 Sustainable pest management of cacao in the Neotropics

Challenges and opportunities

Patricia Morán, Rossana Castro, Dorys T. Chirinos, Luz Cecilia García, Jessenia Castro, and Takumasa Kondo

9.1 Introduction

Cacao, *Theobroma cacao* L. (Malvaceae), is cultivated in tropical countries and is in demand because its beans are the raw material used for the production of chocolate and other products (Armengot et al., 2020). The moderate consumption of chocolate is attributed to its high content of polyphenols, which have been associated with cardiovascular health (Vanderschueren et al., 2019). In 2021, 5,58,0432.37 tons of cacao were harvested from 11,53,5884 ha of cacao crops in the world (Food and Agriculture Organization of the United Nations [FAO], 2023).

Tscharntke et al. (2023) pointed out that the genetic diversification of cacao is in the western Amazon, a region in which it was originally domesticated, which includes southern Ecuador and northern Peruvian Amazon before it was introduced to Central America. Hence, although cacao originated and was initially domesticated in the Amazon rainforest, its cultivation expanded pantropically from Central America (Tscharntke et al., 2023). Currently, the largest cacao production occurs outside its center of origin, with only about 17% of all cacao in the world being harvested in the Neotropics (FAO, 2023).

Cacao has traditionally been produced under shade within agroforestry systems; however, in recent decades, the general trend has been the intensification in management and the use of high-yield cultivars, which have led to losses of biodiversity (Lagneaux et al., 2021; Tscharntke et al., 2023). This intensive system has been justified based on the achievement of greater production; however, this does not necessarily imply greater income rates. In fact, Jezeer et al. (2017) carried out a meta-analysis with 23 studies relating yield with profitability in intensive systems and concluded that an increase in yield does not lead to an increase in profitability due to the increase in production costs derived from the use of agrochemicals. The quality of the native Neotropical cacao varieties of fine aroma that have existed since pre-Colombian times offer a genetic diversity that could be used as ecosystem services for the biological control of pests (Tscharntke et al., 2023). Babin (2018) indicated that in Latin America where the crop is native, pests have a lower impact on yield compared to Asia and Africa where they cause extensive damage. Thus, the combination of being the

DOI: 10.4324/9781003381761-12

center of origin, production in agroforestry systems, and the variety of types of cacao allow the interactions of plants, birds, insects, fungi, and bacteria, among other organisms, to contribute to pest control and consequently provide important environmental services that will improve the sustainability of production (Avelino et al., 2011).

On the other hand, the excessive use of organosynthetic pesticides after World War II negatively affected biodiversity by interrupting food chains, generating imbalances or resistance phenomena that often result in population outbreaks of primary and secondary pests (Avelino et al., 2011). Additionally, pesticides contaminate soil, water, and air, causing negative effects on human health and the environment.

In this chapter, we analyze the pest management information in the main countries that grow cacao in the Neotropics, highlighting the use of ecosystem services offered by natural enemies, plant resistance, cultural practices, as well as the importance of biological control and the phenological selectivity of pesticides in chemical control. Designing strategies for the sustainable management of cacao crop in the Neotropics constitutes an opportunity, taking advantage of the benefits of its location in the center of origin of the crop, for example, rich genetic variability that could constitute important sources of resistance for genetic improvement, and the presence of natural enemies that have coevolved with native phytophagous insect pests, etc. The above can be used as an advantage to counteract the challenges derived from intensive management, as well as the effects of climate change. This chapter discusses the sustainability of cacao pest management in the Neotropics.

9.2 The agroecosystem approach: ecological bases of cacao pest management

Although agriculture is an economic process, it is carried out with living things (animals, plants, among others), developed within a generally complex environment. Agriculture is a *simplification* of what occurs in nature, forced by the *investment* of energy resources to achieve the proposed socioeconomic *objectives*. However, the ecological nature of the agricultural production process makes it dependent on the interactions of autotroph organisms (plants) with their biotic and abiotic components. The abiotic components are those that determine the interrelations of the biotic components, which are determined by the genetic characteristics of each of the species involved.

The base of the food chain is composed of 'green' plants which are organisms that synthesize organic matter from CO_2, air, water, and solar energy, which becomes metabolically more complex with the participation of mineral nutrients found in the soil. Consumers (heterotrophic organisms) obtain these vital mineral nutrients from the soil that they cannot produce by themselves. Consumption chains are formed from the different trophic levels (phytophagous, carnivorous, and saprophagous), which form structurally and functionally complex networks, especially in diversified systems. In this complexity rests the dynamic stability of biotic communities.

For this reason, it is important to review the information related to the cultivation of cacao and the importance of ecosystem services based on agroecosystem diversification, biological, microbial, and ethological control, as well as plant resistance.

9.3 Alternatives based on ecosystem services

9.3.1 Agroecosystem diversification

Ratnadass et al. (2012) indicated that the high vulnerability of crops to pest damage in modern intensive agroecosystems is a consequence of oversimplification. In contrast, these researchers pointed out that diversification may contribute to different regulatory processes of these biological competitors.

Diversification in cacao is based on agroforestry systems, which are considered multifunctional agroecosystems that provide, along with crop production, regulation and support services, including local climate regulation, the maintenance of soil fertility, carbon storage, and maintenance of the genetic diversity (Lagneaux et al., 2021; Rizali et al., 2022). Furthermore, Mortimer et al. (2018) reported that the capacity of agroforestry systems to regulate pests is a function of the local pressure of pests, but also of the composition and spatial distribution of the vegetation, factors that influence species richness (e.g., predators) and contribute to the general resilience of these agroecosystems to pest outbreaks.

Agroforestry systems in cacao have shown differential results regarding the regulation of pest populations. In La Paz, Alto Beni, Bolivia, a three-year study looked into the incidence of mirid bugs and stem borers under monoculture and agroforestry production systems managed in three ways: conventional, organic, and successional, among other parameters (Armengot et al., 2020). These authors determined that in all production systems, the incidence of mirids was low, while the damage caused by stem borers was significantly lower in conventional and organic agroforestry systems compared to monoculture. Similar results were obtained by Argueta et al. (2022) in a study carried out in the department of Atlántida, Honduras. The researchers observed for 22 years that pests and diseases did not increase their incidence in cacao agroforestry systems compared to timber species planted in monoculture.

Negative effects on pests have also been reported in cacao plantations associated with other crops. Ratnadass et al. (2012) reported that cacao was attacked more frequently by Lepidoptera larvae when it was cultivated under the shade of Leucaena, *Leucaena leucocephala* (Lam.) de Wit (Fabaceae), because the caterpillars used this leguminous tree as an alternative food source. In a study conducted in Bahia, Brazil, it was found that the association of cacao with rubber agroforests resulted in low species richness and high abundance of individuals; hence, it appears that the risk of pest increases in rubber agroforests associated with cacao (Novais et al., 2016). Although diversification could have some negative effects on cacao pests, studies should be oriented toward testing the best associations of other plant species with cacao to determine those that are suitable in different geographical areas where the crop is cultivated. The benefits of diversification include pest management and pollination, which affect the productivity of agroecosystems, and are also essential for sustainable crop development (Maas et al., 2020).

9.3.2 Biological control

From an ecological point of view, biological control has been defined as the action of parasitoids, predators, or pathogens that maintain the population densities of other organisms at lower averages than would occur in their absence (DeBach, 1964). Later it was conceptualized as the human manipulation of natural enemies to control pests (Bosch et al., 1982). The concept of biological control can be summarized as the use of living organisms to suppress a pest, reducing its population or its impact, and making them less abundant or less harmful (Eilenberg et al., 2001).

The conservation and augmentation of natural enemies, as well as classical biological control (importation of natural enemies), are the most common types of biological control strategies (Bosch et al., 1982). In conservation biological control, organisms are not released, but the conditions of natural enemies found in the crop are improved to reduce densities and damage by insect pests (Cotes, 2018). Among the main strategies for conservation biological control, habitat modification for the survival, longevity, and reproduction of natural enemies stands out, as well as minimizing the exposure of natural enemies to pesticides (Löhr et al., 2018).

Augmentative biological control involves supplementary releases of natural enemies, and classical biological control involves the intentional introduction of an exotic biological control agent, which has generally coevolved with the pest, for permanent establishment as a long-term control strategy (Eilenberg et al., 2001). Natural enemies include parasitoids (which feed on a single host) and predators (which consume many prey during their lifetime) (Bosch et al., 1982). Microbial agents will be referred to in the next section.

In the Neotropics, natural enemies associated with different cacao pests have been reported. In Colombia, for the control of *Carmenta foraseminis* (Eichlin) (Lepidoptera: Sesiidae) eggs, releases of *Trichogramma pretiosum* Riley (Hymenoptera: Trichogrammatidae) are recommended (Plantwise, 2020). Ant species (Hymenoptera: Formicidae) can become pests in cacao; however, some ant species have high potential as biological control agents. For example, *Azteca chartifex spiriti* Forel (Hymenoptera: Formicidae: Dolichoderinae: Dolichoderini) can prey on red-banded thrips, *Selenothrips rubrocinctus* (Giard) (Thysanoptera: Thripidae) and mirid bugs (Hemiptera: Miridae), while *Ectatomma tuberculatum* Olivier (Hymenoptera: Formicidae: Ectatomminae: Ectatommini) prey on leaf beetles (Coleoptera: Chrysomelidae), leafcutter ants (Hymenoptera: Formicidae: Myrmicinae: Attini), caterpillars (Lepidoptera), and also on red-banded thrips (Delabie et al., 2007). Some polyphagous exotic pests have invaded cacao in the Neotropics. The passion vine mealybug, *Planococcus minor* (Maskell) (Hemiptera: Pseudococcidae) is a species native to Asia, which is characterized by being polyphagous with about 200 recorded host plants in 190 genera distributed in 80 families (Watson & Kondo, 2022), among which cacao stands out, with notorious damages in the Caribbean region (Lenteren & Bueno, 2020). The passion vine mealybug is highly manageable with biological control, with an abundant number of natural enemies in 12 genera in six families (Watson & Kondo, 2022). In Trinidad, a complex of predators and parasitoids keep the mealybug under effective control, including four species of parasitoid wasps, that is, *Anagyrus mangicola* Noyes, *An. mirzai* Agarwal & Alam, *Coccidoxenoides perminutus* Girault and *Leptomastix dactylopii* Howard (Encyrtidae) (Watson & Kondo, 2022). Currently, populations of the mealybug remain at low levels due to the joint action of predators and parasitoids (Lenteren & Bueno, 2020).

The pink hibiscus mealybug, *Maconellicoccus hirsutus* (Green) (Hemiptera: Pseudococcidae) is highly polyphagous, recorded from more than 330 species belonging to 222 genera in 78 families (Harman, 2022). This mealybug species feeds on young shoots, flowers, and fruits on a wide range of host plants, including Malvaceae, and affects cacao, among other crops (Lenteren & Bueno, 2020). Biological control is effective for the management of *M. hirsutus*; numerous natural enemies have been recorded in 25 insect genera belonging to nine families (Harman, 2022). In the Caribbean islands, the predatory ladybeetle *Cryptolaemus montrouzieri* Mulsant (Coleoptera: Coccinellidae) and the hymenopteran endoparasitoids *Anagyrus kamali* Moursi and *Gyranusoidea indica* (Shafee, Alam & Agarwal) (Hymenoptera: Encyrtidae) are effective natural enemies (Harman, 2022). The pest was first reported in Grenada in 1994, and by 2003, it had spread to 28 Caribbean territories (Lenteren & Bueno, 2020). Classical biological control programs were implemented between 1996 and 1998 by introducing different natural enemies to the Caribbean islands, predators such as *C. montrouzieri, Scymnus coccivora* Ayyar, *Nephus regularis* Sicard (Coleoptera: Coccinellidae), and the parasitoids, *A. kamali* and *G. indica* (Hymenoptera: Encyrtidae). After these introductions, the populations of *M. hirsutus* have remained at low densities in the Caribbean region (Lenteren & Bueno, 2020).

A study carried out in Venezuela on a cacao crop planted in an agroforestry system found that the arthropod fauna was made up of 23 families of parasitoids, eight families of predatory beetles, and four families of phytophagous insects (Mazón et al., 2018). Likewise, a literature review on the diversity of native and exotic parasitoids attacking agricultural pests in Ecuador reported the existence of 13 parasitoid taxa attacking six phytophagous insect species associated with cacao in that country (Chirinos et al., 2021).

Spiders also play an important role as generalist biological control agents. An experiment conducted by Pérez-De La Cruz et al. (2007) in Tabasco,

Mexico, in a cacao agroforestry system with several plant species used as shade trees, identified 1,749 individuals belonging to ten insect Orders trapped in the webs of 54 spider species. The Orders Coleoptera, Diptera, and Hemiptera constituted 74% of the identified families. The main conclusion of this study indicates that spiders are promising biological control agents in cacao agroecosystems.

Some of the natural enemies associated with cacao pests reported in several Neotropical countries in the last 25 years are shown in Figure 1. The food chain built with the reported arthropods shows that there are 56 taxa of natural enemies (parasitoids and predators) associated with 11 phytophagous insects (Figure 9.1). This would indicate a greater diversity of natural enemies concerning phytophagous insects. Birds also play an important role in the biological control of arthropods associated with cacao.

9.3.3 Microbial control

The use of pathogens for pest management, which is defined as microbial control and includes bacteria, fungi, protozoa, and viruses, as well as nematodes,

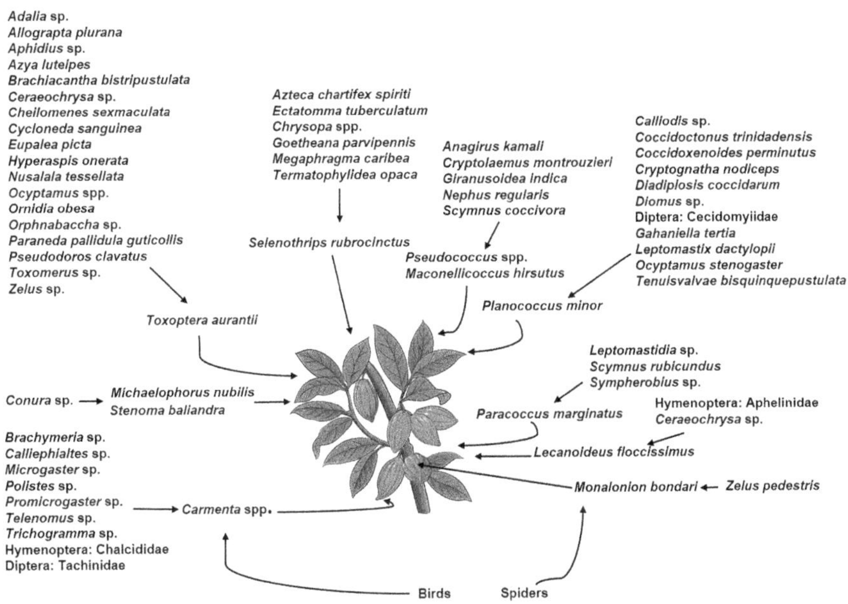

Figure 9.1 Natural enemies associated with pests that attack cacao in the Neotropics. The arrows show the interconnections in the agroecosystem. The arrows that point towards the plant constitute the pests and those that point to the pests represent the natural enemies. The cacao plant was obtained from Canva.

and although the latter are not microorganisms, they are considered as part of this type of control (Bosch et al., 1982).

In cacao, microbial control has been mainly implemented for the management of diseases caused by *Moniliophthora perniciosa* (Stahel) Aime & Phillips-Mora and *Moniliophthora roreri* Cif & Par (Ferraz et al., 2019; Villamil et al., 2015). However, in the Neotropics, 12 entomopathogenic taxa have been recorded associated with ten pest taxa in eight countries, as well as in the Caribbean islands (Table 9.1). Some of these pathogens have been collected within cacao agroecosystems, while others constitute strains obtained from other crops or commercial formulations that were tested to measure their effectiveness (Table 9.1).

Table 9.1 Entomopathogens associated with some cacao pests in Neotropical countries

Pest	*Entomopathogen*	*Control*	*Country*	*Reference*
Aleyrodidae	*Aegerita* sp. aff. *webberi*	Natural	Venezuela	Rojas & Goitía (2004)
	Aschersonia basicystis	Natural	Venezuela	Rojas & Goitía (2004)
	Lecanicillium lecanii	Natural	Venezuela	Rojas & Goitía (2004)
Carmenta spp.	*Paecilomyces* sp.	Natural	Colombia	Muñoz et al. (2017)
	Lecanicillium sp.	Natural	Colombia	Figueroa et al. (2013)
	Beauveria bassiana	Applied	Peru	Cabezas et al. (2017)
	Baccillus thuringiensis	Applied	Peru	Cabezas et al. (2017)
	Steinernema spp.	Applied	Colombia	Cubillos (2014)
Cicadellidae	*Podonectria* sp.	Natural	Venezuela	Rojas & Goitía (2004)
Coccidae	*Torrubiella* sp.	Natural	Venezuela	Rojas & Goitía (2004)
	L. lecanii	Natural	Venezuela	Rojas & Goitía (2004)
Coleoptera: Scarabaeidae	*Beauveria* sp.	Applied	Nicaragua, El Salvador	Argüello-Chávez (2017), Estrada et al. (2011)
	Metarhizium sp.	Applied	Nicaragua, El Salvador	Argüello-Chávez (2017), Estrada et al. (2011)
Hemiptera	*L. lecanii*	Natural	Venezuela	Rojas & Goitía (2004)
Horiola picta	*B. bassiana*	Natural	Brazil	Da Silva et al. (2020)
Monalonion spp.	*B. bassiana*	Applied	USA	Posada et al. (2010)
Selenothrips rubrocinctus	*B. bassiana*	Applied	Caribbean Islands	Lenteren & Bueno (2020)
Toxoptera aurantii	*L. lecanii*	Natural	Mexico	Cortez-Madrigal (2006)

9.3.4 Ethological control

According to Cañedo et al. (2011), ethology is the study of insect behavior in relation to their environment and, consequently, ethological control is the use of insect behavior to control pests. The same researchers reported that the ethological control of pests includes the use of traps based on baits, chromatic attractants (light traps), and pheromones, and that chemical communication also involves semiochemicals (Cañedo et al., 2011). Semiochemicals are substances that mediate relationships between organisms and have been divided into two groups: allelochemicals that include allomones, synomones, kairomones, and pheromones, among which stand out alarm, trail, and sex pheromones (Norin, 2007). Allelochemicals are involved in interspecific relationships. Thus, allomones benefit the organism that emits them; kairomones benefit the receiver, and synomones benefit both the sender and the receiver (Borrero-Echeverry et al., 2018). Pheromones are chemical compounds that mediate intraspecific communication. The following information is summarized by Borrero-Echeverry et al. (2018). Alarm pheromones are secreted by an individual to communicate to other individuals of its species about the presence of natural enemies or if the colony is threatened. Trail pheromones are secreted by social insects to direct individuals toward the food source or the nest. In pest management, allomones are used as repellents, kairomones as attractants, and both sex and aggregation pheromones have been the most studied in pest control.

Semiochemicals have been scarcely used for pest control in cacao, especially in the Neotropics. In Asia, pheromones have been synthesized for the control of an important endemic pest, *Conopomorpha cramerella* (Snellen) (Lepidoptera: Gracillariidae) (Niogret et al., 2022). In America, behavioral studies have been carried out that suggest the importance of some compounds for the management of some cacao pests. The cacao beetle, *Steirastoma breve* (Sulzer) (Coleoptera: Cerambycidae) is considered an important pest in some Neotropical countries (Liendo-Barandiaran et al., 2010). Field trials conducted in Venezuela demonstrated the importance of the kairomones emitted by the injured cacao tree to capture *S. breve* adults and of the male beetle sex pheromone to attract the female (Liendo-Barandiaran et al., 2010; Liendo et al., 2005).

Likewise, Morillo et al. (2009), after studying the flight activity and sexual behavior of the cacao pod borer, *Carmenta theobromae* (Busck) (Lepidoptera: Sesiidae) in Venezuela indicated that both the flight and sexual behavior of the female were observed from 2:30 pm to 7:30 p.m., which coincided with the greatest fluttering of the male moths. Based on these observations, the researchers suggested that, during this time, the female releases the sex pheromone. In Colombia, Montes et al. (2017) conducted a trial to test the effectiveness of white light and black light traps, as well as McPhail traps (with hydrolyzed corn protein) for monitoring *C. foraseminis*. The researchers found out that McPhail traps were able to capture the highest number of adults (four individuals per trap per sample), thus confirming that it is an effective method for monitoring this pest.

Morán-Rosillo and Castillo-Carrillo (2020) carried out studies on the morphological identification, behavior, and natural enemies of *C. foraseminis* in Peru. They suggested that it is important to study, among other aspects, the effectiveness of attractants, either pheromones or kairomones, to generate control strategies with lower environmental impact against the damage caused to cacao pods by this species.

9.3.5 Plant genetic control

Plants have developed defense mechanisms to protect themselves against herbivores' attacks (Johnson et al., 2015); among these, plant-derived secondary metabolites play an important role in the defense against pathogens and herbivores. Additionally, antixenosis and antibiosis were reported by Mahob et al. (2021) as important resistance mechanisms. The term antixenosis was proposed and defined by Kogan and Ortman (1978) as the resistance mechanism used by the plant to discourage colonization by an insect and may include morphological changes, such as subtle variations in the plant surface, color changes, flavor, waxy or hairy leaves, as well as exudations of gums or resins. In antibiosis, the insect feeds on the plant, but this has adverse effects on its biology, preventing it from completing its development (Mahob et al., 2021).

In the Neotropics, genetic improvement to obtain resistant varieties has been carried out mainly in Trinidad, Brazil, Ecuador (Malhotra & Apshara, 2017), and Colombia (Rodriguez-Medina et al., 2019). However, few studies have been conducted to obtain cacao materials that are resistant to insect pests. In Brazil, studies have been conducted to improve the resistance of the plant to attacks by *Toxoptera aurantii* Boyer de Fonscolombe (Hemiptera: Aphididae) by applying potassium silicate (Pinto et al., 2014). In the above study, high levels of total soluble phenols were found, which was considered by researchers as promising for the protection of cacao genotypes because the increase in the synthesis of these secondary metabolites was associated with a decrease in the damage caused by the aphids (Pinto et al., 2014).

The evaluation of cacao genotypes against damage by insect pests is also an alternative to differentiate susceptible clones from tolerant ones. In Colombia, it has been observed that the genotypes ICS-1, ICS-39, ICS-60, ICS-95, TSH-565, and SCC-61 of the 'Trinitario' cultivar group are highly susceptible to *C. foraseminis* unlike the IMC-clones 67, PA-46, PA-150, and CCN-51 which, apparently, have a higher degree of tolerance (Cubillos, 2013).

Studies carried out by Arevalo-Gardini et al. (2019) to understand cacao domestication evaluated 13 cacao germplasms molecularly, including the wild variety Morona Santiago from northern Peru. The results showed the relationship of fine cacaos from Peru and Ecuador with this wild variety whose existing populations are found in a wide region of northern Peru. These populations of Morona Santiago cacao could be used for genetic improvement to increase fruit quality and resistance to diseases (Arevalo-Gardini et al., 2019) that can also be extended to insect pests.

Ceccarelli et al. (2022) researched the conservation and use of cacao genetic materials in two South American countries (Ecuador and Peru) and four Central American countries (Nicaragua, Honduras, El Salvador, and Guatemala). The study found that Ecuador and Peru maintain many local cacao materials, whereas Central American countries maintain international clones (Ceccarelli et al., 2022). On the other hand, they observed that there are a good number of promising genotypes for high productivity, sensory quality, and resistance to insect pests and diseases. For these reasons, the researchers indicated that good conservation of genetic resources in Latin America benefits both those countries and the entire cacao sector.

9.4 Alternatives based on the manipulation of the agroecosystem

Some of the agricultural practices carried out in cacao crops include those that are directed toward pest management, among which are the elimination of alternate pest hosts, shade management, and the planting of plant species that serve as reservoirs for natural enemies, etc. (Babin, 2018). On the other hand, Adu-Acheampong et al. (2015) reported that the use of insecticides in cacao should be considered as a complementary strategy to cultural practices, selection of varieties, and other methods, within an Integrated Pest Management program. In this section, we detail some cultural and chemical control practices applied in some countries in the Neotropics for insect pest management of cacao crops.

9.4.1 Cultural practices

Arvelo et al. (2017) indicated the cultural practices used in the management of *C. foraseminis*, bugs of the genus *Monalonion* (Hemiptera: Miridae), and for the red-banded thrips, *S. rubrocinctus*, the three important pests that affect cacao pods in Mexico. For *C. foraseminis*, they suggested the removal of infested fruits and the protection of fruits from an early age with specialized plastic bags, as well as avoiding the movement of infested fruits. In the case of *Monalonion*, they pointed out the importance of frequent harvesting of fruits, shading regulation, and avoiding very short planting distances in cacao plantations. Similar practices are recommended for *S. rubrocinctus* and are suggested for the control of weeds that are hosts of the insect pest (Arvelo et al., 2017). In the Dominican Republic, biweekly harvest of fruits is suggested to control this thrips species (Batista, 2009).

Additionally, Cabezas et al. (2017) proposed pruning to improve lighting and ventilation to repel *C. foraseminis*, as well as covering the harvest remains with plastic to prevent the exit of adult moths or to add anti-sporulating substances such as lime, 20% salt water, and 15% urea solution. The scolitine, *Xylosandrus compactus* Eichhoff (Coleoptera: Curculionidae), has been reported in Peru as a pest of cacao in nursery plants, for which the removal of affected plants and good aeration and lighting of nurseries are recommended (Delgado & Couturier, 2017).

9.4.2 Chemical control

Botanical insecticides based on the sandbox tree, *Hura crepitans* L. (Euphorbiaceae), mountain garlic, *Gallesia integrifolia* (Spreng.) Harms (Petiveriaceae), and dehydrated tobacco, *Nicotiana tabacum* L. (Solanaceae) have been tested for the management of *Monalonion* in Bolivia, in which plots that were treated with *H. crepitans* had the lowest incidence and damage of the insect pest (Huaycho Callisaya et al., 2017).

A bioassay carried out in Costa Rica that evaluated the efficacy of some insecticides on *S. rubrocinctus* found that, of the insecticides tested, spinetoram was the most effective (Walter et al., 2018). The researchers concluded that, since the thrips species can be resistant to insecticides, to reduce resistance, it is recommended to rotate insecticides with different modes of action, as well as to spray them only when necessary, which implies the determination of economic thresholds for the damage caused by the insect (Walter et al., 2018).

In Colombia, for the control of *C. foraseminis*, spraying of pyrethroids directed at the fruits of approximately four months of age at biweekly intervals has been recommended (Plantwise, 2020).

9.4.3 Main arthropod pests that affect cacao production in the Neotropics

Approximately 71 phytophagous taxa of both primary and secondary importance were found to be associated with cacao in Neotropical countries (Figure 9.3).

Of these pests, due to the damage reported in the countries in Central and South America, two species of the genus *Carmenta* and three of *Monalonion* stand out as primary (Figure 9.2). *Carmenta theobromae* and *C. foraseminis* have been detected in Central and South American countries (Delgado, 2005). For Venezuela, it was reported that *C. theobromae* larvae drill into the cacao pods, but remain in the epicarp of the fruit and occasionally pierce the mesocarp to feed on the seeds (Delgado, 2005), whereas in Colombia they have been detected attacking the stem and inflorescences (Vasquez et al., 2015), and in Peru, it is known to perforate fruits and stems (Morán-Rosillo & Castillo-Carrillo, 2020).

C. foraseminis has been observed feeding on fruits and seeds in Brazil (Benassi et al., 2013), Colombia, Panama, Venezuela (Delgado, 2005), and Peru (Cabezas et al., 2017; Vasquez et al., 2015). Direct damage is attributed to this last species by perforating the fruit and feeding on the seed and by causing indirect damage due to the entry of rainwater, fungi, bacteria, and other insects through the exit holes of the adults (Cabezas et al., 2017). *C. foraseminis* is considered the species that causes the greatest economic losses (del Sánchez

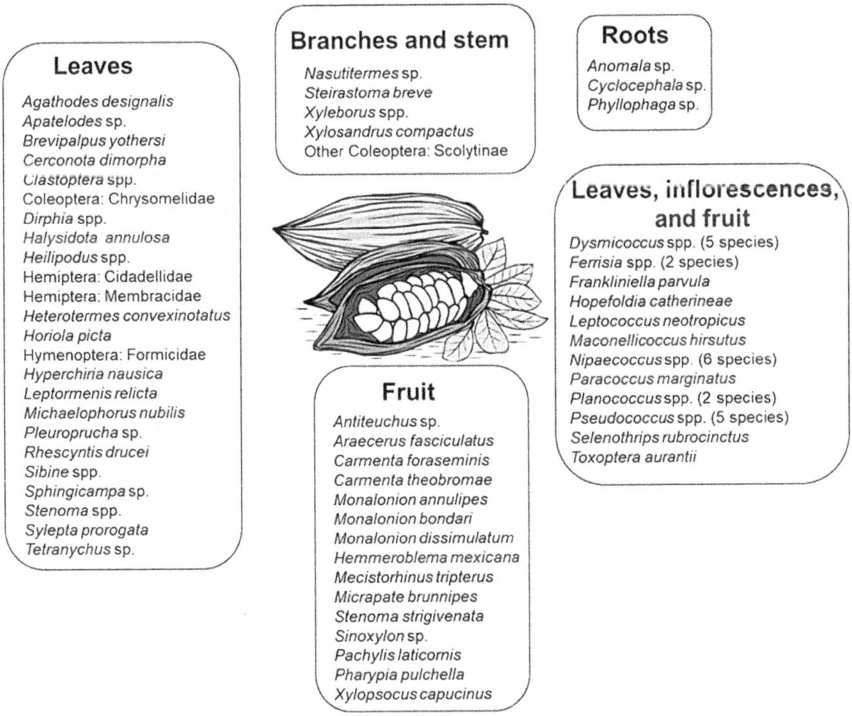

Figure 9.2 Insect pests reported on cacao in Neotropical countries. The cacao fruit was obtained from Canva.

et al., 2011). In Peru, 34.6% damage to fruits has been reported (Cabezas et al., 2017), and in Colombia, grain losses ranging from 23% to 30% have been reported (Cubillos, 2013, 2014; Muñoz et al., 2017). Additionally, in Colombia, Muñoz et al. (2017) reported up to 55.3% incidence of the insect.

Different *Monalonion* species have been reported in different Central and South American countries. In Brazil, although there are seven species associated with cacao, *Monalonion bondari* Costa Lima is the most common species (Babin, 2018; Santos et al., 2021); *Monalonion dissimulatum* Distant has been reported in Bolivia, Colombia, Ecuador, Peru, and Venezuela (Cañarte-Bermúdez & Navarrete-Cedeño, 2021; Quispe et al., 2021; Ramos & Montilla, 2018; Vargas et al., 2005); *Monalonion annulipes* Signoret has been mentioned in Colombia (Arvelo et al., 2017), Costa Rica (Villacorta, 1973), and El Salvador (Sermeño-Chicas et al., 2019) and has been related to the dieback of fruits, young shoots, and cacao tree (Villacorta, 1973). Batista (2009) pointed out an unidentified species of *Monalonion* affecting cacao fruits in the Dominican Republic.

Vargas et al. (2005) reported that the adults and nymphs of *Monalonion* sp. suck the sap from the endocarp of the pods, producing lesions that cause

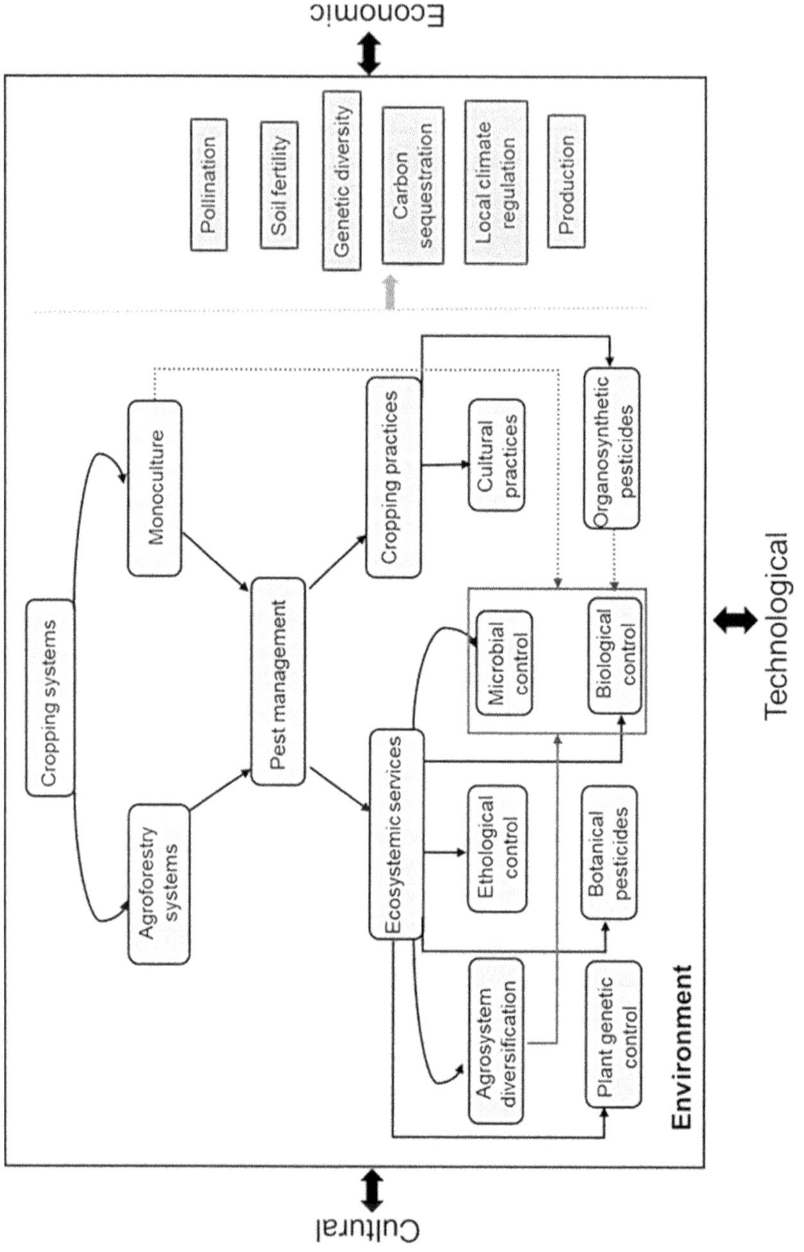

Figure 9.3 Alternatives for the sustainable management of insect pests in cacao. Own source.

malformations, reduction in size, and even abortion of young fruits; bug lesions expose pods to the attack by other insects and fungi. The same researchers mentioned that crop losses due to feeding damage vary from 15% to 80%.

A second group of pests would be made up of *S. rubrocinctus, T. aurantii*, scolitines (Coleoptera: Scolytinae), and mealybugs (Hemiptera: Pseudococcidae). Walter et al. (2018) reported that the feeding of immature and adult *S. rubrocinctus* can cause chlorotic spots and premature leaf drop, while on the pods it causes brown spots that merge in intense infestations, forming a corky brown layer of dead cells. The same researchers indicated that, although it is just cosmetic damage, this damage makes it difficult to estimate the maturity of the pods (Walter et al., 2018). Damage caused by *S. rubrocinthus* has been reported in Brazil (Delabie et al., 2007), Colombia (Arvelo et al., 2017), Ecuador (Cañarte-Bermúdez & Navarrete-Cedeño, 2021), El Salvador (Sermeño-Chicas et al., 2019), the Caribbean islands (Lenteren & Bueno, 2020), Mexico (Capetillo-Concepción et al., 2014), Peru (Castillo, 2013), and the Dominican Republic (Batista, 2009).

Toxoptera aurantii damages young leaves and inflorescences in cacao (Pinto et al., 2014). It has been reported in the main cacao-producing countries in the Neotropics, but there are approximately 18 taxa of natural enemies, both parasitoids (braconids) and predators (coccinellids, lacewings, hoverflies, and assassin bugs) (Figure 9.2), which could be important biological control agents in a sustainable insect pest management program.

Scolitines (Coleoptera: Curculionidae: Scolytinae), known as ambrosia beetles, show different adaptations to certain trees or particular parts of them. Some are monophagous and others are polyphagous with preferences to colonize dead, freshly cut, old, or dying trees and can also attack live trees under certain conditions (Pérez-De La Cruz et al., 2009a). Currently, the main problem with this group of insects is their association with *Ceratocystis cacaofunesta* Engelbrecht & Harrington, a fungus that causes wilting in cacao plants (Paladines-Rezabala et al., 2022). More than 60 species of scolitines have been reported in Neotropical countries. Among these, Colombia (Bernal Zuluaga, 2021), Ecuador (Cañarte-Bermúdez & Navarrete-Cedeño, 2021), Honduras, México (Pérez-De La Cruz et al., 2009b), El Salvador (Sermeño-Chicas et al., 2019), Peru (Delgado & Couturier, 2017; Ramos & Montilla, 2018), and Venezuela (Ramos & Montilla, 2018).

Mealybugs (Hemiptera: Pseudococcidae) through their sucking mouthparts feed on the phloem, which reduces the vigor of the host plants (Puig et al., 2021). Mealybugs attack different plant tissues, and severe leaf infestations cause defoliation; some species inject toxins causing twisted or stunted growth (Puig et al., 2021). On cacao, the main economic impact of mealybugs is the transmission of viruses of the genus Badnavirus. *Planococcus citri* is considered the dominant mealybug species on cacao in Brazil, Trinidad, Grenada, Haiti, and the Dominican Republic (Watson & Kondo, 2022). *P. citri* (Risso) is an important vector of cacao viruses in Trinidad (Puig et al., 2021). In the West Indies, *P. citri* has been associated with physical damage to cacao; in Grenada, the formation of abnormally large cushions that bear high numbers of flowers in some trees but produce few pods (Watson & Kondo, 2022). Twenty-five taxa of mealybugs have been reported in different countries of the Neotropics associated with cacao (García Morales et al., 2020; Kondo et al., 2008; Williams & Granara de Willink, 1992) (Figure 9.2). Other important pest problems are the cacao beetle, *S. breve*, damaging branches of young cacao trees (Liendo-Barandiaran et al., 2010), as well as several species of ants (Hymenoptera: Formicidae) and termites, among others (Figure 9.2).

9.5 Designing sustainable cacao pest management programs

Following the criteria established by Metcalf and Luckmann (1975) for the design of a pest management program in cacao, it is necessary to:

A. Evaluate the development of the crop in the area of interest including the characteristics of genotypes.
B. Identify primary and potential pests, as well as the incidence of environmental factors.
C. Maximize the biotic and abiotic factors of mortality.
D. Frame pest problems within ecological, economic, and social complexity.
E. Where possible diversify the agroecosystem.
F. Consider management alternatives.
G. Search for weak points in the life cycle of pest species and attack them selectively, avoiding broad impacts on the agroecosystem.
H. Structure the management program, providing flexibility to include adjustments if necessary.
I. Ensure adequate technical assistance (e.g., infestation monitoring and management decision-making).

The interaction of factors in the design of a sustainable pest management program is suggested in Figure 9.3 In Neotropical countries, diversification of systems has been practiced by small farmers (García-Briones et al., 2021), which favors the regulation of ecosystem services by natural enemies. The resistance or tolerance of clones, the use of chromatic traps, pheromones, allelochemicals, and botanical or microbial insecticides if available can also contribute to the reduction of pest problems. All are framed within the strict economic feasibility of cacao farmers.

Although monoculture simplifies trophic chains and consequently can have negative effects on natural enemies, they are also present in this type of agroecosystem. For this reason, the use of organosynthetic insecticides must be included with the greatest possible selectivity to favor the conservation of natural enemies and minimize as much as possible the disruption of trophic chains.

9.5.1 Economic aspects

Pests, diseases, and weeds are the main factors limiting crop yields (Allinne et al., 2016). These problems are generally controlled with organosynthetic pesticide sprays. However, control based on the use of pesticides does not necessarily reduce these losses. Rincon et al. (2018), based on data from the FAO, indicated that the use of pesticides in the world during the period 1990–2010 increased by an average of 98,941 tons of active ingredient per year, but the losses caused by pests, diseases, and weeds remained between 30% and 40%. This coupled with the side effects that pesticides can have on crops imposes the need to analyze the use of these products from the economic, ecological, and social perspectives.

In cacao, the use of pesticides should be carefully considered due to possible negative effects on specialized pollinating flies, *Forcipomyia* spp. (Diptera: Ceratopogonidae), as well as the presence of residues in grains for export, disruption of the ecosystem, and alteration of human health, among others (Cabezas et al., 2017).

On the other hand, intensive systems based on monocultures are justified by increasing productivity. However, this does not take into account that by simplifying the systems, the regulation ecosystem services are reduced, and consequently, to guarantee production, the use of agrochemicals is increased (Jezeer et al., 2017). Therefore, it is essential to estimate the relationship between the cost of applying chemical inputs and the value of production to calculate the profitability of cacao grown both in monoculture and agroforestry systems. It is noteworthy that intensive systems which are highly dependent on chemical inputs have contributed to the deterioration of soil and water contamination, which may affect human health (Jezeer et al., 2017).

Castillo et al. (2022) studied in Jamal, Cuba, the association of short-cycle crops with cacao. After the economic analyses, the researchers concluded that associating short-cycle crops with cacao allows harvesting of different crops annually, increasing agricultural production, generating additional income for the farmers, and improving food sovereignty. These factors positively add to the sustainability of the cacao agroforestry system.

9.5.2 Perspectives for sustainable pest management in cacao

Pest management in cacao cultivation in the Neotropics faces some challenges to achieve sustainability, among which are modern intensive systems, the use of high-yield genotypes, the damage caused to the pod by the main insect pests, and the possible change in the status of secondary or potential pests to primary pests. On the other hand, climate change that brings about an increase in atmospheric CO_2, as well as the alteration of patterns of rainy and dry seasons, could impact all the components of the cacao agroecosystem (Cilas & Bastide, 2020).

Therefore, the challenges consist of retaking ancestral forms of cultivation and increasing the implementation of agroforestry systems that promote carbon sequestration, soil fertility management, the use of local materials for genetic improvement, and sustainable pest management. The diversification of the agrosystems and the agroecological management of insect pests constitute the basis for more resilient agriculture, taking advantage of the opportunities that provide cacao cultivation in the center of origin of the crop in the Neotropics.

9.6 Conclusions

The main pests that affect cacao in the Neotropics are those that attack the pod, *Carmenta* spp., and *Monalonium* spp. But there are insect pests that attack other plant parts such as *Selenothrips rubrocinctus, Toxoptera aurantii*, and mealybugs (Hemiptera: Pseudococcidae) on leaves, inflorescences, and fruit and scolitines (Coleoptera: Scolytinae) on tree branches, which can also cause extensive damage if their population is not managed properly. Furthermore, some insect pests may act as vectors of disease, that is, some mealybug species may transfer viruses of the genus Badnavirus and scolitines may transmit *Ceratocystis cacao-funesta*, a fungus that is associated with the wilting of cacao trees. The ecosystem services offered by natural enemies, the diversification of production systems, and the genetic improvement of local cacao materials constitute alternatives that together with cultural practices and chemical pesticides must be used in the most compatible way possible. The challenges for cacao production in the Neotropics consist of retaking ancestral forms of cultivation and adapting them to the new scientific-technological knowledge to achieve sustainable cacao production in the context of climate change. Designing more resilient cacao production is possible in the Neotropics, taking advantage of the opportunities that provide the cultivation of cacao crop in its center of origin.

References

Adu-Acheampong, R., Sarfo, J., Appiah, E., Nkansah, A., Awudzi, G., Obeng, E., Tagbor, P., & Sem, R. (2015). Strategy for insect pest control in cacao. *American Journal of Experimental Agriculture*, *6*(6), 416–423. https://doi.org/10.9734/ajea/2015/12956

Allinne, C., Savary, S., & Avelino, J. (2016). Delicate balance between pest and disease injuries, yield performance, and other ecosystem services in the complex coffee-based systems of Costa Rica. *Agriculture, Ecosystems and Environment*, *222*, 1–12. https://doi.org/10.1016/j.agee.2016.02.001

Arevalo-Gardini, E., Meinhardt, L. W., Zuñiga, L. C., Arévalo-Gardni, J., Motilal, L., & Zhang, D. (2019). Genetic identity and origin of "Piura Porcelana" – A fine-flavored traditional variety of cacao (*Theobroma cacao*) from the Peruvian Amazon. *Tree Genetics & Genomes*, *15*(11).1-11. https://doi.org/10.1007/s11295-019-1316-y

Argüello-Chávez, H. (2017). Aspectos bioecológicos y de manejo de gallina ciega (*Phyllophaga* sp.), en la producción de cacao (*Theobroma cacao*) en Río San Juan, Nicaragua. *La Calera*, *17*(29), 63–67. https://doi.org/10.5377/calera.v17i29.6526

Argueta, O., Orozco-Aguilar, L., Dubón, A., Díaz, F. J., Sánchez, J., & Casanoves, F. (2022). Timber growth, cacao yields, and financial revenues in a long-term experiment of cacao agroforestry systems in northern. *Frontiers in Sustainable Food Systems, 6*, 941743.

Armengot, L., Ferrari, L., Milz, J., Velásquez, F., Hohmann, P., & Schneider, M. (2020). Cacao agroforestry systems do not increase pest and disease incidence compared with monocultures under good cultural management practices. *Crop Protection, 130*, 105047. https://doi.org/10.1016/j.cropro.2019.105047

Arvelo, M. Á., González, D., Delgado, T., Maroto, S., & Montoya, P. (2017). *Estado actual sobre la producción, el comercio y cultivo del cacao en América*. IICA.

Avelino, J., Hoopen, G. M., & DeClerck, F. (2011). Ecological mechanisms for pest and disease control in coffee and cacao agroecosystems of the Neotropics. In J. Le Coq & J. Beer (Eds.), *Ecosystem Services from Agriculture and Agroforestry: Measurement and Payment* (pp. 91–117). Earthscan Publications. https://doi.org/10.4324/9781849775656

Babin, R. (2018). Pest management in organic cacao. In V. Vacante & S. Kreiter (Eds.), *Handbook of Pest Management in Organic Farming* (pp. 502–518). CAB International 2018. https://doi.org/10.1079/9781780644998.0502

Batista, L. (2009). *Guía Técnica El Cultivo del Cacao*. CEDAF. https://repositorio.iica.int/bitstream/11324/6181/1/BVE17089191e.pdf

Benassi, V. L., Souza, C. A., Valente, F., & Lenzi, J. C. (2013). *Carmenta foraseminis* (Lepidoptera: Sesiidae), nova broca de frutos de cacau no Brasil. *Brazilian Journal of Agriculture, 88*(1), 70–75. https://doi.org/10.37856/bja.v88i1.42

Bernal Zuluaga, J. T. (2021). Evaluación de la diversidad de géneros de escolitinos (Coleoptera: Curculionidae: Scolytinae) asociados al agrosistema de cacao (*Theobroma cacao* L.) en Victoria-Caldas. Undergraduate thesis. Universidad de Caldas, Colombia. https://repositorio.ucaldas.edu.co/bitstream/handle/ucaldas/17190/JenyTatiana_BernalZuluaga_2021.pdf?sequence=1

Borrero-Echeverry, F., Barreto-Triana, N., Aragón-Rodríguez, S. M., Rivera-Trujillo, H. F., Oehlschlager, C., & Cotes, A. M. (2018). Las feromonas en el control de insectos. In A. M. Cotes (Ed.), *Control biológico de fitopatógenos, insectos y ácaros: agentes de control biológico. V. 1* (pp. 414–452). Agrosavia. http://hdl.handle.net/20.500.12324/34073

Bosch, R., Messenger, P. S., & Gutierrez, A. P. (1982). *An Introduction to Biological Control*. Plenum Press.

Cabezas, O., Gil, J. L., Gómez, R., Dávila, C., Morón, S., & Ramírez, C. (2017). Estado fitosanitario en la producción de cacao (*Theobroma cacao* L.) en la región de Huánuco (Perú): incremento del impacto de *Carmenta foraseminis* Eichlin. In Laliberté, B. (Ed.) *International Symposium on Cacao Research (ISCR)*.

Cañarte-Bermúdez, E., & Navarrete-Cedeño, B. (2021). *Reconocimiento de artrópodos - plaga y controladores biológicos como herramienta para el manejo ecológico de plagas en cacao* (1 edición). Instituto Nacional de Investigaciones Agropecuarias. http://repositorio.iniap.gob.ec/handle/41000/5750

Cañedo, V., Alfaro, A., & Kroschel, J. (2011). *Manejo Integrado de plagas de insectos en hortalizas Principios y referencias técnicas para la Sierra Central de Perú*. Centro Internacional de la Papa (CIP) https://cipotato.org/wp-content/uploads/2014/08/005739.pdf

Capetillo-Concepción, E., Córdova-Ávalos, V., Sánchez-Soto, S., Romero-Nápoles, J., Pérez-De La Cruz, M., & Mendoza-Hernández, J. H. R. (2014). Fluctuación poblacional de *Selenothrips rubrocinctus* (Giard) en cultivo de cacao en Huimanguillo, Tabasco, México. *Revista Peruana de Entomología, 49*(2), 137–142. https://www.revperuentomol.com.pe/index.php/rev-peru-entomol/article/view/244

Castillo, P. (2013). Insectos plagas y sus enemigos naturales en el cultivo de *Theobroma cacao* L. (cacao) en los valles de Tumbes y Zarumilla, Perú. *Revista Manglar*, *10*(2), 3–16. https://doi.org/10.17268/manglar.2013.002

Ceccarelli, V., Lastra, S., Loor, R. G., Chacón, W. W., Nolasco, M., Sotomayor, I. A., Plaza, L. F., López, D. A., Fernández, F. M., Dessauw, D., Orozco-Aguilar, L., & Thomas, E. (2022). Conservation and use of genetic resources of cacao (*Theobroma cacao* L.) by gene banks and nurseries in six Latin American countries. *Genetic Resources and Crop Evolution*, *69*, 1283–1302. https://doi.org/10.1007/s10722-021-01304-3

Chirinos, D. T., Anchundia, M., Castro, R., Castro, J., Geraud, J. E., & Kondo, T. (2021). Diversity of native and exotic parasitoids attacking agricultural pests in Ecuador: Are Ecuadorian biocontrol programs in decline? *Anartia*, *33*(2), 7–26. https://doi.org/10.5281/zenodo.5812417

Cilas, C., & Bastide, P. (2020). Challenges to cacao production in the face of climate change and the spread of pests and diseases. *Agronomy*, *10*(9), 1–8. https://doi.org/10.3390/agronomy10091232

Cortez-Madrigal, H. (2006). Efecto de coadyuvantes en *Lecanicillium lecanii* (Zimm.) Zare y Gams y su virulencia hacia *Toxoptera aurantii* Boyer. *Revista Mexicana de Fitopatología*, *24*(1), 59–64. https://www.redalyc.org/pdf/612/61224109.pdf

Cotes, A. M. (2018). El concepto de control biológico y sus premisas fundamentales. In: Cotes, A. M. (Ed.) *Control biológico de fitopatógenos, insectos y ácaros Volumen 1* (pp. 42–54). Agrosavia. http://hdl.handle.net/20.500.12324/34057

Cubillos, G. (2013). *Manual del Perforador de la Mazorca del Cacao Carmenta foraseminis (Busck) Eichlin*. Compañía Nacional de Chocolates S.A.S. http://hdl.handle.net/20.500.12324/13314

Cubillos, G. (2014). Introductory assays of the effects of entomo-pathogenic nematodes and B. thuringiensis on the cacao Pod Borer, *Carmenta foraseminis*. *News of the Lepidopterists' Society*, *56*(4), 186–189. https://www.lepsoc.org/content/news

Da Silva, A. F. C., Moreira, S. M. da C. de O., Cunha, R. do S. da S., Moreira, D. A., Teodoro, A., Do, N. A. H. C., & Evangelista, J. W. S. (2020). Effect of isolates of *Beauveria bassiana* pathogenic to leafhopper cacao *Horiola picta* (Hemiptera: Membracidae). *Amazonian Journal of Plant Research*, *4*(1), 456–461. https://doi.org/10.26545/ajpr.2020.b00054x

DeBach, P. (1964). *Biological Control of Insect Pests and Weeds*. 1sted. Chapman and Hall Ltd.

Delabie, J. H. C., Jahyny, B., Do Nascimento, I., Mariano, C. S. F., Lacau, S., Campiolo, S., Philpott, S. M., & Leponce, M. (2007). Contribution of cacao plantations to the conservation of native ants (Insecta: Hymenoptera: Formicidae) with a special emphasis on the Atlantic Forest fauna of southern Bahia, Brazil. *Biodiversity and Conservation*, *16*, 2359–2384. https://doi.org/10.1007/s10531-007-9190-6

Delgado, C., & Couturier, G. (2017). Primer registro de *Xylosandrus compactus* (Coleoptera: Curculionidae: Scolytinae) sobre cacao en Perú. *Revista Colombiana de Entomologia*, *43*(1), 121–124.

Delgado, N. (2005). Caracterización morfológica de los Sesiidae (Insecta: Lepidoptera) perforadores del fruto del cacao (*Theobroma cacao* L.), presentes en la región costera del Estado Aragua, Venezuela. *Entomotropica*, *20*(2), 97–111.

Eilenberg, J., Hajek, A., & Lomer, C. (2001). Suggestions for unifying the terminology in biological control. *BioControl*, *46*, 387–400. https://doi.org/10.1023/A:1014193329979

Estrada, J. M., Romero, G., & Moreno, J. A. (2011). *Guía técnica del cultivo de cacao manejado con técnicas agroecológicas* (p. 18). http://infocafes.com/portal/wp-content/uploads/2015/12/Estrada_et_al_Guia_Tecnica_Cacao.pdf

Food and Agriculture Organization of the United Nations (FAO) (2023). *Food and Agriculture Data*. Datos Sobre Alimentación y Agricultura. http://www.fao.org/faostat/en/#data/QC

Ferraz, P., Cássio, F., & Lucas, C. (2019). Potential of yeasts as biocontrol agents of the phytopathogen causing cacao witches' broom disease: Is microbial warfare a solution? *Frontiers in Microbiology*, *10*, 1766. https://doi.org/10.3389/fmicb.2019.01766

Figueroa, W., Ramirez, J. A., & Sigarroa, A. K. (2013). Efecto de las cepas nativas *Paecilomyces* sp. (Bainier) y *Lecanicillium* sp. (Zimm.) en el control de *Carmenta foraseminis* Eichlin (Lepidoptera: Sesiidae) en cultivos de cacao (*Theobroma cacao* L.). *Acta Agronomica*, *62*(3), 279–286. http://hdl.handle.net/10893/11655

García-Briones, A., Pico-Pico, B., & Jaimez, R. (2021). La cadena de producción del cacao en Ecuador: Resiliencia en los diferentes actores de la producción. *Novasinergia*, *4*(2), 152–172. https://doi.org/10.37135/ns.01.08.10

García Morales, M., Denno, B., Miller, D. R., Miller, G. L., Ben-Dov, Y., & Hardy, N. B. (2020). *ScaleNet: A Literature-Based Model of Scale Insect Biology and Systematics*. Database. http://scalenet.info

Harman, H. M. (2022). 4.2.6. Pink hibiscus mealybug, *Maconellicoccus hirsutus* (Green, 1908) (Hemiptera: Coccomorpha: Pseudococcidae). In T. Kondo, & G. W. Watson (Eds.), *Encyclopedia of Scale Insect Pests* (pp. 149–152). CABI.

Huaycho Callisaya, H., Maldonado Fuentes, C., & Manzaneda Delgado, F. (2017). Control del chinche del cacao (*Monalonion dissimulatum* Dist.) con aplicación de bioinsecticidas en la región de los Yungas de Bolivia. *Revista de Investigación e Innovación Agropecuaria y de Recursos Naturales*, *4*(1), 31–39. http://www.scielo.org.bo/scielo.php?script=sci_arttext&pid=S2409-16182017000100005&lang=es%0A http://www.scielo.org.bo/pdf/riiarn/v4n1/v4n1_a05.pdf

Jezeer, R. E., Verweij, P. A., Santos, M. J., & A Boot, (2017). Shaded coffee and cacao – Double dividend for biodiversity and small-scale farmers. *Ecological Economics*, *140*, 136–145. https://doi.org/10.1016/j.ecolecon.2017.04.019

Johnson, M. T. J., Campbell, S. A., & Barrett, S. C. H. (2015). Evolutionary interactions between plant reproduction and defense against herbivores. *Annual Review of Ecology, Evolution, and Systematics*, *46*, 191–213. https://doi.org/10.1146/annurev-ecolsys-112414-054215

Kogan, M., & Ortman, E. F. (1978). Antixenosis: A new term proposed to define Painter's "nonpreference" modality of resistance. *Bulletin of the Entomological Society of America*, *24*(2), 175–176. https://doi.org/10.1093/%20besa/24.2.175

Kondo, T., Ramos-Portilla, A. A., & Vergara-Navarro, E. V. (2008). Updated list of mealybugs and putoids from Colombia (Hemiptera: Pseudococcidae and Putoidae). *Boletín del Museo de Entomología de la Universidad del Valle*, *9*(1), 29–53.

Lagneaux, E., Andreotti, F., & Neher, C. M. (2021). Cacao, copoazu and macambo: Exploring *Theobroma* diversity in smallholder agroforestry systems of the Peruvian Amazon. *Agroforestry Systems*, *95*, 1359–1368. https://doi.org/10.1007/s10457-021-00610-0

Lenteren, J. C., & Bueno, V. H. P. (2020). Biological control in the remaining Caribbean Islands. In J. C. van Lenteren, V. H. P. Bueno, M. G. Luna, & Y. C. Colmenarez (Eds.), *Biological Control in Latin America and the Caribbean: Its Rich History and Bright Future* (pp. 403–423). CABI Publishing. https://www.cabidigitallibrary.org/doi/book/10.1079/9781789242430.0000

Liendo-Barandiaran, C. V., Herrera-Malaver, B., Morillo, F., Sánchez, P., & Hernández, J. V. (2010). Behavioral responses of *Steirastoma breve* (Sulzer) (Coleoptera: Cerambycidae) to host plant *Theobroma cacao* L., brushwood piles, under field conditions. *Applied Entomology and Zoology*, *45*(3), 489–496. https://doi.org/10.1303/aez.2010.489

Liendo, C., Morillo, F., Sánchez, P., Muñoz, W., Guerra, J., Cabrera, A., & Hernández, J. V. (2005). Olfactory behavior and electroantennographic responses of the cacao beetle, *Steirastoma breve* (Coleoptera: Cerambycidae). *Florida Entomologist*, *88*(2), 117–122. https://doi.org/10.1653/0015-4040(2005)088[0117:OBAERO]2.0.CO;2

Löhr, B., Díaz Niño, M. F., Manzano, M. R., Narváez Vásquez, C. A., Gómez-Jiménez, M. I., Carabalí, A., Vargas, G., Kondo, T., & Bustillo Pardey, A. E. (2018). Uso de parasitoides en el control biológico de insectos plaga en Colombia. In A. M. Cotes (Ed.), *Control biológico de fitopatógenos, insectos y ácaros. Volumen 1. Agentes de control biológico* (pp. 454–485). Agrosavia.

Maas, B., Thomas, E., Ocampo-Ariza, C., Vansynghel, J., & Steffan-Dewenter, I. (2020). Transforming tropical agroforestry towards high socio-ecological standards. *Trends in Ecology & Evolution*, *35*(12), 1049–1052. https://doi.org/10.1016/j.tree.2020.09.002

Mahob, R. J., Ngah, I. M., Feumba, R. D., Mahot, H. C., Bassogog, C. B., Bilong, C. F., Edoun Ebouel, F., Etam, P. B., Taliedje, D. M., Hanna, R., & Babin, R. (2021). Secondary metabolite effects of different cacao genotypes on feeding preference of the mirid *Sahlbergella singularis* Hagl. *Arthropod-Plant Interactions*, *15*, 821–831. https://doi.org/10.1007/s11829-021-09857-x

Malhotra, S. K., & Apshara, S. E. (2017). Genetic resources of cacao (*Theobroma cacao* L.) and their utilization-an appraisal. *Indian Journal of Genetics and Plant Breeding*, *77*(2), 199–213. https://doi.org/10.5958/0975-6906.2017.00027.X

Mazón, M., Sánchez-Angarita, D., Díaz, F. A., Gutiérrez, N., & Jaimez, R. (2018). Entomofauna associated with agroforestry systems of timber species and cacao in the southern region of the Maracaibo Lake Basin (Mérida, Venezuela). *Insects*, *9*(46). https://doi.org/10.3390/insects9020046

Metcalf, R. L., & Luckmann, W. H. (1975). *Introduction to insect pest management*. 1s ed. Wiley.

Montes, M., Carabali, A., & Mira, R. H. (2017). *Evaluación de métodos de monitoreo de Carmenta foraseminis Eichlin (Lepidoptera: Sesiidae) perforador del fruto de Cacao Theobroma cacao* (p. 1). http://hdl.handle.net/20.500.12324/22165

Morán-Rosillo, J. L., & Castillo-Carrillo, P. S. (2020). El barrenador del fruto y tallo del cacao (*Carmenta theobromae*, Lepidoptera: Sesiidae) en el valle de Zarumilla, Tumbes, Perú. *Revista Colombiana de Entomología*, *46*(1), e10165. https://doi.org/10.25100/socolen.v46i1.10165

Morillo, F., Sánchez, P., Herrera, B., Liendo-Barandiaran, C., Muñoz, W., & Hernández, J. V. (2009). Pupal development, longevity and behavior of *Carmenta theobromae* (Lepidoptera: Sesiidae). *Florida Entomologist*, *92*(2), 355–361. https://doi.org/10.1653/024.092.0222

Mortimer, R., Saj, S., & David, C. (2018). Supporting and regulating ecosystem services in cacao agroforestry systems. *Agroforestry Systems*, *92*, 1639–1657. https://doi.org/10.1007/s10457-017-0113-6

Muñoz, J. A., Vásquez, Y., & Muriel, S. B. (2017). Estimación de pérdidas generados por *Carmenta foraseminis* (Busck) Eichlin (Lepidoptera: Sesiidae) en el grano comercial de cacao (*Theobroma cacao* L.) y registro de controladores biológicos en la Granja "Rafael Rivera", San Jerónimo (Antioquia – Colombia). *Boletín Del Museo de Entomología de La Universidad del Valle*, *17*(2), 29–36. http://hdl.handle.net/10893/11655

Niogret, J., Ekayanti, A., & Zhang, A. (2022). Sex pheromone of cacao pod borer, *Conopomorpha cramerella*: Field activity evaluation of pheromone formulations in an Indonesia plantation. *Insects*, *13*(663), 1–9. https://doi.org/10.3390/insects13080663

Norin, T. (2007). Semiochemicals for insect pest management. *Pure and Applied Chemistry*, *79*(12), 2129–2136. https://doi.org/10.1351/pac200779122129

Novais, S. M. A., Macedo-Reis, L. E., Darocha, W. D., & Neves, F. S. (2016). Effects of habitat management on different feeding guilds of herbivorous insects in cacao agroforestry systems. *Revista de Biología Tropical*, *64*(2), 763–777. https://doi.org/10.15517/rbt.v64i2.19100

Paladines-Rezabala, A., Moreira-Morrillo, A. A., Mieles, A., & Garcés-Fiallos, F. R. (2022). Avances en la comprensión de la interacción entre *Ceratocystis cacaofunesta* y *Xyleborus ferrugineus* (Coleoptera: Curculionidae: Scolytinae) en árboles de cacao. *Scientia Agropecuaria, 13*(1), 43–52. https://doi.org/10.17268/sci.agropecu.2022.004

Pérez-De La Cruz, M., Equihua-Martínez, A., Romero-Nápoles, J., Sánchez-Soto, S., & García-López, E. (2009a). Diversidad, fluctuación poblacional y plantas huésped de escolitinos (Coleoptera: Curculionidae) asociados con el agroecosistema cacao en Tabasco, México. *Revista Mexicana de Biodiversidad, 80*, 779–791. https://doi.org/10.22201/ib.20078706e.2009.003.173

Pérez-De La Cruz, M., Equihua-Martínez, A., Romero-Nápoles, J., Valdez-Carrasco, J. M., & De La Cruz-Pérez, A. (2009b). Claves para la identificación de escolítinos (Coleoptera: Curculionidae: Scolytinae) asociados al agroecosistema cacao en el sur de México. *Boletín del Museo de Entomología de la Universidad del Valle, 10*(1), 14–29. http://hdl.handle.net/10893/730

Pérez-De La Cruz, M., Sánchez-Soto, S., Ortíz-García, C. F., Zapata-Mata, R., & De La Cruz-Pérez, A. (2007). Diversidad de insectos capturados por arañas tejedoras (Arachnida: Araneae) en el agroecosistema cacao en Tabasco, México. *Neotropical Entomology, 36*(1), 90–101. https://doi.org/10.1590/S1519-566X2007000100011

Pinto, D. G., Aguilar, M. A. G., Souza, C. A. S., Silva, D. M., Siqueira, P. R., Cao, J. R., & Zanetti, L. V. (2014). Photosynthesis, growth and incidence of insect pest in cacao genotypes sprayed with silicon. *Bioscience Journal, 30*(3), 715–724. https://doi.org/10.1590/S0034-737X2012000300010

Plantwise. (2020). *Guía de Manejo de Plagas: Lista Verde y Amarilla. Carmenta o perforador de los frutos del Cacao Carmenta foraseminis*. CAB International. https://www-products.plantwise.org/KnowledgeBankBeta/pmdg/20207800516

Posada, F. J., Chaves, F. C., Gianfagna, T. J., Pava-Ripoll, M., & Hebbar, P. (2010). Establishment of the fungal entomopathogen *Beauveria bassiana* as an endophyte in cacao pods (*Theobroma cacao* L.). *Revista U.D.C.A Actualidad & Divulgación Científica, 12*(3), 71–78. https://repository.udca.edu.co/handle/11158/2273

Puig, A., Wurzel, S., Suarez, S., Niogret, J., & Marelli, J. P. (2021). Molecular identification of mealybug species (Hemiptera: Pseudococcidae) affecting *Theobroma cacao* for improved pest management. *Proceedings, 68*, 10399. https://doi.org/10.3390/iece-10399

Quispe, H. I., Nova, M. L., & Sánchez, B. M. (2021). Aplicación de hongos entomopatógenos y producto tecsil para el control del chinche de cacao (*Monaloniun dissimulatum* Dist.) en Alto Beni, La Paz. *Acta Nova, 10*(1), 3–21. http://www.scielo.org.bo/scielo.php?pid=S1683-07892021000100002&script=sci_arttext

Ramos, C., & Montilla, R. (2018). Insectos asociados al cultivo del cacao depositados en el Museo de Insectos de Interés Agrícola (MIIA) del INIA - Venezuela. *Agronomía Tropical, 68*(3–4), 163–176.

Ratnadass, A., Fernandes, P., Avelino, J., & Habib, R. (2012). Plant species diversity for sustainable management of crop pests and diseases in agroecosystems: A review. *Agronomy for Sustainable Development, 32*, 273–303. https://doi.org/10.1007/s13593-011-0022-4

Rizali, A., Himawan, T., Yuniasari, N., Yuliastanti, N., Bachtiar, M. A., & Rafid, E. D. R. (2022). Contribution of agricultural landscape composition on shaping the interaction between pests and natural enemies in cacao agroforestry Akhmad. *AGRIVITA Journal of Agricultural Science, 44*(3), 479–489. http://doi.org/10.17503/agrivita.v44i3.3388

Rodriguez-Medina, C., Caicedo Arana, A., Sounigo, O., Argout, X., Alvarado, G. A., & Yockteng, R. (2019). Cacao breeding in Colombia, past, present and future. *Breeding Science, 69*, 373–382. doi:10.1270/jsbbs.19011

Rojas, T., & Goitía, W. (2004). Riqueza y abundancia de hongos entomógenos en tres sistemas de manejo de cacao en Choroní, Venezuela. *Manejo Integrado de Plagas y Agroecología (Costa Rica), 71*, 46–53. http://repositorio.bibliotecaorton.catie.ac.cr/handle/11554/6788

del Sánchez, M. C., Navarro, R., Marín, C., Caseres, R., & Fuentes, V. (2011). Duración de la fase adulta y emergencia de machos y hembras del perforador del fruto de cacao en Choroní y Maracay, Estado Aragua. *Agronomia Tropical*, *61*(3–4), 241–251. http://sian.inia.gob.ve/repositorio/revistas_ci/AgronomiaTropical/at613-4/pdf/at613-4_sanchez_m.pdf

Santos, E. B., Soares, A. M. L., Bahia, B. L., Pereira, R. R. D. C., & Fávaro, C. F. (2021). First record of *Zelus pedestris* (Hemiptera: Reduviidae) preying on the cacao pest *Monalonion bondari* (Hemiptera: Miridae). *Biocontrol Science and Technology*, *32*(4), 511–514. https://doi.org/10.1080/09583157.2021.1990857

Sermeño-Chicas, J. M., Pérez, D., Serrano-Cervantes, L., Parada-Jaco, M. E., Joyce, A. L., Maldonado-Santos, E. J., de los Alvanes-Leiva, Y. Á., Rodríguez-Sibrían, F. M., Girón-Segovia, C. D., García-Sánchez, D. A., Hernández-León, C. E., Rivas-Nieto, F., Rivera-Mejía, F. A., Parada-Berríos, F. Á., Rodríguez-Urrutia, E. A., Vásquez-Osegueda, E. A., & Lovo-Lara, L. M. (2019). Insectos como plagas potenciales del cacao (*Theobroma cacao* L.) en El Salvador. *Revista Minerva*, *2*(2), 117–134. https://revistas.ues.edu.sv/index.php/minerva/article/view/2423

Tscharntke, T., Ocampo-Ariza, C., Vansynghel, J., Ivañez-Ballesteros, B., Aycart, P., Rodriguez, L., Ramirez, M., Steffan-Dewenter, I., Maas, B., & Thomas, E. (2023). Socio-ecological benefits of fine-flavor cacao in its center of origin. *Conservation Letters*, *16*(1), 1–10. https://doi.org/10.1111/conl.12936

Vanderschueren, R., Montalvo, D., De Ketelaere, B., Delcour, J. A., & Smolders, E. (2019). The elemental composition of chocolates is related to cacao content and origin: A multi-element fingerprinting analysis of single origin chocolates. *Journal of Food Composition and Analysis*, *83*, 103277. https://doi.org/10.1016/j.jfca.2019.103277

Vargas, A., Somarriba, E., & Carballo, M. (2005). Dinámica poblacional del chinche (*Monalonion dissimulatum* Dist.) y daño de mazorcas en plantaciones orgánicas de cacao del Alto Beni, Bolivia. *Agroforestería en las Américas*, *43–44*, 72–76. http://www.sidalc.net/repdoc/A2335E/A2335E.PDF

Vasquez Castaneda, Y., Munoz Gutiérrez, J., Muriel Ruiz, S. B., & Hernández-Baz, F. (2015). Ocurrencia de los barrenadores *Carmenta foraminis* Eichlin y *Carmenta theobromae* (Busck) (Lepidoptera: Sesiidae) en *Theobroma cacao* L., en el Departamento de Antioquia-Colombia. *Boletín del Museo de Entomología de la Universidad del Valle*, *16*(1), 34–38. http://hdl.handle.net/10893/11780

Villacorta, A. (1973). Fluctuación anual de las poblaciones de *Monalonion annulipes* Sig. y su relación con la "muerte descendente de *Theobroma cacao*" en Costa Rica. *Revista Peruana de Entomología*, *16*(1), 21–24. https://www.revperuentomol.com.pe/index.php/rev-peru-entomol/article/view/537

Villamil, J. E., Viteri, S. E., & Villegas, W. L. (2015). Aplicación de Antagonistas microbianos para el control biológico de *Moniliophthora roreri* Cif & Par en *Theobroma cacao* L. bajo condiciones de campo. *Revista Facultad Nacional de Agronomía Medellín*, *68*(1), 7441–7450.

Walter, N. T., Adeleye, V. O., Muthomi, P. K., Ortiz, R. J., Strzyzewski, I., Funderburk, J., & Martini, X. (2018). Toxicity of different insecticides against two thrips (Thysanoptera: Thripidae) pests of concern in Central America. *Florida Entomologist*, *101*(4), 627–633. https://doi.org/10.1653/024.101.0426

Watson, G. W., & Kondo, T. (2022). 4.2.11. *Planococcus* spp. (Hemiptera: Coccomorpha: Pseudococcidae). In T. Kondo & G. W. Watson (Eds.), *Encyclopedia of Scale Insect Pests* (pp. 177–192). CABI.

Williams, D. J., & Granara de Willink, M. C. (1992). *Mealybugs of Central and South America*. CAB International.

Part IV

Industrialization and by-product valorization

10 Cacao post-harvest processing

From tradition to modernization and its role to quality diversification

Zoi Papalexandratou, Manuel Kirchmayr, Eugenia Lugo-Cervantes, Dulce Velásquez-Reyes, and Anne Gschaedler

10.1 Introduction

Fermented foods are of high importance as they represent a big part of the daily human food consumption (Tamang et al., 2016; De Roos & De Vuyst, 2018). Fermentation allows for conversion of food properties through a microbial process which may include one or more microbial groups. Yeasts and/or bacteria consume certain substrates present in the food matrix and convert them into several metabolites with distinct organoleptic properties (Tamang et al., 2016).

The seeds of the cacao tree, namely *Theobroma cacao* L., are the raw material for chocolate making, a widely appreciated food. The cacao beans are bitter and astringent when freshly collected and need to undergo a curing process to generate flavor precursors that are important for the chocolate flavor development later (D'Souza et al., 2017; Sari et al., 2023b). The curing process of the cacao beans includes mainly a fermentation step and a drying step which take place at the cacao-producing origin after the harvest of the fresh cacao fruits, namely cacao pods, while the method and duration vary widely among the different cacao-producing regions (Hernández-Hernández et al., 2016; Velásquez-Reyes et al., 2021; Verce et al., 2021; Cortez et al., 2023). The last decades extra emphasis has been put on the importance of cacao post-harvest handling. The market demand for high-quality cacao is continuously increasing and the cacao-producing origins are striving for volume increase but also improvement of their fermentation and drying practices.

While historians are still discovering new evidence about the exact origin of cacao, it is proven that cacao's use dates to pre-Colombian times (Motamayor et al., 2008; Nieves-Orduña et al., 2023); apart from its use as a currency, discoveries have brought in light findings with regard to equipment and methods used to process and brew cacao. As occurred also with other raw materials, for example, grapes and milk, the fact that cacao's organoleptic properties improve through curing was accidentally found in Mesoamericas back in the times of

DOI: 10.4324/9781003381761-14

the Olmecs, the Mayans, and the Mexica/Aztecs (Motamayor et al., 2008; Grivetti & Shapiro, 2009).

Nowadays chocolate is broadly considered as a simple confection, but it is more than that; it must be appreciated as an actual delight. In the Mayan and Mexica/Aztec times, cacao was considered as a real gift from gods and it was only consumed by a few (Arevalo-Gardini et al., 2019). Apart from the spiritual and sociocultural uses, cacao was used in the form of a paste from crushed roasted cacao beans or as frothed beverage for its medicinal and stimulant properties.

Food historians still debate about the real "first" introduction of cacao to Europe, but it took almost 4,000 years to experience cacao as chocolate (Grivetti & Shapiro, 2009; Presilla, 2009). It is reported that Hernán Cortés brought back to Spain cacao beans that got planted in Iberia and the plant was then studied and observed; this helped the Spaniards understand how to produce more pleasant flavorsome drinks. It was the "exciting exotic" dark drink that attracted the attention of Europeans at first (Grivetti, 2005; Grivetti & Shapiro, 2009; Coe & Coe, 2019). Although cacao remained expensive, sugar production became cheaper and its availability increased by the seventeenth century, so eventually more sweetened chocolate recipes could be found, and cacao could charm the palate of everyone who could afford it (Grivetti & Shapiro, 2009).

Research in recent decades has shown the importance of fermentation and drying to obtain the final quality of cacao beans (Rodríguez-Campos et al., 2012; Castro-Alayo et al., 2019; Barrientos et al., 2019; Utrilla-Vázquez et al., 2020; Moreno-Rojas et al., 2023). Cacao is a complex raw matrix; unlike wine, yogurt, beer, or vinegar where fermentation of the raw produce is mostly carried out by a single microbial group and in most cases under an industrialized controlled process, cacao undergoes a spontaneous mixed microbial fermentation. The fresh cacao pods contain the cacao beans embedded into a mucilaginous pulp. The content of nonbroken cacao pods is sterile and once broken open in the plantation the cacao pulp-bean mass (CPBM) gets immediately contaminated from the surrounding microbiota initiating this spontaneous fermentation process (Beckett, 2009; Afoakwa, 2010; De Vuyst & Weckx, 2016; Pereira et al., 2016; Ghisolfi et al., 2023). The viscous sugar-rich pulp is the fermentation substrate for a "zoo" of microorganisms [yeasts, lactic acid bacteria (LAB), acetic acid bacteria (AAB), and in many cases enterobacteria], and this natural process will indirectly initiate physicochemical changes inside the cacao beans, the actual product of interest. Such process when properly done will allow for killing of the bean embryo, reduction of bitterness, and astringency as well as flavor and color enhancement (De Vuyst & Leroy, 2020).

The variety of cacao, the level and type of microbial contamination, and the general practices from harvest to the wet mass collection will affect the speed of this essential spontaneous process. Cacao is primarily fermented at the farm level by smallholder farmers. In most cases, the volume accumulated from one small farm (1–2 ha) is inadequate for proper fermentation conditions (below 80–100 kg wet mass), and it can cause underprocessing and inconsistency in the quality of the final product. Also, small farmers tend to pay low/no attention to the proper monitoring of this important process, either because the market does not pay differentials for improved quality or because of lack of knowledge. The concept of centralized fermentation facilities dates to post-colonial times of Brazil, where farmers got organized and brought cacao to one central location for processing in bigger volumes under continuous monitoring (Wood & Lass, 2001; Beckett, 2009). The methods used for CBF vary and they usually reflect traditional practices passed from generation to generation. With time, science has given insides of how to better steer and control the CBF process.

On another note, CBF helps with the elimination of the sticky mucilage which permits easier drying of the seeds. Drying is the next step of cacao curing, and it is of equal importance. Biochemical reactions inside the beans continue during this process and will contribute further to flavor, color, and aroma development (Nielsen et al., 2014; Pereira et al., 2016; De Vuyst & Leroy, 2020). Drying methods and techniques vary as well, and climate plays a key role in the evolution of this process (Rodríguez-Campos et al., 2012). The cases of underestimation of the importance of this step are many.

In this chapter, an in-depth description of the microbial process of CBF is given along with a broad introduction to fermentation and drying methods and practices applied for cacao in Latin America and beyond, as experienced by the authors while explaining the pros and cons of each practice. In the past few years the intense labor work that cacao requires has forced processors to find and implement modernized techniques that can help them process cacao in a more efficient and less time-consuming way. In that sense, a comparative discussion of the impact of these practices on cacao quality is made. As cacao quality is of high importance, a linkage between science and the approaches for market access is discussed.

10.2 Post-harvest processing of cacao beans

10.2.1 Cacao bean fermentation

The cacao bean fermentation (CBF) process is initiated immediately after manual or mechanical opening of the cacao pods at the farm. The freshly collected mass' inoculation comes from the surrounding environment and mainly from the surface of pods and (post-)harvest equipment (Verce et al., 2021; Ghisolfi et al., 2023). Also, biological vectors, such as fruit flies, beetles, bees, and wasps, may help inoculate the freshly collected mass (De Vuyst & Weckx, 2016). The state of the pods (healthy, damaged, or diseased), the handling practices, the hygiene of the workers, or the spatial separation of the place where the fermentation takes place determines the number of microorganisms initially present in the wet mass. The fermentation method, the volume of the fresh mass, and the duration of the fermentation differ between cacao-producing regions and even within a region or among neighboring farms. CBF can be carried out in heaps, boxes, baskets, sacks, trays, platforms, or in traditional Mexican canoe-like toyas (Camu et al., 2007; Nielsen et al., 2014; Rodríguez-Campos et al., 2011, 2012; Arana-Sánchez et al., 2015; Utrilla-Vázquez et al., 2020; Velásquez-Reyes et al., 2021) (Figure 10.1), and it may take place on farm level or at centralized locations (collection points or fermentation centers) when wet mass from several farms is consolidated for bigger volume handling under better monitored conditions.

10.2.2 Microbial and metabolic dynamics during CBF

The fresh and ripe CPBM contains mainly water and high amounts of glucose and fructose, creating the perfect environment for carbohydrate-consuming microorganisms. The low initial pulp pH (3.0–3.8), which relates to the concentration of citric acid, is an inhibitory factor for the growth of non-acid-tolerant microbes. The initial microbial load may vary from 10^4 to 10^8 cells/g of CPBM depending on practices between cacao pod breaking and collection of the fresh mass (Lima et al., 2011). Figure 10.2 describes the microbial dynamics reported in different fermentations in Latin America for the groups of yeasts, LAB, AAB, and spore-forming bacteria, enterobacteria and filamentous fungi (Arana-Sánchez et al., 2015; Papalexandratou et al., 2019; Serra et al., 2019; Velásquez-Reyes et al., 2021; Mota-Gutierrez et al., 2021). In addition, the general metabolism of the first three groups and examples of microorganisms that have been reported to produce certain compounds in spontaneous CBFs are integrated. In Latin America, low initial total microbial counts have been reported (Arana-Sánchez et al., 2015; Papalexandratou et al., 2019; Velásquez-Reyes et al., 2021). This may relate to the better hygienic conditions and careful separation of healthy from infected pods in comparison to other origins (Nielsen et al., 2014; Papalexandratou et al., 2019), thanks to extensive training and coaching of farmers. The growth rate of the different microbial groups depends strongly on the implemented fermentation practices (fermentation method, mixing technique, etc.).

The evolution of the different microbial groups during CBF is considered somehow well-defined and is independent of the cacao-producing origin (Pereira et al., 2016; Papalexandratou et al., 2019). Yeasts dominate the total microbial counts during the first stage of the process, while LAB species surpass the yeasts usually after 24 h of fermentation. The yeasts start to decline after approximately 72 h, due to the steep growth of AAB. In the last stage, spore-forming aerophilic bacteria (SFB) may appear especially when the temperature of the fermentation is maintained above 48°C for a long period (Lima et al., 2011). Besides, at the onset of the CBF, the presence of diverse non-pathogenic environment- and plant-associated bacteria belonging mainly to the families *Yersiniaceae*, *Erwiniaceae*, and *Enterobacteriaceae* has been reported (Janda & Abbott, 2021). These species play an important role in vegetative processes and they are widely spread by insect vectors. *Enterobacteriaceae*

Figure 10.1 Cacao bean fermentation methods: (a) heap, (b) horizontal-type box, (c) tray, (d) platform, (e) basket, (f) rotary drum, (g) bag, and (h) toya.

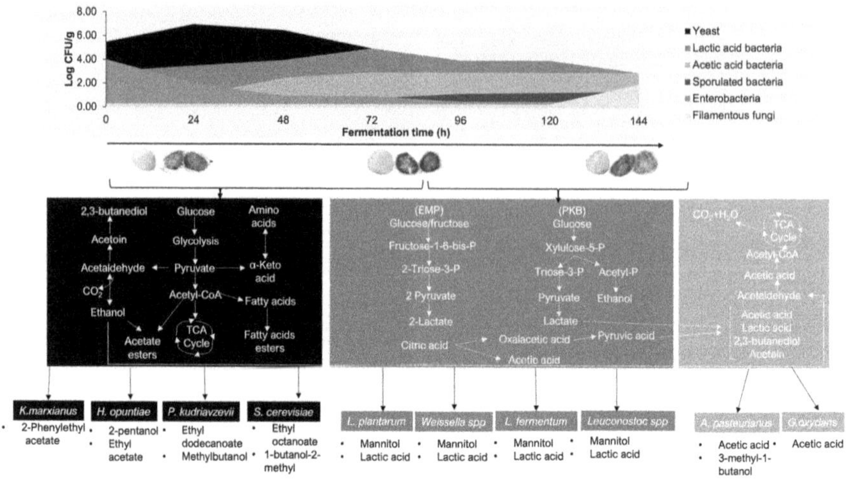

Figure 10.2 Microbial dynamics of CBFs and the general metabolism of the three main groups of microorganisms [yeasts, lactic acid bacteria (LAB), and acetic acid bacteria (AAB)] as reported for Latin America.

at the start as well as occasional mold growth at the end of fermentation may also impact the fermentation process and therefore the sensory quality of the fermented cacao beans (Ozturk & Young, 2017).

Diverse authors have described the evolution of microbial groups during CBF; however, several factors may influence the sequence of their appearance, the species diversity of each microbial group, their respective maximum populations, as well as the duration of their involvement in the process. In Figure 10.2 a schematic representation of the most frequently reported dynamic changes of the different microbial groups during CBF is shown. There are several suggestions of how to characterize/divide the different phases of a CBF process based on physical parameters such as temperature or pH (Gutiérrez-Ríos et al., 2022), the absence/presence of oxygen (De Vuyst & Leroy, 2020), the different microbial population peaks (Viesser et al., 2021), or the implicated highest concentrations of main metabolites that are commonly observed along the process (Hirko et al., 2023). The initial microbial diversity is considered essential for the metabolic process of CBF. The semi-solid matrix of the fresh pulp creates a unique semi-anaerobic environment.

The predominant yeast consumes high amounts of sugars in the fresh pulp converting them efficiently into ethanol, CO_2, and other (aromatic) metabolites leading to a rapid increase of their population within the first 24 h (De Vuyst & Leroy, 2020). Starting under environmental temperatures of 28–30°C, the yeasts' metabolism is flexible; aerobic conditions favor their growth by respiration while anaerobic conditions favor their fermentative metabolism. In the past, cacao-related yeasts have been reported to produce and secrete pectinolytic enzymes causing drainage of the pulp (Lagunes-Gálvez et al., 2007;

De Vuyst & Leroy, 2020), though latest studies report on lack/no expression of these enzymes under the micro-environment of CBF (Samagaci et al., 2014; Díaz-Muñoz et al., 2021).

In parallel, the population of LAB species increases; they are also capable of growing under high osmotic pressure and in the absence of oxygen while consuming glucose and fructose (Ho et al., 2018; Viesser et al., 2020). Also, the low initial pH favors their growth compared to other bacterial groups. Moderate temperatures between 30°C and 37°C are reported to be optimal and most LAB prefer anaerobic conditions to proliferate and convert pulp sugars into organic acids, ethanol, and mannitol, among other metabolites. It is interesting to mention that the pulp pH starts increasing (reaching values up to 4.0) due to the assimilation of citric acid and conversion into lactic and acetic acid (causing again pH drop to values of 3.7–3.8) by certain species of this microbial group during the first days of fermentation (Meersman et al., 2017; Mota-Gutierrez et al., 2021).

Once most of the fermentable pulp sugars are consumed by yeasts and LAB and a certain concentration of ethanol has been produced, the population of AAB starts increasing. Though AAB are present from the onset of the CBF, their growth is boosted under moderate temperatures (30–37°C) and aerobic conditions, initiating oxidation of the ethanol into acetic acid (De Roos & De Vuyst, 2018). Oxygen penetrates the fermenting mass thanks to pulp drainage and the aeration of the mass caused during turning. Their metabolic activity is highly exothermic and therefore monitoring the fermentation temperature is a good indicator for their action; when temperature rises to 40–46°C the AAB metabolism is intense and their further dominance in the CBF process can cause overoxidation of the acetic acid into CO_2 and water pushing temperatures to 48–50°C and reaching pulp pH of 4.2–4.5 toward the end of the fermentation process (Adler et al., 2014; Ho et al., 2018; Papalexandratou et al., 2019).

Also, SFB that are present in the surrounding environment and contaminate the freshly collected CPBM stay inactive for most of the duration of the CBF; they are highly thermotolerant, but their exact role is not yet clarified. *Bacillus* spp. may get activated during the later stages of the fermentation when inside the fermenting CPBM pH and aeration increase and the temperature rises above 48°C for a prolonged period (Schwan & Wheals, 2004), allowing for their spores break open and get activated. Such aerobic bacteria can produce multiple chemical substances, including pyrazines, organic acids, and 2,3-butanediol but also off-flavors relating to C_3-C_5 fatty acids (Lopez & Quesnel, 1971). These compounds may have an impact on the acidity of fermented cacao beans but potentially negatively affect their final flavor (Schwan, 1998; Magagna et al., 2017). According to Ouattara et al. (2008), the degradation of the pulp during CBF may be caused by a combined action of enzymes produced by both yeast and *Bacillus* strains; however, characterization of extracellular pectate lyases produced by these strains shows no activity in laboratory experiments at common cacao pulp temperature and pH (29–45°C; pH 3.9–4.5), perhaps because these enzymes exhibit optimal activity at 60°C and pH 7.5.

Filamentous fungi prefer environments with low water activity and aerobic conditions (Delgado-Ospina et al., 2021). Their involvement in CBF mainly tends to be toward the end of the process when temperature starts dropping and enough oxygen can circulate in the mass. Mold usually appears at spots that are more exposed to air, thus on the surface or bottom layer of the fermenting mass or at wide openings or corners of the fermentation boxes. They can sporulate in adverse conditions, for example at high temperatures, and they can survive the heat generated during ethanol oxidation by AAB. Molds are considered undesirable in the fermentation of cacao beans because they may affect the flavor and physical aspects of the cacao beans and they may produce mycotoxins, which can be poisonous for the final consumer. Good hygiene and handling of the raw material as well as proper maintenance of the fermentation equipment are strictly required to avoid the contamination by fungi and buildup of mycotoxins (Copetti et al., 2012).

Thanks to the microbial activity in the pulp, the cacao bean testa starts gradually becoming permeable allowing heat, ethanol, acids, and other microbial metabolites to penetrate. This causes eventually killing of the embryo inhibiting its further germination (De Vuyst & Leroy, 2020); the cotyledon cells break due to osmosis and enzymes get activated and start cutting down chains of complex sugars, proteins, and polyphenols into smaller size molecules. Upon fermentation, sucrose is hydrolyzed by cacao bean invertase into glucose and fructose. The level of sucrose in well-fermented cacao beans should be almost zero. The remaining reducing carbohydrates, such as glucose and fructose, are important flavor precursors and are of high importance for the roasting step of the dry cacao (Beckett, 2009). The polyphenols of cotyledons are responsible for the astringent flavor of cacao. They consist of anthocyanins, responsible for the purple color of unfermented cacao beans, colorless catechin and epicatechin, and their oligomers, the proanthocyanidins, as well as polymers (Wollgast & Anklam, 2000). The anthocyanins rapidly break down to anthocyanidins and carbohydrates (galactose and arabinose) early in the fermentation by glycosidase activity, bleaching the purple color of the cotyledons through exudation and releasing and reducing monosaccharides that participate in flavor formation during roasting (Wollgast & Anklam, 2000). Epicatechin and small procyanidins are soluble and, therefore, contribute to the astringency of cacao. During fermentation and oxygen penetration, oxidation by polyphenol oxidase and polymerization of these compounds into insoluble high-molecular-mass compounds (tannins) result in the reduction of the astringency and bitterness, which can be further reduced by exudation and/or enzymatic reactions. Polyphenol oxidase converts the main cacao bean polyphenol epicatechin and free anthocyanidins into quinones. Polyphenols and quinones can form complex molecules with other polyphenols, proteins, peptides, and amino acids (tannins), giving rise to brown coloration (Afoakwa, 2010). The theobromine and caffeine contents slightly decrease during fermentation because of diffusion out of the beans and hence further reduce bitterness. During fermentation, in general, the cacao bean proteins are degraded through enzymatic hydrolysis (aspartic endoprotease and carboxypeptidase activities), resulting in the production of hydrophobic free amino acids, as well as hydrophobic peptides, all of major importance for flavor development. The globulin-type protein is degraded during fermentation, but the albumin-type is not (Afoakwa, 2010).

10.2.3 Microbial species diversity of CBF

Several studies have unraveled the cacao microbial species diversity through culture-dependent techniques; the limitations of such techniques, for example, agar media selectivity or inability of certain species to grow on certain agar media, may have biased the outcome of these studies (Miguel et al., 2017; Arana-Sánchez et al., 2015; Velásquez-Reyes, 2021). Since the application of culture-independent methods and more precisely the Next Generation Sequencing (NGS) the detected microbial diversity associated with CBFs has significantly expanded (Viesser et al., 2021; Fernández-Niño et al., 2021; Serra et al., 2019; Verce et al., 2021). In addition, frequency and time points of sampling can impact the outcome of the study. This section reports on the most frequently encountered species related to the process.

10.2.3.1 Yeasts

The most often encountered yeast genera involved in CBF, in decreasing order of relative abundance, are *Pichia*, *Hanseniaspora*, *Saccharomyces*, and *Candida*, and they represent an even wider species diversity. *Saccharomyces cerevisiae, Pichia kudriavzevii, Hanseniaspora opuntiae, Hanseniaspora uvarum, Hanseniaspora guilliermondii, Pichia manshurica, Pichia kluyveri*, and *Candida tropicalis* are the most prevalent yeast species found in spontaneous CBF processes. *Wickerhamomyces*, *Kazachstania*, *Kluyveromyces*, *Torulaspora*, and *Rhodotorula* are additional yeast genera that have been found in spontaneous fermentations (Gutiérrez-Ríos et al., 2022; Díaz-Muñoz & De Vuyst, 2022). *Hanseniaspora* species are frequently present at the start of the fermentation process but gradually vanish in favor of yeast species better adapted to stress conditions such as higher temperatures and concentrations of ethanol, including *S. cerevisiae, P. kudriavzevii, P. manshurica, or P. kluyveri* (Papalexandratou et al., 2013; Díaz-Muñoz et al., 2021). Several yeast species have been reported for production of esters and alcohols that relate with floral and fruity notes (De Vuyst & Leroy, 2020). The genus *Pichia* has been identified as a producer of phenylethanol, methylbutanol, and 1-butanol as well as esters such as ethyl acetate, isoamyl acetate, ethyl dodecanoate, and 2-hydroxy-4-methyl-ethyl (Koné et al., 2016; Ouattara et al., 2020). Species of the genus *Hanseniaspora* such as *H. opuntiae* have been described as capable of producing volatile compounds in spontaneous CBFs such as ethyl acetate, ethyl lactate, 2-pentanol, 2-phenylethyl acetate, ethyl hexanoate and acetophenone (Ho et al., 2018; Mota-Gutierrez et al., 2018; Verce et al., 2021). On the other hand, *Saccharomyces cerevisiae* has been positively correlated with esters such as ethyl dodecanoate, ethyl octanoate, and ethyl decanoate as well as higher alcohols such as 1-butanol-2-methyl and 1-propanol-2-methyl (Díaz-Muñoz et al., 2023). CBFs carried out in Latin countries are often dominated by several yeast genera and species that tend to be active in parallel (Papalexandratou & De Vuyst, 2011; Saerens & Swiegers, 2013) or in succession (Arana-Sánchez et al., 2015; Papalexandratou & De Vuyst, 2011; Papalexandratou et al., 2019). Non-*Saccharomyces* species have attracted attention as they are often involved in fine cacao processing of Latin America, probably adding on to the aromatic character of these cacao types (Arana-Sánchez et al., 2015; Papalexandratou & Nielsen, 2016; Pereira et al., 2017; Chagas-Junior et al., 2021; de Ferreira et al., 2022).

10.2.3.2 Lactic acid bacteria

In the group of lactic acid bacteria (LAB), the most abundant genus associated with CBFs corresponds to *Lactobacillus* (recently reclassified, Zheng et al., 2020), but other genera such as *Fructobacillus*, *Leuconostoc*, *Weissella*, *Pediococcus*, and *Lactococcus* are also commonly reported (De Vuyst & Leroy, 2020). At the beginning of the process, microaerophilic, acid-tolerant, ethanol-tolerant, and fructophilic LAB species (citrate-converting or not), in particular *Leuconostoc pseudomesenteroides, Fructobacillus pseudoficulneus, F. tropaeoli, Liquorilactobacillus cacaonum, Lactiplantibacillus fabifermentans*, and *Lactiplantibacillus plantarum* are present. This diverse group of LABs prevails mainly at the beginning of the CBF process, followed by strictly heterofermentative *Limosilactobacillus fermentum* that persists upon further fermentation (De Vuyst & Weckx, 2016). Its predominance may be ascribed to its citrate-converting, mannitol-producing and ethanol-, acid- and heat-tolerant properties (Adler et al., 2013). It is important to mention that the LAB species diversity and abundance varies widely between different geographic locations with diverse cacao types, fermentation conditions, and the presence of other microbes. The *Leuconostoc* genus was positively correlated with lactate, mannitol, myo-inositol, acetoin, and alcohols (Díaz-Muñoz et al., 2023), while *L. fermentum* capacity to produce an increased concentration of mannitol and 2,3-butanediol (Romanens et al., 2019). Latin and Caribbean CBFs have often revealed atypical LAB growth in comparison to African or Asian processes (Lagunes-Gálvez et al., 2007; Papalexandratou et al., 2013, 2011, 2019; Verce et al., 2021). Delayed appearance of LAB and more specifically of *L. fermentum* has been reported in fermentations of the Dominican Republic and Nicaragua, probably relating to low initial pulp pH and/or dominance of yeasts. Also, *Fructobacillus* spp. have been found to dominate Ecuadorian fine CBFs reflecting a high mannitol production (Papalexandratou et al., 2011b). Inadequate separation of healthy cacao pods from infected ones in Brazil has shown a very wide LAB diversity (mainly compost-related LAB species, such as *Liquorilactobacillus nagelii*) predominating the whole fermentation process (Garcia-Armisen et al., 2010, Papalexandratou et al., 2011c).

10.2.3.3 Acetic acid bacteria

Aerobic acetic acid bacteria (AAB) are ubiquitous in the environment and they are strongly associated with plant material (e.g., fruits, flowers and pollen) and insects (e.g., honeybees) (Hata et al., 2023). The diversity of AAB present in CBFs is limited to two genera, namely *Acetobacter* and *Gluconobacter*. At species level, *Acetobacter pasteurianus* is the most identified AAB during CBFs around the world presumably because of its ethanol-, mannitol-, and lactate-oxidizing capacity and acid and heat tolerance (Farrera et al., 2021). At the beginning of the process, however, other species such as *A. ghanensis, A. senegalensis, A. syzygii, A. tropicalis, A. persici*, and *A. fabarum*, among others, have been identified (Papalexandratou et al., 2011a, 2011b, 2011c; Pereira et al., 2012, 2016; Gutiérrez-Ríos et al., 2022). *Gluconobacter* spp. have frequently been encountered in early stages of Latin CBFs (Papalexandratou et al., 2011b, 2019); *G. oxydans* is a main representative, but in general, it is a less prevalent genus. The growth of glucose-oxidizing *Gluconobacter* spp. at early stages of the fermentation process may lead to the production of gluconic acid, when ethanol and lactic acid are depleted (De Vuyst & Leroy, 2020). Gluconic acid can positively impact organoleptic properties by adding on mild herbal flavor and/or balancing sourness and bitterness, as described also during fermentations of fine cacao varieties in Nicaragua (Papalexandratou et al., 2019). From the *Acetobacter* genus, a positive correlation between the production of 3-methyl-1-butanol, phenylacetaldehyde and acetic acid has been previously reported (Mota-Gutierrez et al., 2018).

10.2.3.4 Spore-forming bacteria

The spore-forming bacteria (SFB) found in CBF are of the genus *Bacillus*. These bacteria are ubiquitously found in soil, air, and water. Due to their ability to abolish vegetative life cycle by forming endospores, they can survive in adverse conditions such as absence of water or high temperatures (De Vuyst & Leroy, 2020; Ouattara et al., 2020). At species level, there are numerous representatives that are associated with CBFs, among which are *B. subtilis, B. licheniformis, B. firmus, B. coagulans, B. polymyxa, B. circulans, B. pasteurii, B. megaterium, B. brevis*, and *B. cereus* (Schwan et al., 1986, Schwan, 1998, Nielsen et al., 2014). In recent studies, where NGS approaches have been applied, other genera, such as *Paenibacillus, Fontibacillus, Brevibacillus, Lysinibacillus*, or *Rummeliibacillus*, have also been reported (Serra et al., 2019). The production of L-phenylalanine and 3-hydroxy-2-butanone by strains of *B. subtilis* has been reported (Castro-Alayo et al., 2019, Ouattara et al., 2020). The appearance of *Bacillus* spp. during the last days of CBFs of Latin origins has always been related to the presence of off-flavors to the final dry cacao (Papalexandratou et al., 2019). *Bacillus* species are occasionally reported during Mexican and Nicaraguan CBFs (Papalexandratou et al., 2019; Velásquez-Reyes et al., 2021), probably relating to the very high fermentation temperatures for prolonged period in the respective processes.

10.2.3.5 Enterobacteria

As mentioned above, the bacterial families belonging to *Enterobacterales* are *Yersiniaceae*, *Erwiniaceae*, and *Enterobacteriaceae*, and they are present frequently but transiently during CBFs (De Vuyst & Weckx, 2016). The diversity of these facultative anaerobic enterobacteria is broad and the most often cacao-related genera are *Tatumella, Pantoea, Klebsiella, Serratia*, and *Erwinia* (Lefeber et al., 2011a; Viesser et al., 2021). At species level, isolates of *T. saanichensis, T. punctata, T. terrea, T. morbirosei, T. ptyseos, P. gaviniae, K. pneumoniae, S. marcescens, E. soli, E. billingiae*, and *E. tasmaniensis* have been identified (Garcia-Armisen et al., 2010, Papalexandratou et al. 2011a, 2011b, 2011c; Illeghems et al., 2012; Binh et al., 2017; Serra et al., 2019; Pacheco-Montealegre et al., 2020). Their exact role in CBF is not yet fully understood. *Tatumella* has been positively associated with gluconate production in cacao pulp and with ethyl isovalerate and benzaldehyde in cacao beans (Díaz-Muñoz et al., 2023). Enterobacteria may also produce pectinolytic enzymes and assimilate citric acid (Illeghems et al., 2012; Papalexandratou et al., 2019). Their concentration is relatively low in the CPBM ($<10^4$ CFU/g), and once the anaerobic yeasts and LAB start to take over the process, they are no longer detected (often within 24–72 h).

10.2.3.6 Filamentous fungi (molds)

Airborne contamination is the primary source by which fungal spores can be transmitted from different media to the final product during storage. The diversity of this group is high: *Aspergillus, Penicillium, Fusarium, Colletotrichum, Trichosporon*, among several others (Delgado-Ospina et al., 2021). It is likely that filamentous fungi have a significant effect on the growth of other microorganisms as well as on the quality of the cacao since they can produce extracellular enzymes and aid in the degradation of pulp components. With regard to food safety, however, considering their capacity of producing mycotoxins (e.g., ochratoxin A), fermentation practices should be conducted toward the enhancement of acetic acid production, which leads to inhibition of fungal growth (Copetti et al., 2012).

10.2.4 Pre-fermentation practices

In certain cacao-producing countries of Latin America (e.g., Brazil, Ecuador, Central America) and beyond (e.g., the Philippines, Malaysia), it is common to encounter additional practices that are carried out before the use of the fermentation equipment. Such practices influence the onset inoculation through changes in the physicochemical properties of the pulp.

10.2.4.1 De-pulping

De-pulping of the fresh wet mass refers to partial or full removal of mucilage. This is a practice for moisture, pectin, and sugar reduction. Such practice results in acceleration of cacao trading (through shorter fermentation duration or direct cacao drying). Some de-pulping methods are mentioned below:

10.2.4.1.1 RINSING WITH WATER

This practice is mostly applied when cacao is intended to be traded as non-fermented cacao (adequate for traditional beverage markets, bulk market, cosmetics, cacao butter extraction). The fresh wet mass is placed in perforated buckets/baskets, and it is rinsed with water (preferably applied with pressure). The mass is stirred periodically and then it is left standing for a few hours so that most water can drain away. Afterward, the mass is spread as a thin layer in direct sunlight for drying. Washed cacao must be dried quickly during the first days for moisture reduction to avoid random contamination and non-controlled fermentation.

10.2.4.1.2 MECHANIC/HYDRAULIC

Cacao types with high pulp content may be mechanically spun/centrifuged for pulp reduction (Figure 10.3a). High pulp content can cause fermentation delay and/or high acidity production. Alternatively fresh mass may be pressed hydraulically to reduce the amount of pulp (Figure 10.3b). Proper equipment maintenance is required to avoid severe contamination of the collected mass.

10.2.4.1.3 SUN EXPOSURE (PRE-DRYING OF FRESH MASS)

In the case of very pulpy cacao or big bean size, fresh wet mass may be spread on drying areas immediately after harvest or after few hours/days of pulp pre-drainage (see Section 10.2.4.2) for brief sessions of direct exposure to sun (Moreno-Rojas et al., 2023). Due to friction against the drying surface in combination with the drying effect of direct sunlight, the pulp amount will be reduced or partially dried.

10.2.4.2 Pre-drainage

Pre-drainage of the fresh CPBM allows for partial pulp reduction but also a more intense yeast activity. This could be done in two ways.

10.2.4.2.1 STRICTLY ANAEROBIC CONDITIONS

Keeping the fresh pulp in closed recipients or well-sealed PP sacks overnight (approximately 12 h) that permit dripping may result in quicker ethanol production and therefore in an accelerated fermentation process and higher cut-test results (>% well-fermented beans).

Figure 10.3 (a) De-pulping equipment and appearance of the mass after partial pulp removal (Mexico) and (b) hydraulic-pressed fresh CPBM in wooden recipients for pulp juice reduction (Brazil).

10.2.4.2.2 BOOST OF YEAST INOCULATION

The freshly collected wet mass may be spread on elevated surfaces made of net or wood or on fresh banana leaves, preferably with slight inclination for better dripping for a period of 8 to 12 h (Schwan et al., 2023). Aerobic exposure of fresh mass on a big surface permits more uniform yeast inoculation under oxygen conditions that enhance fruity notes to final product.

10.2.5 Fermentation methods

The CBF method varies between cacao-producing origins, and it has evolved with time. Most common methods are heap and box fermentations, while tray or basket fermentations are often applied by remote smallholder farmers (Nielsen et al. 2014; Santos et al., 2020; Streule et al., 2022). In the last decades, a common practice in Latin countries has been fermenting in polypropylene bags while several countries have been experimenting with drum fermentations.

Heap fermentation is the simplest method and the most common in West Africa and particularly in Ghana and Ivory Coast (Camu et al., 2007; Guehi et al., 2010; Nielsen et al., 2014; Mota-Gutierrez et al., 2018). It is a low-cost method, and it can result in high-quality cacao when properly monitored. Fresh CPBM of 25–1,000 kg (though ideally better handled in volume that does not exceed 300 kg) is piled into a heap on a mattress made of banana or plantain leaves laid on the ground (Figure 10.2a). The heap is covered with extra leaves, a practice that prevents the produced heat from escaping and protects the fermenting CPBM from undesirable contamination. Though not traditionally done, it is advised to cover the heap with plastic to protect from rain and animals, better conserve the temperature, and keep the fermenting mass moist. The liquefied pulp or sweating produced during the fermentation can drain away and penetrate the ground; in many cases the heap is installed on an inclined surface and a hole is made on the ground for recuperation of the sweating during the first day of the process. The heap fermentation lasts for four to seven days, and a periodical mixing of the fermenting mass is required to add oxygen and homogenize the mass. The mixing technique varies from origin to origin. Inadequate mixing will cause overfermentation in certain layers and underfermentation in others; when farmers do not apply any mixing, the fermenting mass does not generate sufficient heat and the superficial layer gets moldy or the mass gets a slimy texture. Changing the banana/plantain leaves at every mixing guarantees better hygiene and does not allow for contamination from the decomposing vegetal material. The heap is mainly carried out in the plantation but lately farmers tend to move it near their houses or even indoors because of increased risk of theft. Though not frequently encountered in Latin countries, the heap method could be a solution for remote cacao farmers with no means for quick transport of the fresh mass or without special installations.

An increasingly popular method is **fermentation in wooden boxes** (Nielsen et al., 2014; De Vuyst & Weckx, 2016) (Figure 10.2b). The use of wooden boxes is a good option for fermentation of 100 to 1,500 kg of fresh CPBM. It is applied widely in Latin America as it is the ideal system for centralized cacao post-harvest handling. It must be mentioned that the design of the box is crucial for successful fermentation; well-designed boxes allow the liquefied pulp to properly drain away and air to enter, thanks to their construction with wooden planks that form narrow gaps between them (2–3 mm). While common practice in the past, perforation of the wood does not contribute much to additional aeration or better drainage of the sweating; in contrary, perforations get quickly blocked and if not frequently cleaned, they become hubs of mold growth risking contamination of the fermenting mass. The box can be a single unit, but it is recommended to have several compartments. Several compartments allow for proper inoculation of the fermentation equipment for the different fermentation stages. It is important to establish adequate microbiology at each compartment for each fermentation phase. The escalator-type boxes allow mixing of the beans by moving the containing mass from an upper box into a lower one. The boxes are usually protected from bad weather conditions by placing them under a roof or in fermenter rooms. The fresh mass is placed inside the boxes and covered with banana or plantain leaves and jute sacks to retain the heat and prevent drying of the surface. Use of banana/plantain leaves as a thin layer at the bottom of the box helps control the speed of drainage of the sweating and it also works as a barrier for possible contaminants (egg-laying insects, worms, etc.). Changing the banana/plantain leaves at every mixing guarantees better hygiene and does not allow for contamination from the decomposing vegetal material. Mixing of the fermenting CPBM is advised to be done periodically to achieve homogeneity and add oxygen to the mass. The duration of a box fermentation is normally five to seven days. This method provides relatively increased aeration of the fermenting CPBM and, hence, too frequent mixing is not desirable as it will lead to increased acidity. The exact mixing pattern is defined according to the targeted flavor profile.

In the tray system, the raw cacao beans are placed in 10-cm-deep wooden trays, 4 to 10 of which are stacked on top of each other. While the method was originally developed in Ghana to ensure aeration of the fermenting CPBM without mixing the beans, in Latin American countries the mass of each tray is still mixed, and the positioning of the trays is rotated to guarantee better uniformity of the mass. The trays have slatted split bottoms for pulp drainage. Using the tray system approximately 80–120 kg of fresh CPBM can be fermented at each layer. Every tray or just the top tray may be covered with banana/plantain leaves and/or sacks. A tray system allows uniform fermentation, but it can be very labor-intensive when mixing/rotating of trays is applied. Recent studies have shown high herbal and spicy notes in cacao fermented under the tray system in Central America while an enhanced cacao flavor was found in tray-fermented cacao from Ghana (Figure 10.4 data not peer reviewed; Nielsen et al., 2014).

(a)

Figure 10.4 Innovative marketing tool for promotion of cacaos with distinct sensory profiles: (a) cacao Flavor Guide for 32 cacao bean profiles developed in Central America (Honduras, Nicaragua, El Salvador) and their respective geographic origin. Partial Least Squares regression (PLS, software: LatentiX 2.12, latentix.com) using aroma data as the X-matrix and sensory data as the Y-matrix. All values were autoscaled (PC#1 (22.774%); PC#2 (18.896%). Petersen M.A., Seguine E.S., Papalexandratou Z., Sukha D., De Winne T., Camacho A., Jakobsen R.R., & Nielsen D.S., 2017 (data not peer-reviewed; © Lutheran World Relief).

(Continued)

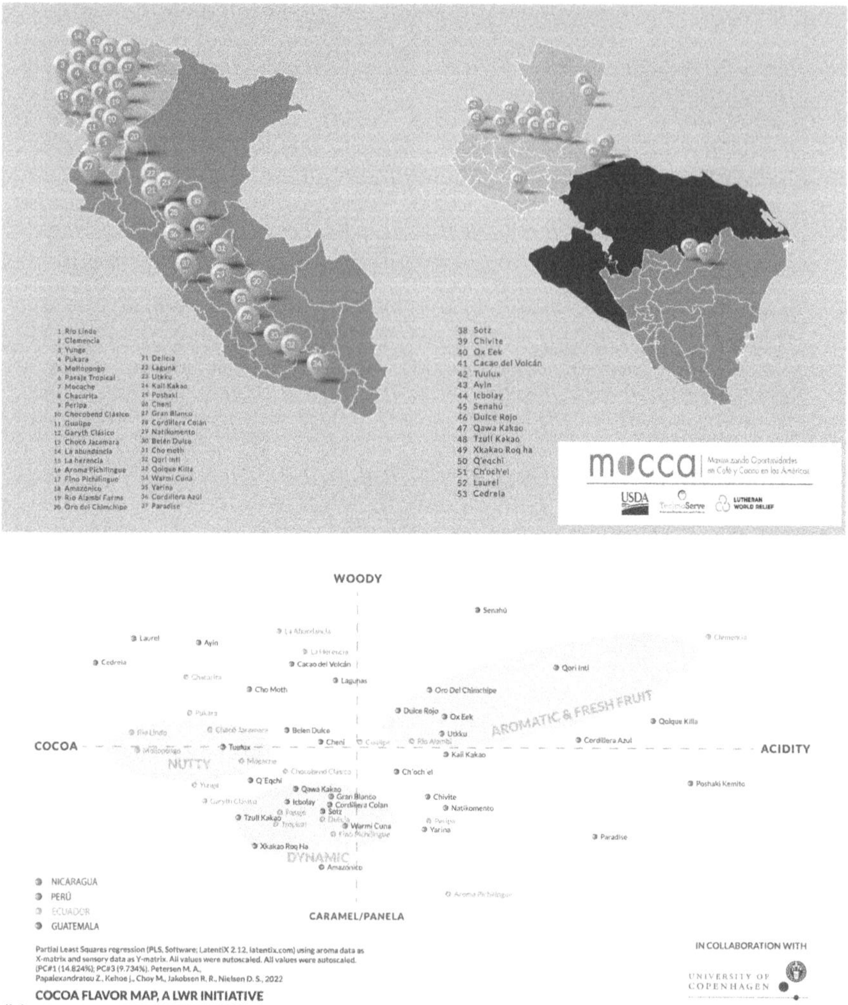

(b)

Figure 10.4 (Continued). Innovative marketing tool for promotion of cacaos with distinct sensory profiles: (b) cacao Flavor Guide for 53 cacao bean profiles developed in Peru, Ecuador, Guatemala, and Nicaragua and their respective geographic origin. Partial Least Squares regression (PLS, software: LatentiX 2.12, latentix.com) using aroma data as the X-matrix and sensory data as the Y-matrix. All values were autoscaled. All values were autoscaled. (PC#1 (14.824%); PC#3 (9.734%). Petersen M.A., Papalexandratou Z., Kehoe J., Choy M., Camacho A., Jakobsen R.R., & Nielsen D.S., 2022 (data not peer-reviewed; © Lutheran World Relief).

In Ecuador, the traditional technique of the 1980s to produce *Arriba* cacao was a two- to three-day fermentation on drying **platforms** (Papalexandratou et al., 2011b; De Vuyst & Leroy, 2020). According to Lehrian & Patterson (1983), the cacao beans were customarily spread out for 2–3 h during the day on bamboo or wooden platforms and heaped during the night. As of late, platform equipment is only available in the older Ecuadorian plantations. Instead, farmers may spread the freshly collected mass for 5–6 h and heap the pre-dried mass for further fermentation by covering with polypropylene bags (Figure 10.2). The process may be repeated for 2–4 days before being spread for full drying.

When farmers do not collect sufficient volume to ferment under one of the above methods, they tend to carry out very small-scale fermentations (5–25 kg) in traditional **woven baskets** or **plastic recipients** (Nielsen et al., 2014). The recipient is layered with banana leaves and the wet mass is poured and wrapped completely. The woven baskets allow for some pulp drainage while the plastic buckets are not always perforated. The mass is sometimes gently stirred, and fermentation will progress for 2–7 days. High importance must be put on the hygiene of the baskets as the natural material gets humid and moldy, releasing unpleasant odors that can impact the quality of the beans.

The last decade effort has been put into the mechanization of the fermentation process. The purpose is to make the work less labor-intensive without compromising quality. One of the first models introduced is **drum/barrel-shaped wooden recipients**. Inspired by the wine or whisky barrels, such design is claimed to have less "dead" corners that cause inconsistent temperatures. In most cases the drum/barrel is made of hard wood narrow planks with openings that permit pulp drainage. The design varies from cylindric to hexagonal, while a central axis is connected to a steering wheel that helps rotate the barrel but also homogenizes/stirs the fermenting mass inside the single compartment. Though this design was at first suggested for more frequent mixing and oxygenation of the mass for quick temperature increase, very frequent rotation usually results in increased acidity, disturbance of microbial activity, and therefore, no balanced mellow chocolate profiles. The drum/barrel design comes with several flaws; the central axis is in most cases unable to mix properly the mass, leaving non-mixed layers stuck on the interior wooden surface or near the axis. Also, there are plenty of metal nails and screws at the interior of the recipient and they get oxidized once in contact with the acidic pulp. Stirring is also not as handy; a full recipient is very heavy and when stirring is not mechanized it needs effort to achieve proper and sufficient rotation. For that reason, the capacity of such device is kept small (maximum 250 kg wet mass). (Un)loading and maintenance of the device are challenging too.

The need for quick cash and no price differential according to quality pushes farmers to quick alternatives. It is now very common to encounter CBFs carried out **in polypropylene sacks** at the farm level. It is a practice to mainly allow for some drainage due to gravity and it lasts not more than 3–4 days. The sack may be shaken to achieve some level of uniformity and a little aeration. The partially fermented mass is then spread for sun drying.

10.2.6 (Pre-) Drying

A five- to seven-day CBF is succeeded by a drying process of well-fermented beans (Wood & Lass, 2001). The main purposes of drying are termination of the microbial fermentation, reduction of moisture from 40% to 60% (wt/wt) to 6–7% (wt/wt), and elimination of acetic acid (Afoakwa, 2010; Dzelagha et al., 2020). The removal of water prevents mold growth during storage and allows exudation of not only acetic acid but also lactic acid from the bean to the shell. The duration of the drying normally depends on the method and conditions applied. Dry beans weigh on average 1.0–2.0 g.

The common drying practice of small farmers is **sun-drying** (Figure 10.5a), which is supposed to be the most efficient way and usually leads to cacao beans of high quality with low cost (Lehrian & Patterson, 1983; Wood & Lass, 2001; Thompson et al., 2007; Rodríguez-Campos et al., 2011; Dzelagha et al., 2020). Briefly, the fermented mass is spread as a thin layer on cane, bamboo, wooden platforms, shallow trays or, in some cases, concrete terraces and mixed regularly while clumps are broken up during the day for uniform drying. In case of bad weather conditions and increased rainfall, plateaus or retractable roofs can be pulled above the drying platform to prevent bean soaking. The duration of sun-drying (normally 7–10 days) is weather-dependent and can last up to 25 days. An extended drying period, incomplete and slow drying, or rain soaking may result in undesirable mold contamination of the fermented cacao beans. Heavy growth of molds could lead to

Figure 10.5 Drying techniques for cacao: (a) natural and direct sun-drying on movable wooden platforms in el Salvador; (b) natural sun-drying in solar tunnel with polycarbonate roofing in Nicaragua; and (c) mechanical drying in rotating stainless-steel drum with gas as energy source in Nicaragua.

penetration of the testa and production of off-flavors with detrimental effects on the quality and flavor of fermented dry cacao beans.

A common practice in Latin countries is the gradual drying process (Rodríguez-Campos et al., 2011, 2012). The fermented mass is at first spread for drying as a thick layer of 6–25 cm or alternatively drying starts under shade with gradual exposure to direct sun. Such practices may help with acidity reduction, no shrinkage, improved brown color development, and better fissuring of the dry cacao beans. Also, cacao is often handled under a very humid microclimate. In such cases, the drying platforms are often installed in a greenhouse-like installation, the so-called solar tunnel made of UV plastic or transparent (polycarbonate) plastic sheets (Figure 10.5b). This installation helps farmers for less labor-intensive handling; cacao is protected from rain, and dust while drying is carried out under moderate sunlight and relative humidity. Attention must be put to the design of such equipment guaranteeing sufficient air flow and control of the relative humidity as it can often cause accumulation of bad smells and/or intense condensation, especially during the early morning hours which can stimulate mold growth (Dzelagha et al., 2020).

Mechanical drying is used for fast drying of large amounts of fermented cacao beans requiring reduced manual labor (Wood & Lass, 2001; Thompson et al., 2007; Dzelagha et al., 2020). However, this method requires energy supply and well-maintained equipment for successful drying. Numerous designs have been developed, but indirect heating using heat exchangers is preferred, fueled by wood, gas, or oil (Figure 10.5c). Technologies like microwave and freeze-drying have been studied as well; however, they seem unsuitable for rural cacao regions and are highly costly (Dzelagha et al., 2020).

The initial drying rate must be slow and with frequent mixing to achieve uniform water reduction and desirable flavor and aroma precursor-producing biochemical reactions. This is achieved by keeping the temperature at ≥60°C and drying for at least 48 h. The main problem of artificially dried cacao is their contamination with smoky/hammy undesirable flavors, due to poorly maintained equipment or its inappropriate use. In smoky beans and cacao powder, compounds such as phenol, methyl-phenols, and dimethoxy-phenols have been found (Beckett, 2009; Rodríguez-Campos et al., 2012; Dzelagha et al., 2020). Prolonged drying normally leads to a moisture decrease under the optimal levels, making further processing of these beans difficult. Rapid drying may cause case hardening, meaning fast drying of the upper layers of the beans with moisture retention in the center of the beans, which may prevent oxidative changes from being completed and result in excessive acidity.

10.3 Cacao post-harvest innovations

In the meantime, the industry focuses on investigation of the industrialization options of cacao post-harvest processing while cacao is still processed in a spontaneous manner to a great extent by smallholder farmers; the benefits and challenges for industrialization of post-harvest processing are discussed to shorten the gap between farmers and industry.

10.3.1 Development and application of starter cultures in CBF

A starter culture (SC) is defined as a preparation of living microorganisms (single or consortium) which are deliberately used to assist the beginning of a fermentation process (Pereira et al., 2020a). The principal aim of this practice is to make specific changes in the chemical composition and the sensory properties of the substrate and finally obtain a more homogeneous product. With an increasing demand for consistency in cacao quality, researchers have focused on the development of microbial SC as *initiators* of the CBF too. Inspired by other food matrices, it is believed that the application of a known microbial consortium at the onset/during the CBF may allow for a shorter, more uniform fermentation with desired organoleptic characteristics ruling out the complex variables of the process (Pereira et al., 2020a, 2020b; Romanens et al., 2020; Díaz-Muñoz & De Vuyst, 2022). Unlike dairy, bakery, and alcoholic products, CBF remains a spontaneous process and no commercial SC are yet broadly used, while several SCs are under research.

The main properties that are targeted during the selection of strains for a cacao SC are (Pereira et al., 2020a):

- faster removal of the mucilaginous pulp that surrounds the cacao beans, by the activity of pectin-degrading enzymes
- hydrolysis of residual sugars, especially sucrose when cacao is underripe by yeast invertases
- generation of volatile organic compounds of interest enhancing the organoleptic properties of cacao (more cacao, fruity, or floral notes; bitterness/astringency reduction)
- assimilation of citrate, which is associated with LAB metabolism and, recently, some specific strains of yeasts
- acceleration of fermentation process by early ethanol production and earlier conversion into acetic acid

Pichia, *Hanseniaspora*, *Saccharomyces*, and *Candida* are the four most common yeast genera involved in CBF, and several respective strains (cacao isolates or other) have been investigated as potential SCs (Lefeber et al., 2011b, 2012; Menezes et al., 2016; Visintin et al., 2017; Chagas-Junior et al., 2021; Viesser et al., 2021). From bacterial species, the main applied genera are *Lactiplantibacillus*, *Limosilactobacillus*, *Acetobacter*, and *Gluconobacter* (Lefeber et al., 2011b; Magalhães da Veiga Moreira et al., 2017; Díaz-Muñoz et al., 2021; Alvarez-Villagomez et al., 2022). Depending on the desired characteristics of the fermented cacao beans, the most suitable SC shall be selected, thus there is no universal "recipe". Table 10.1 gives an overview of

Table 10.1 Effects of Different Inoculation Cultures on Cocoa Fermentation and Final Chocolate Product in Different Latin American Countries

Country *Variety of cocoa* *Reference*	*Composition of starter culture*	*Volume of fermentation (material of the vessel)*	*Quantity of microorganisms used for inoculation*	*Strategy of culture starter preparation and of inoculation*	*Effects on fermented cocoa beans*	*Effects on chocolate and/or liquors*
Costa Rica Trinitario Van de Voorde et al., 2023	- *Saccharomyces cerevisiae* - *Hansenispora opuntiae* - *Pichia kudriavzevii* - *Limosilactobacillus fermentum* - *Acetobacter pasteurianus* In different combinations.	36 kg of cocoa pulp-bean mass fermented in 49 L plastic vessel.	Initial cell densities of 6.0, 6.0, and 5.0 log of colony-forming units (CFU) per g of cocoa pulp-bean mass for yeasts, LAB, and AAB, respectively.	Yeasts and bacteria were grown in a specific liquid media and then lyophilized together (depending on each specific starter). To inoculate the cocoa pulp-bean mass, the lyophilized biomass of the strains was resuspended in saline solution.	Different results depending on the starter culture used. *S. cerevisiae* is necessary for successful fermentation (high production of alcohols and esters). With *H. opuntiae* underfermentation and poor VOC profiles are detected.	Differences in VOC profiles because of the starter culture mixtures applied were most distinguishable among the cocoa liquors instead of the final chocolates.
Costa Rica Trinitario Díaz-Muñoz et al., 2023	- *Saccharomyces cerevisiae* - *Hansenispora opuntiae* - *Limosilactobacillus fermentum* - *Acetobacter pasteurianus* In different combinations.	36 kg of cocoa pulp-bean mass fermented in plastic vessel during 96h.	Initial cell densities: 5.0 x10^5 CFU/g of cocoa pulp-bean mass.	Biomass was produced for each strain separately, then was centrifuged and the cell paste was lyophilized and conserved at 4°C. Before inoculation the lyophilized powders were rehydrated in sterile saline solution.	Reproducibility in the post-harvest processing chain.	*S. cerevisiae* positively influences aroma profiles of cocoa liquors and chocolates. *H. opuntiae* has no clear effects.
Dominican Republic Criollo Korcari et al., 2023	- *Lactiplantibacillus fabifermentans* - *Furfurilactibacillus rossiae* Different inoculation strategies: each strain separately with one inoculation (t 0) or two (t 0 and t 48h); the two bacteria together.	60 kg of cocoa pulp-bean mass fermented in in closed plastic containers perforated to allow pulp drainage.	Final cell concentration for each strain, ranging between 10^6 and 10^7 CFU/g.	LAB cultures grown in MRS broth were centrifuged and the cellular pellet was resuspended in a 50 g/L sucrose solution to a final concentration between 10^8 and 10^9 CFU/mL.	The two strains did not significantly impact fermentation parameters (temperature and pH) but they gave different profiles when considering the final product. The inoculated LAB had a protective effect against mould contamination.	Inoculated LABs seemed to increase the complexity of the flavour profile of the chocolate (increasing honey and red fruit notes).

(*Continued*)

Table 10.1 (Continued)

Country *Variety of cocoa* *Reference*	*Composition of starter culture*	*Volume of fermentation (material of the vessel)*	*Quantity of microorganisms used for inoculation*	*Strategy of culture starter preparation and of inoculation*	*Effects on fermented cocoa beans*	*Effects on chocolate and/or liquors*
Brazil Forastero De Sousa Ferreira et al., 2022	- *Saccharomyces cerevisiae* - *Pichia manshurica* The two species were used separately.	Wooden fermentation boxes (1.0 m length × 0.4 m width × 0.3 m height). 45 kg of pulp-bean mass in each box. 7 days of fermentation.	10^6 cells/g of cocoa pulp-bean mass.	After growing in the bench bioreactor, the cells were centrifuged, and the biomass was diluted in 1 L of sterile saline peptone water and stored up under refrigeration (4°C) until use.	*P. manshurica* performed better than the *S. cerevisiae* culture in terms of aldehydes, ketones, esters, and alcohols productions as well as low acidity and high phenolic component concentration.	Not mentioned.
Colombia Not mentioned. Garcia Gonzalez et al., 2022	- *Saccharomyces cerevisae* - *Hanseniaspora* spp	Wooden fermentation boxes (12 × 12 × 12 cm; capacity 1728 cm^3) loaded with 500 g of cocoa beans.	Initial concentration adjusted to 10^6 CFU/g of pulp bean mass.	Not mentioned.	The starter culture fermentation process showed a higher production of organic acids than the spontaneous fermentation process.	Not mentioned.
Mexico Trinitario Alvarez-Villagomez et al., 2022	- *Saccharomyces cerevisiae* - *Lactiplantibacillus plantarum (Lactobacillus plantarum)* - *Limosilactobacillus fermentum (Lactobacillus fermentum)* - *Acetobacter pasterianus* - *Acetobacter tropicalis*	Wooden (cedar) fermentation boxes of 0.08 m3 (120 kg of pulp-bean mass for each fermentation).	The starter culture was extended to 40 cm from the upper surface and the rest was added on the surface, with total concentration of 10^9 –10^{10} CFU /kg of cocoa beans.	Each microorganism was cultured independently in a specific medium. The starter culture was prepared in 1L of steril peptone water with 70% of yeast (7 g), 15% of lactic acid bacteria (1.5 g) and acetic acid bacteria (1.5 g).	Inoculated samples fermented with the starter culture produce a high content of alcohol or aldehydes/ketones, which could be converted into pyrazines (which generate interesting notes in the final product).	Not mentioned.
Brazil Cocoa hybrids CEPEC2002 and FA13 Moreira et al., 2021	- *Pichia kluyveri* - *Saccharomyces cerevisiae* (commercial yeast CA11) combined or *S. cerevisiae* alone.	Wooden fermentation boxes of 0.08 m^3 (100 kg of pulp-bean mass for each fermentation batch).	Initial concentration adjusted to 10^5 CFU/g of pulp bean mass.	*S. cerevisiae* was assessed according to the manufacturer's instructions. *P. kluyveri* was grown on YPD.Cells were recovered and re-suspended in 1 L of sterile peptone water.	*P. kluyveri* and *S. cerevisiae* influence the composition of volatile organic compounds.	Inoculation process improved the chocolate sensorial qualities (result of sensorial analysis).

Costa Rica Trinitario Díaz-Muñoz et al., 2021	- *Saccharomyces cerevisiae* - *Pichia kudriavzevii* - *Limosilactobacillus fermentum* - *Acetobacter pasteurianus* in different combinations.	28 L plastic vessels containing 20 kg of cocoa pulp-bean mass.	Initial cell densities of 4.0 and 5.0 log [(CFU)/g] in pulp-bean mass for yeasts and both LAB and AAB, respectively.	Yeasts and bacteria were grown in a specific liquid media and then lyophilized together (depending on each specific starter).	All fermentations with starter cultures showed an enhanced production of desired metabolites, in particular VOCs. Each starter has some specific effects.	Not mentioned.
Honduras Trinitario hybrids mix with forastero Romanens et al., 2020	- *Saccharomyces cerevisiae* - *Limosilactobacillus fermentum* (*Lactobacillus fermentum*); 2 different strains; 2 different combinations.	Wooden fermentation wooden boxes (measuring 0.6 m x 0.6 m x 0.5 m) filled with 180 kg of pulp-bean mass.	*Lb. fermentum*: 10^6 CFU/g of pulp bean mass *S. cerevisiae*: 10^5 CFU/g of pulp bean mass.	Lyophilised co-cultures were produced in Switzerland and used to inoculate the fermentation.	Inoculate fermentation: 10x higher concentration of mannitol in cotyledons, slower fermentation, less beans well fermented. Starter to improve cocoa bean quality compared to Starter 1: less acetic acid, higher % of well-fermented beans, lower astringency, lower off-flavours.	Not mentioned.

(*Continued*)

Table 10.1 (Continued)

Country Variety of cocoa Reference	*Composition of starter culture*	*Volume of fermentation (material of the vessel)*	*Quantity of microorganisms used for inoculation*	*Strategy of culture starter preparation and of inoculation*	*Effects on fermented cocoa beans*	*Effects on chocolate and/or liquors*
Brazil Scavina variety Cocoa Sande Santos et al., 2020	- *Rhodotorula mucilaginosa* - *Torulaspora delbrueckii* - *Candida parapsilosis* - *Pichia galeiformis* - *Pichia kluyveri* - *Issatchenkia orientalis* - *Saccharomyces cerevisiae* - *Pichia membranifaciens*	Wooden Fermentation boxes with 45 kg of pulp-bean mass.	The boxes were inoculated with 3.26 x 10^8 yeast cells.kg^{-1} of cocoa bean	All strains were grown previously and separately, overnight in 250 mL of YEPD. This culture was then used as a pre-inoculum for 3 L of YEPD. After centrifugation, the biomass obtained was resuspended in 40 mL YEPD medium and use as inoculum in the fermentations.	*Candida parapsilosis*, *Torulaspora delbrueckii* and *Pichia kluyveri* were the strains that accumulated the most physicochemical changes, (free amino acids, titirable acidity and reducing sugar analysis) that are relevant to the flavor.	Not mentioned.
Brazil Cocoa hybrids PS1319 and SJ02 resistant to	- *Saccharomyces cerevisiae* - *Torulaspora delbrueckii* Different combinaciones.	Fermentation boxes loaded with 300 kg of pulp-bean mass.	Initial concentration: 10^6 cells/g and 10^5 cells/g of cocoa beans for *S. cerevisiae* and *T. delbrueckii*,	The inoculum was prepared in YPD at 30 °C and replicated every 24 h for five days. The cells were recovered by centrifugation and re-suspended in 1 L	*S. cerevisiae* and *T. delbrueckii* consumed sugars and citric acid faster. They accelerated the increase in temperature and pH.	The starter culture influenced the PS1319 hybrid end-product quality, changing analytic profile and sensory perception of chocolates.
Brazil Cocoa hybrids PH16 Da Cruz Pedrozo Miguel et al., 2017	- *Saccharomyces cerevisiae* - *Limosilactobacillus fermentum* (Lactobacillus fermentum).	Wooden fermentation boxes of 0.08 m^3 (100 kg of pulp-bean mass for each fermentation batch).	*S. cerevisiae* (105 cells/ml) and L. fermentum (10^4 cells/ml).	Not mentioned.	The spontaneous cocoa bean fermentation process showed a greater variety of species compared to the inoculated fermentations.	Spontaneous fermentative process: dominance of bitter flavour; inoculated fermentation presented bitter, astringent and acid as dominant flavours.

SCs implemented in CBFs in Latin America referring to preparation method as well as their impact on the process and quality of final product. To the knowledge of the authors, no publications are available with the implementation of commercial SCs.

Six bacterial and 18 different yeast species have been reported. *Saccharomyces cerevisiae*, *A. pasteurianus*, and *L. fermentum* are the most applied species thanks to their importance to CBF (Papalexandratou et al., 2013; Lefeber et al., 2011b, 2012; Díaz-Muñoz et al., 2021). The cacao origin with the most studies on cacao SCs is Brazil, followed by Colombia and Costa Rica. The detection method for the natural as well as the inoculated microorganisms varies; massive sequencing techniques are applied in the most recent studies for more accurate identification and differentiation on genus, species, or strain level, as it is essential to prove the predominance and role of the inoculated strains to the CBF process (Figure 10.6) (Viesser et al., 2021; Gutiérrez-Ríos et al., 2022). Most studies targeted improvement of the process using a consortium of yeasts and bacteria while Korcari et al. 2023 have recently reported the use of LAB (*L. fabifermentans* and *Furfurilactobacillus rossiae*) and their impact against mold contamination referring to the organoleptic changes to the final product.

Considering the high competition on pulp substrates between the natural cacao-related microbiota and the inoculated microbial consortium, the preparation, conservation, and inoculation methods are vital to the application setup for the SCs (Marwati et al., 2018; Figueroa-Hernández et al., 2019; de Ferreira et al., 2022; Ooi et al., 2022). The physiological state of the microorganisms at the time of inoculation becomes critical. Most of the publications report on the application of inoculum with fresh microorganisms; however, they do not mention their physiological state. It would be important to share additional information about the preparation of the previous culture. Some articles refer to "overnight growth" for the yeasts, but they do not specify in what phase of the growth they were collected as this is critical for the viability of the culture (Crafack et al., 2013; Aßkamp et al., 2018; Meersman et al., 2016); it should be considered therefore as a crucial factor in future studies. Another critical point is the conservation phase of the SC before inoculation. Ideally, a SC is applied directly after the culture phase, but in practice this is almost impossible. So, it is necessary to go through a preservation phase of the SC. Two methods are broadly used: after cultivation, the microorganisms are either concentrated and stored under refrigeration in a suitable medium or the conservation is obtained through lyophilization (Figure 10.7). The latter method is very interesting because a freeze-dried culture is very practical to handle, but this technology does not allow for a satisfactory preservation of all microorganisms. In the case of yeasts, *S. cerevisiae* can be freeze-dried without any problem, but non-*Saccharomyces* yeasts show a very low % viability after freeze-drying (Casas-Godoy et al., 2021). This could explain why in the case of

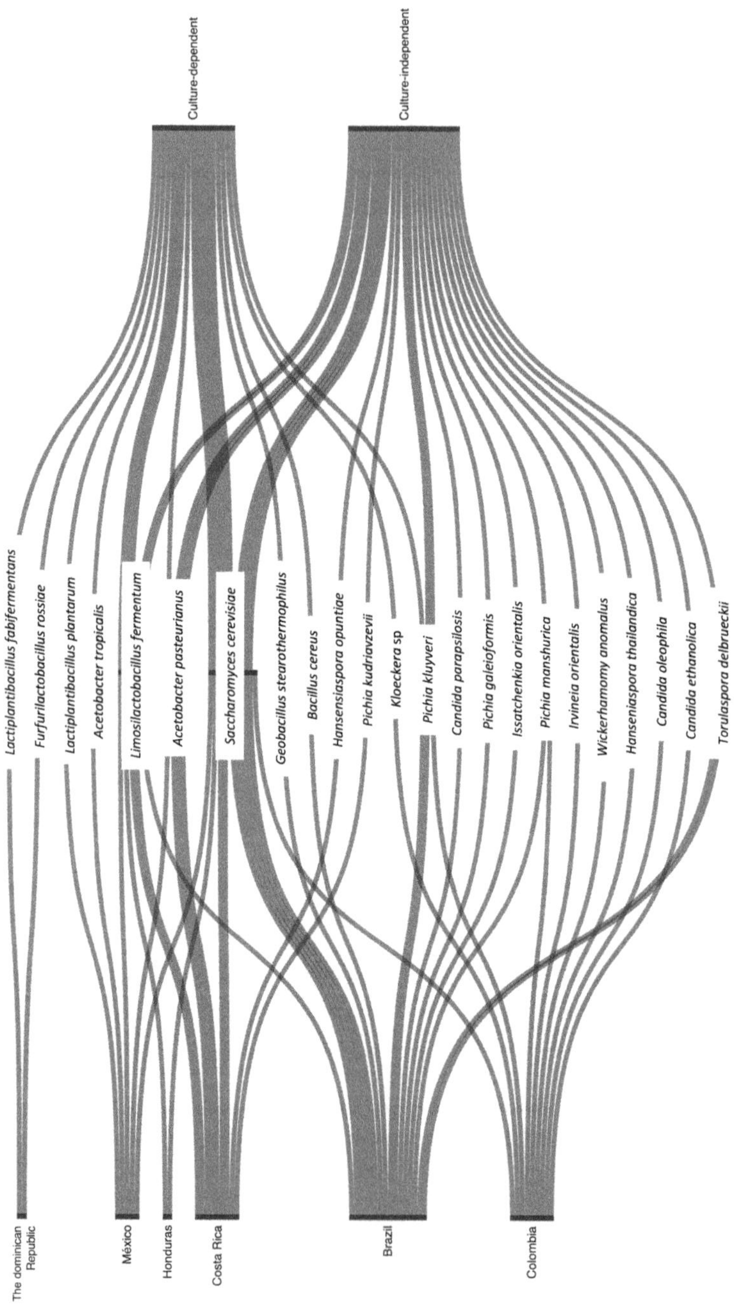

Figure 10.6 Main microorganisms identified in CBFs through culture-dependent and culture-independent techniques in Latin American countries.

Figure 10.7 Application of lyophilized bacterial and fresh yeast starter cultures in Ghanaian cacao bean heap fermentations.

Díaz-Muñoz et al. 2023 no conclusive effect is detected when using *H. opuntiae* as SC, relating probably to the preservation method used. Some of the work reported in Table 10.1 demonstrated that non-*Saccharomyces* spp., like *P. manshurica* (De Sousa Ferreira et al., 2022), *Candida parapsilosis*, *Torulaspora delbrueckii*, and *P. kluyveri* (Santos et al., 2020), have a positive and interesting effect on the fermentations.

It must be mentioned that working in the tropics and mostly at the farm level poses challenges in selection and preservation of affordable and viable SCs. Lack of basic utilities (water and electricity) and the low education and income of many cacao farmers are limitation factors to the wide and routine use of initiators. The commercial use of SCs in cacao is still under investigation; Lalcoc by Lallemand (https://lalcacaoyeast.com/en/) as well as Christian Hansen aromatic wine yeast cultures (Saerens & Swiegers, 2013) have been used in the field. Barry Callebaut has launched several chocolate products made with cacao fermented with in-house SCs (e.g., Terra Cacao, Q Fermentation) though these SCs were not available for sale but rather distributed to their associated farmers. Compost/fertilizer-like products with Effective Microorganisms (EM) of the brand EMRO or Blue & Green Solutions have been promoted in the last years in Latin America as initiators for CBF, but no official studies have been published yet.

10.3.2 Modern handling practices

While most of the food fermentations are carried out in an industrialized environment using stainless steel recipients, cacao is still handled in a traditional manner making use of natural materials (wood, banana leaves). Several research groups have been experimenting with pilot designs for (non) automatized stainless-steel or plastic cacao fermenters (Koffi et al., 2017; Pereira et al., 2013, 2016) (Figure 10.8). One advantage of such material is the easy maintenance though it could be better when combined with SCs. The wooden surfaces usually carry the necessary inoculum for spontaneous fermentation, while working with clean stainless-steel recipients could cause a delay in the fermentation process. Several models have been promoted within Peru,

Figure 10.8 Plastic non-perforated recipients for collection and fermentation of CPBM (up to 1 MT wet mass) as a modern alternative to classic wooden box in Ecuador (left); a semi-automatized stainless-steel cacao fermenter (up to ~250 kg wet mass) in Peru (right).

Ecuador, Brazil, and also Malaysia and Ivory Coast, but the average capacity is below 300 kg wet mass.

The big chocolate industry also looks for laboratory methods to treat and cure fresh CPBM as this may shorten the process while artisan chocolate makers look for modern solutions with flavor-steering results. Treatment of cacao beans with ethanol, enzymes, or acid during incubation under specific temperatures, microwave/UV processing, addition of fruit juices and spices at different stages of the fermentations process (in the lab or the field), or aging of the dry cacao in wooden barrels of several alcoholic spirits (whisky, bourbon, rum) are some of the reported and marketed approaches (Bernaert & Camu, 2009; Schnell et al., 2014; Lemarcq et al., 2022; Tamimi et al., 2023; Wirth & de Bortoli, 2023).

While many chocolate consumers do not know how chocolate is made, the recent launch of Ruby chocolate by chocolate company Barry Callebaut woke up their interest in cacao (Dumarche et al., 2012; Tuenter et al., 2021). As explained in the respective patent, un(der)-fermented cacao of West Africa or Ecuador is treated at the industrial level with citric or phosphoric acid to preserve the natural purple colorants before being pulverized and added into a white chocolate recipe, skipping thus fermentation and roasting steps that cause browning of the beans. This also explains that lack of chocolate flavor and the perception of fruity berry notes from the combination of fruity acid treatment and the purple color of the product. Also, with the interest of preserving cacao polyphenols and enhancing its healthy potential, craft chocolate industry has been promoting "raw" cacao products, meaning products handled below 46–47°C. This translates to fermentation of short duration or with few/no mixings, drying at low temperatures and no/light roasting of the product. Extra attention must be given to the total microbial counts of such products as it is not always guaranteed that handling kills pathogens and inactivates toxins.

10.4 Cacao post-harvest and market access

Seeing how competitive the cacao market has become with most cacao origins elevating their national and regional standards, the need for cacao quality improvement and diversification is a must (Papalexandratou et al., 2019; Jaimez et al., 2022; Sari et al., 2023a; Fanning et al., 2023; Ramón et al., 2023). Access to exquisite markets with unique identities is crucial for sustainable cacao production. Several initiatives try to gain visibility for the efforts put into quality focusing on a marketing approach without in-depth research, as the latter is not always affordable. Product lists, business to business events, and participation in fairs and cacao/chocolate competitions help cacao farmers transmit the message to the international markets. A drawback though in some cases is that qualities are presented without any scientific background and/or with limited exploration of actual potential.

Several cacao geographic maps have been developed pinpointing locations and their respective organoleptic profile (e.g., https://www.chocolateroads.com/images/MAPA.jpg). However, they tend to generalize flavor and quality potential per country or region despite the numerous publications that show that small tweaks in the fermentation and drying practices have a big impact on final quality. To overcome such drawbacks, and as an initiative by Lutheran World Relief (LWR), an interesting marketing tool for visualization of ~80 cacao profiles from Latin and Central America was developed between 2016 and 2022 (Figure 10.4a and b). This Regional Flavor Map was inspired by marketing tools of other foods, like wine, whisky, and gin (e.g., https://www.scotchmaltwhisky.co.uk/whiskyflavourmap.pdf), and it was based on a statistical correlation (Partial Least Squares regression) of fermentation, sensory, and volatile compound data (data not published). Each cacao product was tailored with the most adequate fermentation protocol, considering genetic/phenotypic pool, equipment, and volume available. The products that were mapped under this innovative platform were fully monitored during post-harvest processing and roasting and in parallel their full aromatic composition was identified as an objective support to subjective sensory evaluations (Figure 10.4a and b; data not peer-reviewed). All samples marked fermentation index >1 (FI) indicating complete fermentation, while sensory appreciation was independent of cut-test and FI results (data not shown). This tool helps cacao producers better connect with buyers of fine-specialty cacao, who will be able to quickly identify the sources for the cacao flavors, profiles, and characteristics they are seeking for.

Further, flavor networks are nowadays core research for food technologists, but they are also used in the culinary world for the perfect recipe development (Ahn et al., 2011). Understanding the raw material composition and its relation to the flavor of the final product applies also to cacao. Fermentation practices can be tweaked to enhance certain flavor notes detected in the fresh pulp for an indirect impact on the final dry cacao, as already reported (Eskes et al., 2012; Bickel Haase et al., 2021). As an empirical example, in the Philippines building a flavor network between the diverse regional cacao clones and the organoleptic quality of the available products in the market, helped expand the diversity of qualities offered (https://aurochocolate.com/blogs/community-programs/from-tree-to-bar-exploring-the-unique-flavours-of-auro-chocolate; Figure 10.9; data not peer-reviewed). Some of these cacao types had intense specific flavor characteristics that could be overruled when worked in a general cacao blend. Discovering the potential of single varietals helped apply tailored post-harvest practices for full expression of the flavor potential on these cacaos.

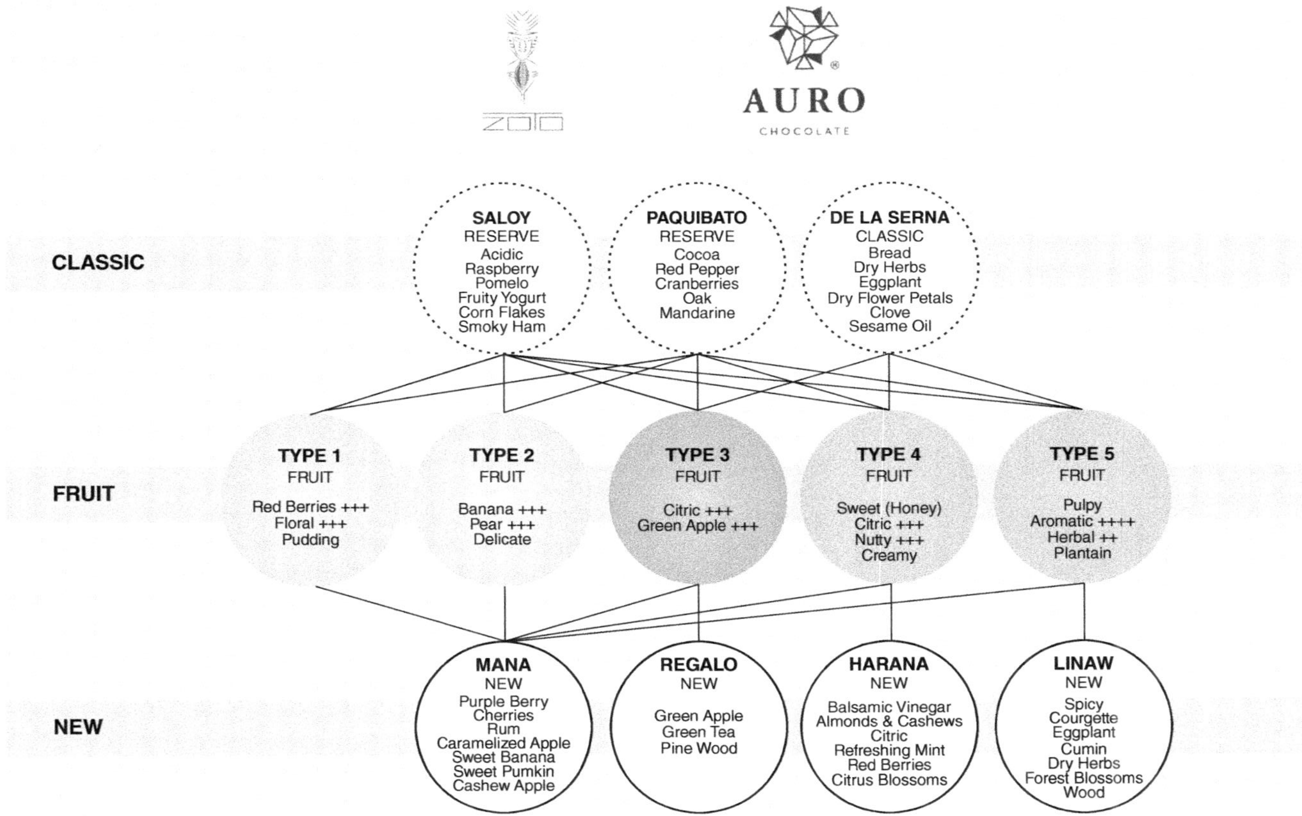

Figure 10.9 Flavor network between sensory characteristics of fresh cacao pulp of different Filipino cacao varieties and the organoleptic profiles of dry cacaos (single varieties or blends) (Papalexandratou Z., OCampo M., Go K., 2017; data not peer-reviewed).

References

Aßkamp, M., Bergounhou, A., D'Souza, R.N., Kumari, N., John, W.A., Ho, P.-W., Ullrich, M.S., Böttcher, N.L. & Nevoigt, E. (2018). Forcing fermentation: profiling proteins, peptides and polyphenols in lab-scale cacao bean fermentation. *Food Chemistry*, *278*, 786–794. Doi: 10.1016/j.foodchem.2018.11.108.

Adler, P., Bolten, C.J., Dohnt, K., Hansen, C.E. & Wittmann, C. (2013). Core fluxome and metafluxome of lactic acid bacteria under simulated cacao pulp fermentation conditions. *Applied and Environmental Microbiology*, *79*(18), 5670–5681. Doi: 10.1128/AEM.01483-13.

Adler, P., Frey, L.J., Berger, A., Bolten, C.J., Hansen, C.E. & Wittmann, C. (2014). The key to acetate: metabolic fluxes of acetic acid bacteria under cacao pulp fermentation-simulating conditions. *Applied and Environmental Microbiology*, 80(15), 4702–4716. Doi: 10.1128/AEM.01048-14.

Afoakwa, E.O. (2010). *Chocolate Science and Technology*. John Wiley & Sons Ltd, Chichester.

Ahn, Y., Ahnert, S., Bagrow, J. & Barabási, A.-L. (2011). Flavor network and the principles of food pairing. *Scientific Reports*, *1*, 196. Doi: 10.1038/srep00196.

Alvarez-Villagomez, K.G., Ledesma-Escobar, C.A., Priego-Capote, F., Robles Olvera, V.J. & García-Alamilla, P. (2022). Influence of the starter culture on the volatile profile of processed cacao beans by gas chromatography–mass spectrometry in high resolution mode. *Food Bioscience*, *47*, 101669. Doi: 10.1016/j.fbio.2022.101669.

Arana-Sánchez, A., Segura-García, L.E., Kirchmayr, M., Orozco-Ávila, I., Lugo-Cervantes, E. & Gschaedler-Mathis, A. (2015). Identification of predominant yeasts associated with artisan Mexican cacao fermentations using culture-dependent and culture independent approaches. *World Journal of Microbiology and Biotechnology*, *31*, 359–369. Doi: 10.1007/s11274-014-1788-8.

Arevalo-Gardini, E., Meinhardt, L.W., Zuñiga, L.C., Arévalo-Gardni, J., Motilal, L. & Zhang, D. (2019). Genetic identity and origin of "Piura Porcelana"—a fine-flavored traditional variety of cacao (*Theobroma cacao*) from the Peruvian Amazon. *Tree Genetics and Genomes*, *15*(1), 1–11. Doi: 10.1007/s11295-019-1316-y.

Barrientos, L.D.P., Oquendo, J.D.T., Garzón, M.A.G. & Álvarez, O.L.M. (2019). Effect of the solar drying process on the sensory and chemical quality of cacao (*Theobroma cacao* L.) cultivated in Antioquia, Colombia. *Food Research International*, *115*(52), 259–267. Doi: 10.1016/j.foodres.2018.08.084.

Beckett, S.T. (2009). *Industrial Chocolate Manufacture and Use* (4th edition). John Wiley & Sons, Ltd., Chichester.

Bernaert, H. & Camu, N. (2009). *Method for Flavouring Cacao Beans*. Patent, W.I.P. (US20090324798A1).

Bickel Haase, T., Schweiggert-Weisz, U., Ortner, E., Zorn, H. & Naumann, S. (2021). Aroma properties of cacao fruit pulp from different origins. *Molecules*, *26*(24), 7618. Doi: 10.3390/molecules26247618.

Binh, P.T., Tru, N.V., Dung, V.T.T., Thoa, N.T., Thao, P.V., Tham Ha, T.T. & Thang, V.V. (2017). Bacteria in wooden box fermentation of cacao in Daklak, Vietnam. *Journal of Microbiology & Experimentation*, *5*(7), 00176. Doi: 10.15406/jmen.2017.05.00176.

Camu, N., De Winter, T., Verbrugghe, K., Cleenwerck, I., Vandamme, P., Takrama, J.S., Takrama, J.S., Vancanneyt, M. & De Vuyst, L. (2007). Dynamics and biodiversity of populations of lactic acid bacteria and acetic acid bacteria involved in spontaneous heap fermentation of cacao beans in Ghana. *Applied and Environmental Microbiology*, *73*, 1809–1824. Doi: 10.1128/AEM.02189-06.

Casas-Godoy, L., Arellano-Plaza, M., Kirchmayr, M., Barrera-Martínez, I. & Gschaedler-Mathis, A. (2021). Preservation of non-*Saccharomyces* yeasts: current technologies and challenges. *Comprehensive Reviews in Food Science and Food Safety*, *20*(4), 3464–3503. Doi: 10.1111/1541-4337.12760.

Castro-Alayo, E.M., Idrogo-Vásquez, G., Siche, R. & Cardenas-Toro, F.P. (2019). Formation of aromatic compounds precursors during fermentation of Criollo and Forastero cacao. *Heliyon*, *5*(1), e01157. Doi: 10.1016/j.heliyon.2019.e01157.

Chagas-Junior, G.C.A., Ferreira, N.R., Gloria, M.B.A., da Martins, L.H. S. & Lopes, A.S. (2021). Chemical implications and time reduction of on-farm cacao fermentation by *Saccharomyces cerevisiae* and *Pichia kudriavzevii*. *Food Chemistry*, *338*, 127834. Doi: 10.1016/j.foodchem.2020.127834.

Coe, S.D. & Coe, M.D. (2019). *The True History of Chocolate* (3rd edition). Thames and Hudson Ltd., London.

Copetti, M.V., Iamanaka, B.T., Mororó, R.C., Pereira, J.L., Frisvad, J.C. & Taniwaki, M.H. (2012). The effect of cacao fermentation and weak organic acids on growth and ochratoxin A production by *Aspergillus* species. *International Journal of Food Microbiology*, *155*(3), 158–164. Doi: 10.1016/j.ijfoodmicro.2012.01.026.

Cortez, D., Quispe-Sanchez, L., Mestanza, M., Oliva-Cruz, M., Yoplac, I., Torres, C. & Chavez, S.G. (2023). Changes in bioactive compounds during fermentation of cacao (*Theobroma cacao*) harvested in Amazonas-Peru. *Current Research in Food Science*, *6*, 100494. Doi: 10.1016/j.crfs.2023.100494.

Crafack, M., Mikkelsen, M.B., Saerens, S., Knudsen, M., Blennow, A., Lowor, S., Takrama, J., Swiegers, J.H., Petersen, G.B., Heimdal, H. & Nielsen, D.S. (2013). Influencing cacao flavour using *Pichia kluyver*i and *Kluyveromyces marxianus* in a defined mixed starter culture for cacao fermentation. *International Journal of Food Microbiology*, *167*(1), 103–116. Doi: 10.1016/j.ijfoodmicro.2013.06.024.

De Sousa Ferreira, O., Chagas-Junior, G.C.A., Campos Chisté, R., da Silva Martins, L.H., de Aguiar Andrade, E.H., Do Nascimento, L.D. & Santos Lopes, A. (2022). *Saccharomyces cerevisiae* and *Pichia manshurica* from Amazonian biome affect the parameters of quality and aromatic profile of fermented and dried cacao beans. *Journal of Food Science*, *87*(9), 4148–4161. Doi: 10.1111/1750-3841.16282.

D'Souza, R.N., Grimbs, S., Behrends, B., Bernaert, H., Ullrich, M.S. & Kuhnert, N. (2017). Origin-based polyphenolic fingerprinting of *Theobroma cacao* in unfermented and fermented beans. *Food Research International*, *99*(1), 550–559. Doi: 10.1016/j.foodres.2017.06.007.

De Roos, J. & De Vuyst, L. (2018). Acetic acid bacteria in fermented foods and beverages. *Current Opinion in Biotechnology*, *49*, 115–119. Doi: 10.1016/j.copbio.2017.08.007.

De Vuyst, L. & Weckx, S. (2016). The cacao bean fermentation process: from ecosystem analysis to starter culture development. *Journal of Applied Microbiology 121*(1), 5–17. Doi: 10.1111/jam.13045.

De Vuyst, L. & Leroy, F. (2020). Functional role of yeasts, lactic acid bacteria and acetic acid bacteria in cacao fermentation processes. *FEMS Microbiology Reviews*, *44*(4), 432–453. Doi: 10.1093/femsre/fuaa014.

Delgado-Ospina, J., Molina-Hernández, J.B., Chaves-López, C., Romanazzi, G. & Paparella, A. (2021). The role of fungi in the cacao production chain and the challenge of climate change. *Journal of Fungi (Basel)*, *7*(3), 202. Doi: 10.3390/jof7030202.

Díaz-Muñoz, C., Van de Voorde, D., Comasio, A., Verce, M., Hernandez, C.E., Weckx, S. & De Vuyst, L. (2021). Curing of cacao beans: fine-scale monitoring of the starter cultures applied and metabolomics of the fermentation and drying steps. *Frontiers in Microbiology*, *11*, 616875. Doi: 10.3389/fmicb.2020.616875.

Díaz-Muñoz, C. & De Vuyst, L. (2022). Functional yeast starter cultures for cacao fermentation. *Journal of Applied Microbiology*, *133*(1), 39–66. Doi: 10.1111/jam.15312.

Díaz-Muñoz, C., Van de Voorde, D., Tuenter, E., Lemarcq, V., Van de Walle, D., Soares Maio, J.P., Mencia, A., Hernandez, C.E., Comasio, A., Sioriki, E., Weckx, S., Pieters, L., Dewettinck, K. & De Vuyst, L. (2023). An in-depth multiphasic analysis of the chocolate production chain, from bean to bar, demonstrates the superiority of *Saccharomyces cerevisiae* over *Hanseniaspora opuntiae* as functional starter culture during cacao fermentation. *Food Microbiology*, *109*, 104115. Doi: 10.1016/j.fm.2022.104115.

Dumarche, A., Troplin, P., Bernaert, H., Lechevalier, P., Beerens, H. & Landuyt, A. (2012). Process for producing cacao-derived material. Patent, European patent (EP2237677B1).

Dzelagha, B.F., Ngwa, N.M. & Nde Bup, D. (2020). A review of cacao drying technologies and the effect on bean quality parameters. *International Journal of Food Science*, Volume 2022, 8830127. Doi: 10.1155/2020/8830127.

Eskes, A., Ahnert, D., Garcia Carrion, L., Seguine, E., Assemat, S., Guarda, D. & Garcia, R.P. (2012). Evidence on the effect of the cacao pulp flavour environment during fermentation on the flavour profile of chocolates. Improving the profitability of small and medium-sized farms: the principal key to a global sustainable cacao economy. In *Proceedings of the 17th International cacao Research Conference (COPAL)*, Yaoundé, Cameroun.

Fanning, E., Eyres, G., Frew, R. & Kebede, B. (2023). Linking cacao quality attributes to its origin using geographical indications. *Food Control*, *151*, 109825. Doi: 10.1016/j.foodcont.2023.109825.

Farrera, L., de la Noue, A.C., Strub, C., Guibert, B., Kouame, C., Grabulos, J., Montet, D. & Teyssier, C. (2021). Towards a starter culture for cacao fermentation by the selection of acetic acid bacteria. *Fermentation*, *7*(1), 42. Doi: 10.3390/fermentation7010042.

Fernández-Niño, M., Juliana, M., Cubillos, R., Rocha, F.H., Anzola, J.M., Lucia, M., Hernández, C., Lorena, J., Mejía, A., Chica, M.J., Olarte, H.H., López, C.R., Calderón, D., Rojas, A.R., Portillo, P., Restrepo, S., Fernando, A. & Barrios, G. (2021). Dissecting industrial fermentations of fine flavour cacao through metagenomic analysis. *Scientific Reports*, *11*, 8638. Doi: 10.1038/s41598-021-88048-3.

de Ferreira, O. S., Chagas-Junior, G.C.A., Chisté, R.C., da Martins, L.H. S., de Andrade, E.H. A., Nascimento, L.D. & Lopes, A.S. (2022). *Saccharomyces cerevisiae* and *Pichia manshurica* from Amazonian biome affect the parameters of quality and aromatic profile of fermented and dried cacao beans. *Journal of Food Science*, *87*(9), 4148–4161. Doi: 10.1111/1750-3841.16282.

Figueroa-Hernández, C., Mota-Gutierrez, J., Ferrocino, I., Hernández-Estrada, Z.J., González-Ríos, O., Cocolin, L. & Suárez-Quiroz, M.L. (2019). The challenges and perspectives of the selection of starter cultures for fermented cacao beans. *International Journal of Food Microbiology*, *301*, 41–50. Doi: 10.1016/j.ijfoodmicro.2019.05.002.

Garcia Gonzalez, E., Mendez Orejuela, J.H., Sierra Banguera, J.S., Chamorro Moreno, D.L., Ordoñez Narváez, G.A., Ochoa Muñoz, A.F. & Montalvo Rodriguez, C. (2022). Ecology and population dynamics of yeast starter cultures in cocoa beans fermentation. *BioTechnologia*, *103*(4), 343–353. Doi: 10.5114/bta.2022.120704.

Garcia-Armisen, T., Papalexandratou, Z., Hendryckx, H., Camu, N., Vrancken, G., De Vuyst, L. & Cornelis, P. (2010). Diversity of the total bacterial community associated with Ghanaian and Brazilian cacao bean fermentation samples as revealed by a 16S rRNA gene clone library. *Applied Microbiology and Biotechnology*, *87*(6), 2281–2292. Doi: 10.1007/s00253-010-2698-9.

Ghisolfi, R., Bandini, F., Vaccari, F., Bellotti, G., Bortolini, C., Patrone, V., Puglisi, E. & Morelli, L. (2023). Bacterial and fungal communities are specifically modulated by the cacao bean fermentation method. *Foods*, *12*(10), 2024. Doi: 10.3390/foods12102024.

Grivetti, L.E. (2005). From aphrodisiac to health food: a cultural history of chocolate. *Karger Gazette*, *68*(1).

Grivetti, L.E. & Shapiro, H.Y. (2009). *Chocolate: History, Culture, and Heritage*. John Wiley and Sons, Inc., Hoboken.

Guehi, S.T., Dabonne, S., Ban-Koffi, L., Kedjebo, D.K. & Zahouli, G.I.B. (2010). Effect of turning beans and fermentation method on the acidity and physical quality of raw cacao beans. *Advance Journal of Food Science and Technology*, *2*(3), 163–171.

Gutiérrez-Ríos, H.G., Suárez-Quiroz, M.L., Hernández-Estrada, Z.J., Castellanos-Onorio, O.P., Alonso-Villegas, R., Rayas-Duarte, P., Cano-Sarmiento, C., Figueroa-Hernández, C.Y. & González-Rios, O. (2022). Yeasts as producers of flavor precursors during cacao bean fermentation and their relevance as starter cultures: a review. *Fermentation*, *8*(7), 331. Doi: 10.3390/fermentation8070331.

Hata, N.N.Y., Surek, M., Sartori, D., Serrato, R.V. & Spinosa, W.A. (2023). Role of acetic acid bacteria in food and beverages. *Food Technology and Biotechnology, 61*(1), 85–103. Doi: 10.17113/ftb.61.01.23.7811

Hernández-Hernández, C., López-Andrade, P.A., Ramírez-Guillermo, M.A., Guerra Ramírez, D. & Caballero Pérez, J.F. (2016). Evaluation of different fermentation processes for use by small cacao growers in Mexico. *Food Science and Nutrition, 4*(5), 690–695. Doi: 10.1002/fsn3.333.

Hirko, B., Mitiku, H. & Getu, A. (2023). Role of fermentation and microbes in cacao fermentation and their impact on cacao quality. *Systems Microbiology and Biomanufacturing. 3*, 509–520. Doi: 10.1007/s43393-023-00160-9.

Ho, V.T.T., Fleet, G.H. & Zhao, J. (2018). Unravelling the contribution of lactic acid bacteria and acetic acid bacteria to cacao fermentation using inoculated organisms. *International Journal of Food Microbiology, 279*, 43–56. Doi: 10.1016/j.ijfoodmicro.2018.04.040.

Illeghems, K., De Vuyst, L., Papalexandratou, Z. & Weckx, S. (2012). Phylogenetic analysis of a spontaneous cacao bean fermentation metagenome reveals new insights into its bacterial and fungal community diversity. *PLoS ONE, 7*, e38040. Doi: 10.1371/journal.pone.0038040.

Jaimez, R.E., Barragan, L., Fernández-Niño, M., Wessjohann, L.A., Cedeño-Garcia, G., Cantos, I.S. & Arteaga, F. (2022). *Theobroma cacao* L. cultivar CCN 51: a comprehensive review on origin, genetics, sensory properties, production dynamics, and physiological aspects. *PeerJ, 10*, e12676. Doi: 10.7717/peerj.12676.

Janda, J.M. & Abbott, S.L. (2021). The changing face of the family *Enterobacteriaceae* (Order: "Enterobacterales"): new members, taxonomic issues, geographic expansion, and new diseases and disease syndromes. *Clinical Microbiology Reviews, 34*(2), e00174–20. Doi: 10.1128/CMR.00174-20.

Koffi, A.S., Yao, N., Bastide, P., Bruneau, D. & Kadjo, D. (2017). Homogenization of cacao beans fermentation to upgrade quality using an original improved fermenter. *International Journal of Nutrition and Food Engineering, 11*(7), 32431. Doi: 10.5281/zenodo.1132573.

Koné, M.K., Guéhi, S.T., Durand, N., Ban-Koffi, L., Berthiot, L., Tachon, A.F., Brou, K., Boulanger, R. & Montet, D. (2016). Contribution of predominant yeasts to the occurrence of aroma compounds during cacao bean fermentation. *Food Research International, 89*(2), 910–917. Doi: 10.1016/j.foodres.2016.04.010.

Korcari, D., Fanton, A., Ricci, G., Rabitti, N.S., Laureati, M., Hogenboom, J., Pellegrino, L., Emide, D., Barbiroli, A. & Fortina, M.G. (2023). Fine cacao fermentation with selected lactic acid bacteria: fermentation performance and impact on chocolate composition and sensory properties. *Foods, 12*(2), 340. Doi: 10.3390/foods12020340.

Lagunes-Gálvez, S., Loiseau, G., Paredes, J.L., Barel, M. & Guiraud, J.P. (2007). Study on the microflora and biochemistry of cacao fermentation in the Dominican Republic. *International Journal of Food Microbiology, 114*(1), 124–130. Doi: 10.1016/j.ijfoodmicro.2006.10.041.

Lefeber, T., Gobert, W., Vrancken, G., Camu, N. & De Vuyst, L. (2011a). Dynamics and species diversity of communities of lactic acid bacteria and acetic acid bacteria during spontaneous cacao bean fermentation in vessels. *Food Microbiology, 28*(3), 457–464. Doi: 10.1016/j.fm.2010.10.010.

Lefeber, T., Janssens, M., Moens, F., Gobert, W. & De Vuyst, L. (2011b). Interesting starter culture strains for controlled cacao bean fermentation revealed by simulated cacao pulp fermentations of cacao-specific lactic acid bacteria. *Applied and Environmental Microbiology, 77*(18), 6694–6698. Doi: 10.1128/AEM.00594-11.

Lefeber, T., Papalexandratou, Z., Gobert, W., Camu, N. & De Vuyst, L. (2012). Onfarm implementation of a starter culture for improved cacao bean fermentation and its influence on the flavour of chocolates produced thereof. *Food Microbiology*, *30*(2), 379–392. Doi: 10.1016/j.fm.2011.12.021.

Lehrian, D.W. & Patterson, G.R. (1983). Cacao fermentation. In: G. Reed (Ed.), *Biotechnology, a Comprehensive Treatise, Volume 5* (pp. 529–575). Chemie, Basel, Switzerland.

Lemarcq, V., Monterde, V., Tuenter, E., Van de Walle, D., Pieters, L., Sioriki, L. & Dewettinck, K. (2022). Flavor diversification of dark chocolate produced through microwave roasting of cacao beans. *LWT*, *159*, 113198. Doi: 10.1016/j.lwt.2022.113198.

Lima, L.J.R., Almeida, M.H., Rob Nout, M.J. & Zwietering, M.H. (2011). *Theobroma cacao* L., "The Food of the Gods": quality determinants of commercial cacao beans, with particular reference to the impact of fermentation. *Critical Reviews in Food Science and Nutrition*, *51*(8), 731–761. Doi: 10.1080/10408391003799913.

Lopez, A. & Quesnel, V.C. (1971). An assessment of some claims relating to the production and composition of chocolate aroma. *Revue Internationale de la Chocolaterie*, *26*, 19–21.

Magagna, F., Guglielmetti, A., Liberto, E., Reichenbach, S.E., Allegrucci, E., Gobino, G., Bicchi, C. & Cordero, C. (2017). Comprehensive chemical fingerprinting of high-quality cacao at early stages of processing: effectiveness of combined untargeted and targeted approaches for classification and discrimination. *Journal of Agricultural and Food Chemistry*, *65*(30), 6329–6341. Doi: 10.1021/acs.jafc.7b02167.

Magalhães da Veiga Moreira, I., de Figueiredo Vilela, L., da Cruz Pedroso Miguel, M.G., Santos, C., Lima, N. & Freitas Schwan, R. (2017). Impact of a microbial cocktail used as a starter culture on cacao fermentation and chocolate flavor. *Molecules*, *22*(5), 766. Doi: 10.3390/molecules22050766.

Marwati, T., Palan Lamadoken, I.P., Djaafar, T.F., Utami, T. & Rahayu, E.S. (2018). Production of dried *Lactobacillus plantarum* HL-15 culture for inhibition growth of mycotoxin producing fungi. *Journal of Agricultural Science and Technology B*, *8*(6), 396–403. Doi: 10.17265/2161-6264/2018.06.007.

Meersman, E., Steensels, J., Struyf, N., Paulus, T., Saels, V., Mathawan, M., Allegaert, L., Vrancken, G. & Verstrepen, K.J. (2016). Tuning chocolate flavor through development of thermotolerant *Saccharomyces cerevisiae* starter cultures with increased acetate ester production. *Applied and Environmental Microbiology*, *82*(2), 732–746. Doi: 10.1128/AEM.02556-15.

Meersman, E., Struyf, N., Kyomugasho, C., Jamsazzadeh Kermani, Z., Santiago, J.S., Baert, E., Hemdane, S., Vrancken, G., Verstrepen, K.J., Courtin, C.M., Hendrickx, M. & Steensels, J. (2017). Characterization and degradation of pectic polysaccharides in cacao pulp. *Journal of Agricultural and Food Chemistry*, *65*(44), 9726–9734. Doi: 10.1021/acs.jafc.7b03854.

Menezes, A.G.T., Batista, N.N., Ramos, C.L., Silva, A.R.A., Efraim, P., Pinheiro, A.C.M. & Schwan, R.F. (2016). Investigation of chocolate produced from four different Brazilian varieties of cacao (*Theobroma cacao* L.) inoculated with *Saccharomyces cerevisiae*. *Food Research International*, *81*, 83–90. Doi: 10.1016/j.foodres.2015.12.036.

Miguel, M.G.D.C.P., Reis, L.V.D.C., Efraim, P., Santos, C., Lima, N. & Schwan, R.F. (2017). Cacao fermentation: microbial identification by MALDI-TOF MS, and sensory evaluation of produced chocolate. *LWT*, *77*, 362–369. Doi: 10.1016/j.lwt.2016.11.076.

Moreira, I., Costa, J., Vilela, L., Lima, N., Santos, C. & Schwan, R. (2021). Influence of *S. cerevisiae* and *P. kluyveri* as starters on chocolate flavour. *Journal Science Food Agriculture*, *101*(10), 4409–4419. Doi: 10.1002/jsfa.11082.

Moreno-Rojas, J.M., Erazo Solorzano, C.Y., Tuárez García, D.A., Pereira-Caro, G., Ordóñez Díaz, J.L., Muñoz-Redondo, J.M. & Rodríguez-Solana, R. (2023). Impact of the pre-drying process on the volatile profile of on-farm processed Ecuadorian bulk and fine-flavour cacao varieties. *Food Research International*, *169*, 112938. Doi: 10.1016/j.foodres.2023.112938.

Mota-Gutierrez, J., Botta, C., Ferrocino, I., Giordano, M., Bertolino, M., Dolci, P., Cannoni, M. & Cocolin, L. (2018). Dynamics and biodiversity of bacterial and yeast communities during fermentation of cacao beans. *Applied and Environmental Microbiology*, *84*(19), 1–17. Doi: 10.1128/AEM.01164-18.

Mota-Gutierrez, J., Ferrocino, I., Giordano, M., Suarez-Quiroz, M.L., Gonzalez-Ríos, O. & Cocolin, L. (2021). Influence of taxonomic and functional content of microbial communities on the quality of fermented cacao pulp-bean mass. *Applied and Environmental Microbiology*, *87*(14), e0042521. Doi: 10.1128/AEM.00425-21.

Motamayor, J.C., Lachenaud, P., da Silva e Mota, J.W., Loor, R., Kuhn, D.N., Brown, J.S. & Schnell, R.J. (2008). Geographic and genetic population differentiation of the Amazonian chocolate tree (*Theobroma cacao* L). *PLoS ONE*, *3*(10), 1–8. Doi: 10.1371/journal.pone.0003311.

Nielsen, D.S., Arneborg, N. & Jespersen, L. (2014). Mixed microbial fermentations and methods for their characterization. In: Schwan, R.F., & Fleet, G. (Eds.), *Cacao and Coffee Fermentations* (pp. 1–42). CRC Press, Boca Raton.

Nieves-Orduña, H.E., Krutovsky, K.V. & Gailing, O. (2023). Geographic distribution, conservation, and genomic resources of cacao *Theobroma cacao* L. *Crop Science*, *63*, 1750–1778. Doi: 10.1002/csc2.20959.

Ooi, T.S., Ting, A.S.Y. & Siow, L.F. (2022). Volatile organic compounds and sensory profile of dark chocolates made with cacao beans fermented with *Pichia kudriavzevii* and *Hanseniaspora thailandica*. *Journal of Food Science and Technology*, *59*(7), 2714–2723. Doi: 10.1007/s13197-021-05292-1.

Ouattara, H.G., Koffi, B.L., Karou, G.T., Sangare, A., Niamke, S.L. & Diopoh, J.K. (2008). Implication of *Bacillus* sp. in the production of pectinolytic enzymes during cacao fermentation. *World Journal of Microbiology and Biotechnology*, *24*, 1753–1760. Doi: 10.1007/s11274-008-9683-9.

Ouattara, H.G., Elias, R.J. & Dudley, E.G. (2020). Microbial synergy between *Pichia kudriavzevii* YS201 and *Bacillus subtilis* BS38 improves pulp degradation and aroma production in cacao pulp simulation medium. *Heliyon*, *6*(1), e03269. Doi: 10.1016/j.heliyon.2020.e03269.

Ozturk, G. & Young, G.M. (2017). Food evolution: the impact of society and science on the fermentation of cacao beans. *Comprehensive Reviews in Food Science and Food Safety*, *16*(3), 431–455. Doi: 10.1111/1541-4337.12264.

Pacheco-Montealegre, M.E., Dávila-Mora, L.L., Botero-Rute, L.M., Reyes, A. & Caro-Quintero, A. (2020). Fine resolution analysis of microbial communities provides insights into the variability of cacao bean fermentation. *Frontiers in Microbiology*, *11*, 650. Doi: 10.3389/fmicb.2020.00650.

Papalexandratou, Z., Camu, N., Falony, G. & De Vuyst, L. (2011a). Comparison of the bacterial species diversity of spontaneous cacao bean fermentations carried out at selected farms in Ivory Coast and Brazil. *Food Microbiology*, *28*(5), 964–973. Doi: 10.1016/j.fm.2011.01.010.

Papalexandratou, Z., Falony, G., Romanens, E., Jimenez, J.C., Amores, F., Daniel, H.M. & De Vuyst, L. (2011b). Species diversity, community dynamics, and metabolite kinetics of the microbiota associated with traditional Ecuadorian spontaneous cacao bean fermentations. *Applied and Environmental Microbiology*, *77*(21), 7698–7714. Doi: 10.1128/AEM.05523-11.

Papalexandratou, Z. & De Vuyst, L. (2011). Assessment of the yeast species composition of cacao bean fermentations in different cacao-producing regions using denaturing gradient gel electrophoresis. *FEMS Yeast Research*, *11*(7), 564–574. Doi: 10.1111/j.1567-1364.2011.00747.

Papalexandratou, Z., Vrancken, G., De Bruyne, K., Vandamme, P. & De Vuyst, L. (2011c). Spontaneous organic cacao bean box fermentations in Brazil are characterized by a restricted species diversity of lactic acid bacteria and acetic acid bacteria. *Food Microbiology*, *28*(7), 1326–1338. Doi: 10.1016/j.fm.2011.06.003.

Papalexandratou, Z., Lefeber, T., Bahrim, B., Lee, O.S., Daniel, H.M. & De Vuyst, L. (2013). *Hanseniaspora opuntiae, Saccharomyces cerevisiae, Lactobacillus fermentum,* and *Acetobacter pasteurianus* predominate during well-performed Malaysian cacao bean box fermentations, underlining the importance of these microbial species for a successful cacao bean fermentation process. *Food Microbiology*, *35*(2), 73–85. Doi: 10.1016/j.fm.2013.02.015.

Papalexandratou, Z. & Nielsen, D.S. (2016). It's gettin' hot in here: breeding robust yeast starter cultures for cacao fermentation. *Trends in Microbiology*, *24*(3), 168–170. Doi: 10.1016/j.tim.2016.01.003

Papalexandratou, Z., Kaasik, K., Kauffmann, L.V., Skorstengaard, A., Bouillon, G., Espensen, J.L., Hansen, L.H., Jakobsen, R.R., Blennow, A., Krych, L., Castro Mejía, J.L. & Nielsen, D.S. (2019). Linking cacao varietals and microbial diversity of Nicaraguan fine cacao bean fermentations and their impact on final cacao quality appreciation. *International Journal of Food Microbiology*, *304*, 106–118. Doi: 10.1016/j.ijfoodmicro.2019.05.012.

Pereira, G.V.M., Miguel, M.G.C.P., Ramos, C.L. & Schwan, R.F. (2012). Microbiological and physicochemical characterization of small-scale cacao fermentations and screening of yeast and bacterial strains to develop a defined starter culture. *Applied and Environmental Microbiology*, *78*(15), 5395–5405. Doi: 10.1128/AEM.01144-12.

Pereira, G.V.M., Magalhães, K.T., de Almeida, E.G., da Silva Coelho, I. & Schwan, R.F. (2013). Spontaneous cacao bean fermentation carried out in a novel-design stainless steel tank: influence on the dynamics of microbial populations and physical-chemical properties. *International Journal of Food Microbiology*, *161*(2), 121–133. Doi: 10.1016/j.ijfoodmicro.2012.11.018.

Pereira, G.V.M., Soccol, V.T. & Soccol, C.R. (2016). Current state of research on cacao and coffee fermentations. *Current Opinion in Food Science*, *7*, 50–57. Doi: 10.1016/j.cofs.2015.11.001.

Pereira, G.V.M., Alvarez, J.P., de Carvalho Neto, D.P., Soccol, V.T., Tanobe, V.O.A., Rogez, H., Góes-Neto, A. & Soccol, C.R. (2017). Great intraspecies diversity of *Pichia kudriavzevii* in cacao fermentation highlights the importance of yeast strain selection for flavor modulation of cacao beans. *LWT*, *84*, 290–297. Doi: 10.1016/j.lwt.2017.05.073.

Pereira, G.V.M., de Carvalho Neto, D.P., Junqueira, A.C.D.O., Karp, S.G., Letti, L.A.J., Magalhães Júnior, A.I. & Soccol, C.R. (2020a). A review of selection criteria for starter culture development in the food fermentation industry. *Food Reviews International*, *36*(2), 135–167. Doi: 10.1080/87559129.2019.1630636.

Pereira, A.P.M., Stellaria, H.A., Leonardo, F., Vilela, L.F., Schwan, R.F. & Sant'Ana, A.S. (2020b). Dynamics of *Geobacillus stearothermophilus* and *Bacillus cereus* spores inoculated in different time intervals during simulated cacao beans fermentation. *LWT*, *120*, 108941. Doi: 10.1016/j.lwt.2019.108941.

Presilla, M. (2009). *The New Taste of Chocolate: A Cultural and Natural History with Recipes*. Ten Speed Press, Berkeley.

Ramón, V., Hernández, H.E., Polania, P. & Suárez, J.C. (2023). Spatial distribution of cacao quality: relationship between physicochemical, functional and sensory attributes of clones from Southern Colombia. *Agronomy*, *13*(1), 15. Doi: 10.3390/agronomy13010015.

Rodríguez-Campos, J., Escalona-Buendía, H.B., Orozco-Avila, I., Lugo-Cervantes, E. & Jaramillo-Flores, M.E. (2011). Dynamics of volatile and non-volatile compounds in cacao (*Theobroma cacao* L.) during fermentation and drying processes using principal components analysis. *Food Research International*, *44*(1), 250–258. Doi: 10.1016/j.foodres.2010.10.028.

Rodríguez-Campos, J., Escalona-Buendía, H.B., Contreras-Ramos, S.M., Orozco-Avila, I., Jaramillo-Flores, E. & Lugo-Cervantes, E. (2012). Effect of fermentation time and drying temperature on volatile compounds in cacao. *Food Chemistry*, *132*(1), 277–288. Doi: 10.1016/j.foodchem.2011.10.078.

Romanens, E., Freimüller Leischtfeld, S., Volland, A., Stevens, M.J.A., Krähenmann, U., Isele, D., Fischer, B., Meile, L. & Miescher Schwenninger, S. (2019). Screening of lactic acid bacteria and yeast strains to select adapted anti-fungal cocultures for cacao bean fermentation. *International Journal of Food Microbiology*, *290*, 262–272. Doi: 10.1016/j.ijfoodmicro.2018.10.001.

Romanens, E., Pedan, V., Meile, L., & Miescher Schwenninger, S. (2020). Influence of two anti-fungal *Lactobacillus fermentum-Saccharomyces cerevisiae* cocultures on cacao bean fermentation and final bean quality. *PloS ONE*, *15*, e0239365. Doi: 10.1371/journal.pone.0239365.

Samagaci, L., Ouattara, H.G., Goualié, B.G. & Niamke, S.L. (2014). Polyphasic analysis of pectinolytic and stress-resistant yeast strains isolated from Ivorian cacao fermentation. *Journal of Food Research* *4*(1), 124–134. Doi: 10.5539/jfr.v4n1p124.

Sari, A.B.T., Fahrurrozi, Marwati, T., Djaafar, T.F., Hatmi, R.U., Purwaningsih, Wanita, Y.P., Lisdiyanti, P., Perwitasari, U., Juanssilfero, A.B. & Rahayu, E.S. (2023a). Chemical composition and sensory profiles of fermented cacao beans obtained from various regions of Indonesia. *International Journal of Food Science*, 5639081, 1–13. Doi: 10.1155/2023/5639081.

Sari, A.B.T., Belgis, M., Amilia, W. & Murtadlo, F.M. (2023b). The sensory profile of unfermented cacao beans and its changes after moisture treatments. *AIP Conference Proceedings*, 2583. Doi: 10.1063/5.0117087.

Saerens, S. & Swiegers, J.H. (2013). Enhancement of cacao quality and flavor by using *Pichia kluyveri* yeast starter culture for cacao fermentation. Patent, W.I.P. (PCT/EP2012/071787).

Santos, D.S., Passos Rezende, R., Ferreira dos Santos, T., de Lima Silva Marques, E., Reis Ferreira, A.C., Barros de Cerqueira e Silva, A., Romano, C.A., da Cruz Santos, D.W., Teixeira Dias, J.C. & Dias Tavares Bisneto, J. (2020). Fermentation in fine cacao type Scavina: change in standard quality as the effect of use of starters yeast in fermentation. *Food Chemistry*, *328*, 127110. Doi: 10.1016/j.foodchem.2020.127110.

Schnell, R.J., Seguine, E.S., Dias, C., Bizotto, C., Marelli, J.-P., Mills, D. & Motamayor, J.C. (2014). Methods of processing unfermented fruit seeds such as cacao beans or cupuacu beans. Patent, W.I.P. (W O 2014/130539 Al).

Schwan, R.F., Vanetti, M.C.D., Silva, D.O., Lopez, A. & de Moraes, C.A. (1986), Characterization and distribution of aerobic, spore-forming bacteria from cacao fermentations in Bahia. *Journal of Food Science*, *51*(6), 1583–1584. Doi: 10.1111/j.1365-2621.1986.tb13872.x.

Schwan, R.F. (1998). Cacao fermentations conducted with a defined microbial cocktail inoculum. *Applied and Environmental Microbiology*, *64*(4), 1477–1483. Doi: 10.1128/AEM.64.4.1477-1483.1998.

Schwan, R.F. & Wheals, A.E. (2004). The microbiology of cacao fermentation and its role in chocolate quality. *Critical Reviews in Food Science and Nutrition*, *44*(4), 205–221. Doi: 10.1080/10408690490464104.

Schwan, R.F., Bressani, A.P.P., Martinez, S.J., Batista, N.N. & Dias, D.R. (2023). The essential role of spontaneous and starter yeasts in cacao and coffee fermentation. *FEMS Yeast Research*, *23*, foad019. Doi: 10.1093/femsyr/foad019.

Serra, J.L., Moura, F.G., Pereira, G.V.M., Soccol, C.R., Rogez, H. & Darnet, S. (2019). Determination of the microbial community in Amazonian cacao bean fermentation by Illumina-based metagenomic sequencing. *LWT*, *106*, 229–239. Doi: 10.1016/j.lwt.2019.02.038.

Streule, S., Freimüller Leischtfeld, S., Galler, M. & Miescher Schwenninger, S. (2022). Monitoring of cacao post-harvest process practices on a small-farm level at five locations in Ecuador. *Heliyon*, *8*(6), e09628. Doi: 10.1016/j.heliyon.2022.e09628.

Tamang, J.P., Shin, D.H., Jung, S.J. & Chae, S.W. (2016). Functional properties of microorganisms in fermented foods. *Frontiers in Microbiology*, *7*, 578. Doi: 10.3389/fmicb.2016.00578.

Tamimi, K.A., Hidayat, C., Utami, T. & Witasari, L.D. (2023). Flavor precursor formation of non-fermented forastero cacao beans after flavourzyme® and glucose treatment. *LWT*, *184*, 114910. Doi: 10.1016/j.lwt.2023.114910.

Tuenter, E., Sakavitsi, M.E., Rivera-Mondragón, A., Hermans, N., Foubert, K., Halabalaki, M. & Pieters, L. (2021). Ruby chocolate: a study of its phytochemical composition and quantitative comparison with dark, milk and white chocolate. *Food Chemistry*, *343*, 128446. Doi: 10.1016/j.foodchem.2020.128446.

Thompson, S.S., Miller, K.B. & Lopez, A.S. (2007). cacao and coffee. In: M.P. Doyle, L.R. Beuchat, & T.J. Montville (Eds.), *Food Microbiology: Fundamentals and Frontiers* (2nd Edition, pp. 837–849). ASM Press, Washington, DC.

Utrilla-Vázquez, M., Rodríguez-Campos, J., Avendaño-Arazate, C.H., Gschaedler, A. & Lugo-Cervantes, E. (2020). Analysis of volatile compounds of five varieties of Maya cacao during fermentation and drying processes by Venn diagram and PCA. *Food Research International*, *129*, 108834. Doi: 10.1016/j.foodres.2019.108834.

Van de Voorde, D., Díaz-Muñoz, C., Hernandez, C.E., Weckx, S. & De Vuyst, L. (2023). Yeast strains do have an impact on the production of cured cacao beans, as assessed with Costa Rican Trinitario cacao fermentation processes and chocolates thereof. *Frontiers in Microbiology*, 14, 1232323.

Velásquez-Reyes, D., Gschaedler, A., Kirchmayr, M., Avendaño-Arazate, C.H., Rodríguez-Campos, J., de Calva-Estrada, S. J. & Lugo-Cervantes, E. (2021). cacao bean turning as a method for redirecting the aroma compound profile in artisanal cacao fermentation. *Heliyon*, *7*(8), e07694. Doi: 10.1016/j.heliyon.2021.e07694.

Verce, M., Schoonejans, J., Hernandez Aguirre, C., Molina-Bravo, R., De Vuyst, L. & Weckx, S. (2021). A combined metagenomics and metatranscriptomics approach to unravel Costa Rican cacao box fermentation processes reveals yet unreported microbial species and functionalities. *Frontiers in Microbiology*, *12*, 1–24. Doi: 10.3389/fmicb.2021.641185.

Viesser, J.A., de Melo Pereira, G.V., de Carvalho Neto, D.P., de Vandenberghe, L.P. S., Azevedo, V., Brenig, B., Rogez, H., Góes-Neto, A. & Soccol, C.R. (2020). Exploring the contribution of fructophilic lactic acid bacteria to cacao beans fermentation: isolation, selection and evaluation. *Food Research International*, *136*, 109478.

Viesser, J.A., Pereira, G.V.M., de Carvalho Neto, D.P., Favero, G.R., de Carvalho, J.C., Goés-Neto, A., Rogez, H. & Soccol, C.R. (2021). Global cacao fermentation microbiome: revealing new taxa and microbial functions by next generation sequencing technologies. *World Journal of Microbiology and Biotechnology*, *37*(7), 118. Doi: 10.1007/s11274-021-03079-2.

Wirth, R. & de Bortoli, V. (2023). Fermentation chamber and method for fermenting cacao fruit. Patent, W.I.P. (EP3782472A1).

Visintin, S., Ramos, L., Batista, N., Dolci, P., Schwan, F. & Cocolin, L. (2017). Impact of *Saccharomyces cerevisiae* and *Torulaspora delbrueckii* starter cultures on cacao beans fermentation. *International Journal of Food Microbiology*, *257*, 31–40. Doi: 10.1016/j.ijfoodmicro.2017.06.004.

Wollgast, J. & Anklam, E. (2000). Review of polyphenols in *Theobroma cacao*: changes in composition during the manufacture of chocolate and methodology for identification and quantification. *Food Research International*, *33*(6), 423–447. Doi: 10.1016/S0963-9969(00)00068-5.

Wood, G.A.R. & Lass, R.A. (2001). *Cacao* (4th edition). Blackwell Science, Oxford.

Zheng, J., Wittouck, S., Salvetti, E., Franz, C.M.A.P., Harris, H.M.B., Mattarelli, P., O'Toole, P.W., Pot, B., Vandamme, P., Walter, J., Watanabe, K., Wuyts, S., Felis, G.E., Gänzle, M.G. & Lebeer, S. (2020). A taxonomic note on the genus *Lactobacillus*: description of 23 novel genera, emended description of the genus *Lactobacillus* Beijerinck 1901, and union of *Lactobacillaceae* and *Leuconostocaceae*. *International Journal of Systematic and Evolutionary Microbiology*, *70*(4), 2782–2858. Doi: 10.1099/ijsem.0.004107.

11 Novel approaches in processing of cacao byproducts and their applications for a sustainable future

R. Suganthi, R. Saravanan, A. Brundha, T. Bowya, A. Lakshmi, P. Abirami, and M. Ashwini

11.1 Introduction

Residual biomass from agriculture and allied sectors resulted in an upsurge of large volumes of waste that can disturb the resilience of the ecosystem. To reduce the quantum of waste generated, various alternative technologies were adopted to transform the underutilized waste into valuable bio-products. Zero waste management is the main objective in the cacao industry of Latin American countries to reduce, recycle, and reuse this waste through various technologies (Hettiarachchi *et al.*, 2018; Mariatti et al., 2021). Cultivation of cacao has been tremendously increasing over the past decades in developing countries. Cacao production was around 4.9 million tons in 2021–2022 as per the International Cacao Organization (ICCO n.d.). Cote d'Ivoire and Ghana were reported to be the largest producing countries, representing 60% of the global production followed by Ecuador with 7%. Farmers are discarding 70–80 % of the cacao biomass as waste without any proper disposal method leading to environmental degradation (Shahbendaz, 2023). The underutilized biomass discarded as waste includescacao pod husk (CPH), cacao bean shells, sweating, etc. These byproducts were found to be rich in lignocelluloses, hemicelluloses, sugars, etc. (Mendoza-Meneses et al., 2021). Nowadays, researchers focus on the valorization of these wastes besides the extraction of high-valued biomolecules in different spectra of sectors such as fuel, medicine, cosmetics, and food due to their physicochemical and structural properties as illustrated in Figure 11.1 (de Souza Vandenberghe *et al.*, 2022; Orejuela-Escobar *et al.*, 2021).

Various chemical and biotechnological approaches were tested for the pretreatment process to extract different biomolecules to increase their market potential. This chapter will concentrate on zero waste management for the cacao processing industries as well as the farming communities. Value-added products from cacao will empower the livelihood of farmers and reduce the input of synthetic chemicals and overexploitation of natural resources (Vasquez *et al.*, 2019).

DOI: 10.4324/9781003381761-15

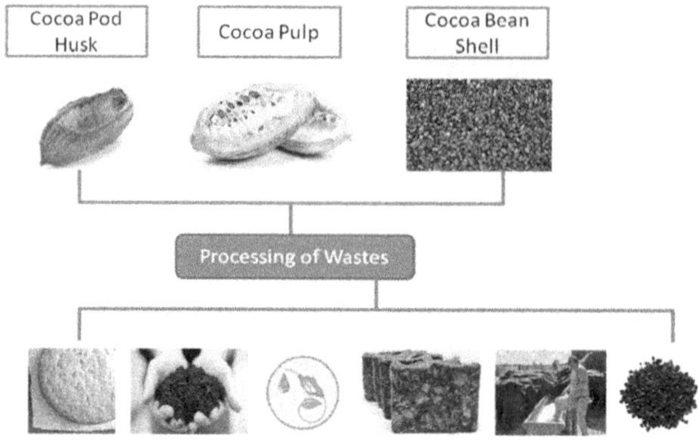

Figure 11.1 Application of Cacao wastes.

11.2 *Theobroma cacao*: 'food of the gods'

The cacao tree is used extensively in the production of chocolate. It is an umbrophilic crop that comes under the Order Malvales, Family Malvaceae, and Genus*Theobroma*. It is native to Central and South America in contrast to their higher production in tropical areas in Asia, Africa, and South America. It is derived from Greek words'Theos' and 'Bromos' meaning 'the food of the gods'. As per the data obtained from ICCO (2018–2019), Cote d'Ivoire was reported to be the largest producer with 2,154 tons, preceded by Ghana and Ecuador. Cacao byproducts contain lignin, cellulose, hemicelluloses, terpenoids, carboxylic acids, amino acids, etc. (Soares &Oliveira, 2022), in addition to basic elemental composition, which are listed in Tables 11.1 and Table 11.2, respectively.

11.2.1 Cacao byproducts

11.2.1.1 Cacao pod husk

CPH is the outer part of the fruit and is the main byproduct generated from cacao processing industries, which constitutes 80% of the dry weight of the fruit. Husk is composed of three different layers,that is, endocarp, mesocarp, and epicarp (Campos-vega *et al.*, 2018). For each ton of cacao bean production, it was reported that 10 tons of waste were generated. It is commonly used as a fertilizer, but its usage is restricted due to pests and fungal infestations. The chemical and structural composition of pod husk makes it applicable in various sectors (Donkoh *et al.*,1991, Meza-sepulveda et al., 2021, Murthy & Naidu, 2012). CPH is found to be rich in lignin, nonstarch polysaccharides such as cellulose, hemicelluloses, and pectin, terpenoids such as chrysophanol, phenolic, and carboxylic acids such as protocatechuic, salicylic, citric, and tartaric acid along with amino acids such as glutamine, asparagine, serine, and lysine,

Table 11.1 Nutritive composition of various cacao byproducts (g/100 g dw)

Compounds (g/100 g dw)	*Cacao Husk*	*Cacao Pulp*	*Cacao Bean Shell*	*Reference*
Carbohydrates	7.5–57.6	10.7–70.1	7.8–70.2	Rojo-poveda *et al.*, 2020 Vasquez *et al.*, 2019 Pinzon-Nuñez *et al.*, 2022
Cellulose	19.7–35.00	20.80–57.50	15.10	Rojo-Poveda *et al.*, 2020 Colovic *et al.*, 2019 Oussou et al., 2023
Hemicellulose	8.72–12.80	7.00–28.5	–	Rousseaux *et al.*, 2023
Lignin	14.60–28.00	12.00–15.00	14–32.41	Cornejal *et al.*, 2023
Pectin	6.10–12.6	0.57–1.33	0.57–15.59	Boateng *et al.*, 2019
Total dietary fiber	24.00–56.10	16.75–16.89	17.40–66.33	Rojo-Poveda et al., 2020 Botella-Martínez et al., 2021 Agus *et al.*, 2018
Total protein	4.21–10.74	0.74–5.56	10.30–27.4	Rojo-Poveda et al., 2020
Lipids	1.50–2.34	0.45–3.54	1.50–8.49	Botella-Martínez et al., 2021
Ash	6.4–10.02	3.76–7.68	5.96–11.40	Nsor-Atindana et al., 2012
Total organic acids	–	17.52	–	Soares *et al.*, 2022
Total phenolics	4.6–6.9	–	1.32–94.95	Mellinas *et al.*, 2020
Anthocyanin	–	–	0.40	Barbosa-Pereira et al., 2018 Lessa et al., 2018
Tannin	5.20	–	3.30–25.30	Joseph *et al.*, 2020
Flavonols	11.88	–	1.50–40.72	Botella-Martínez *et al.*, 2021

Table 11.2 Elemental composition of cacao byproducts

Parameters	*Value*	*Reference*
Proximate analysis (wt%)		
Moisture	3.2–10.2	Bello *et al.*, 2012
Volatile matter	61.17–75.3	Tsai *et al.*, 2018
Fixed carbon	19.78	Syamsiro *et al.*, 2011
Ash	2.2–10.92	Jaimes *et al.*, 2017 Castillo *et al.*, 2018
Ultimate analysis (wt%)		
C	44.3	Bello *et al.*, 2012
H	6.5	Tsai *et al.*, 2018
N	0.1–1.67	Syamsiro *et al.*, 2011
O	45.98	Jaimes *et al.*, 2017
S	0.10	Castillo *et al.*, 2018
Others	49.1	

fibers, and minerals such as selenium, calcium, magnesium, sodium, potassium, zinc, iron, copper, and manganese (Soares *et al.*, 2022).

11.2.1.2 Cacao bean shell

Cacao bean shell is obtained from the bean roasting or pretreatment process. Lignocellulosic content was found to be higher in the bean shell occupying 10–17% of the total weight and dietary fiber ranging from 18 to 60% (Rojo-Poveda et al., 2020). During fermentation, the conversion of waste into biomolecules occurs which found its way towardthe generation of new nutritional products as well as in food, pharmaceutical, cosmetic, and agricultural industries (Lessa et al., 2018). Their application is extended to the environment due to their adsorption capacity, high porosity, and low ash content, and also in the production of carbon monoliths and low-cost carbon adsorbents for the removal of various pollutants (Rojo-Poveda et al., 2020). Furthermore, bean shells contain polysaccharides, phenolic compounds, methylxanthines, catechin, epicatechin, procyanidin, caffeine, theophylline, theobromine, and fiber. Its calorific value was reported to be higher than fuel wood (Grillo et al., 2019).

11.2.1.3 Cacao mucilage

Cacao pulp or sweating is known as mucilage which is the white mass surrounding the cacao beans. The quantity of mucilages depends on various attributes such as the type of cacao, maturation, and physical integrity of the fruits (Soares *et al.*, 2022). This white mass produces a white cloudy liquid due to the microbial activity or pretreatment process which is commonly referred to as cacao pulp juice and it is widely used in the preparation of soft drinks, alcohol, vinegar, citric acid, and cacao jelly (Figueroa et al., 2020, Nunes *et al.,* 2020); it can also be used as a culture medium for industrial microbial culture (Awuah & Frimpong, 2002) or processes due to the presence of fermentable sugars, pectin, pentosans, citric acid, salts, cellulose, hemicelluloses, lignin, and essential minerals without any toxic alkaloids or any other toxic substances (Bastidas et al., 2022).

11.3 Technological intervention in biomass conversion

With the discovery of bioactive compounds and structural properties of cacao waste, the application of processed cacao biomass has been rapidly escalating and got the attention of the scientific community as well as the farmers. Industrial waste processing and its treatment along with byproduct conversion will improve the economy as well as safeguard the environment. For example, biofuel production from biomass will reduce overexploitation of fossil fuels from the earth (Belwal *et al.*, 2022; Mandari & Devarai, 2022).

11.3.1 Physicochemical conversion

It is a mechanical extraction process which is used to generate biofuel which can be subjected to either esterification or transesterification. The main objective of this process will result in the formation of glycerol (Mandari & Devarai, 2022).

11.3.2 Biological conversion

It involves microbes and their associated mechanisms in the decomposition or transformation of the waste materials which will result in bioethanol and biogas production. *Saccharomyces cerevisiae* is commonly used. Wastes are subjected to fermentation and anaerobic digestion to produce biofuel (Igbinadolor & Onilude, 2013). Waste materials rich in organic matter on anaerobic fermentation result in the production of biogas along with methane, carbon dioxide, and traces of hydrogen sulfide *via* hydrolysis, acidogenesis, acetogenesis, and methanogenesis. In the case of fermentation, sugars will transform into ethanol with carbon dioxide (Deena *et al.*,2022). In certain cases, if the waste materials are rich in lignocellulosic compounds, pretreatment is advised before the conversion process (Jönsson & Martín, 2016).

11.3.3 Thermochemical conversion

It involves the breakdown of chemical bonds binding the waste materials resulting in the production of fuels. Different types of thermochemical processes include pyrolysis, combustion, gasification, and liquefaction in which pyrolysis forms the base. The reaction time and the transformation rate were higher in thermochemical processes rather than in biological processes. Pyrolysis is defined as the conversion of biomass into solid, liquid, and gaseous products which is carried out at higher temperatures in the absence of oxygen. The final products after pyrolysis are hydrogen, methane, carbon monoxide, and carbon dioxide in the form of gas, tar, and oil as liquid along with charcoal as solid products (Shahbaz *et al.*, 2020). Even though the production of biochar or charcoal was the focus of the pyrolysis process, now it is diverted into the production of the biofuel sector. Based on the temperature, pyrolysis is classified into various types which include carbonization between 300 and 500°C; slow pyrolysis around 500°C, fast pyrolysis between 500 and 650°C, and flash pyrolysis at temperatures above 650°C (Tursi *et al.*, 2019).

Liquefaction is the process of extraction of biofuel in a liquid medium under high pressure and temperatures ranging from 280–370°C. It is also known as hydrothermal liquefaction and is best suited for algal biomass which neglects the drying of biomass as in pyrolysis. Bio-oil is obtained along with gaseous products (Mendoza-Meneses *et al.*, 2021). Gasification is the process of conversion of biomass into gaseous products with the help of a gasifier. When the waste materials are subjected to gasification above 1200°C, it will result in synthetic gas which consists of hydrogen and carbon monoxide, and under low-temperature results in methane and carbon dioxide. The calorific value of the biomass largely depends upon the raw material used. Gasification is identified as a suitable and reliable method to produce hydrogen gas (Gunasekaran *et a*l., 2021). Combustion refers to the process of conversion of thermal energy to a usable form of energy, that is,thermal energy. It involves the burning of biomass which releases energy in the form of heat. Elements present in the biomass such as hydrogen get converted into water, carbon into carbon dioxide, sulfur into sulfur dioxide, and nitrogen into nitrogen dioxide (Syamsiro *et al.*, 2011).

11.4 Different applications of cacao byproducts

Based on their physicochemical and structural compositions, they are employed in various sectors which are listed in Table 11.3.

11.4.1 Food

The cacao biomass wastes are intensively extracted and processed for nonfood purposes such as the generation of biofuel, biofertilizer, enzymes, briquettes, cosmetics, etc. Exploitation of cacao wastes has been tested in manufacturing various food products which are listed in Table 11.4 (Netania *et al.*, 2022). Cacao bean shell residue is rich in proteins, carbohydrates with low-fat

Table 11.3 Cacao waste products and their utilization

S. No.	*Byproduct*	*Purpose*	*Reference*
1	Cacao pod	Biofuel production, fertilizers	Rodriguez-Garcia *et al.*, 2018
2	Mucilage	Alcohol, vinegar, soft drinks, citric acid, cacao jelly, jam, and pectin extraction	Oddoye *et al.*, 2013
3	Pod ash	Black soap	Ajongbolo, 2020, Ikotun *et al.*, 2018
4	Cacao bean meal, bean shell, and pod husk	Livestock feed	Makinde *et al.*, 2019
5	Pod husk flour	Starch replacement, wheat flour	Delgado-Ospina et al., 2021, Amir *et al.*, 2013
6	Cacao shell	Baking and confectionery products, ethanol, feed, and corn-extruded snacks	Hernandez-mendoza *et al.*, 2021, Díaz-González *et al.*, 2022, Netania *et al.*, 2022
7	Cacao nibs	Flavor enhancer, candy, and activated carbon	Jeong et al., 2022 Delgado et al., 2018 Nikman, 2018

Table 11.4 Alleviation of cacao wastes

S. No.	*Wastes*	*Product*
1	Pulp	Wine, alcohol, syrup, jam
2	Shell	Gluten-free flour, bread mixes, confectionary, and beverages
3	Cacao water	Alternative to plain water
4	Shells and nibs	Caffeine-free alternative for coffee and tea

content, alkaloids such as theobromine, methylxanthine, fiber, phenolic compounds, anthocyanins, and flavonols (Martinez *et al.,* 2012; Okiyama *et al.*, 2017; Lessa *et al.*, 2018; Souza *et al.*, 2022). It was attempted as an intermediate ingredient for functional foods in cereal bars and other bakery products such as biscuits, bread, and cakes (Souza *et al.*, 2022; Martinez *et al.*, 2012; de Barros *et al.*, 2021; de Barros *et al.*, 2020; Handojo *et al.*, 2019). Powdered cacao bean shell is used as flour instead of wheat flour for cake (Souza *et al.*, 2022).

The pulp and beans of a cacao fruit are extracted when it is broken. The mucilaginous pulp is then subjected to the fermentation process. The fermentation process proceeds as a result of the pectinolytic enzymes degrading the pectin content while stimulating the growth and proliferation of acetic bacteria. Cacao honey is a sweet and sour liquid made from pulp liquefaction that shares the same chemical and sensory properties as the original pulp. It has a high carbohydrate and total sugar content (11–19.5 %), a low fat (<3.5%) and ash content (0.2 %), less than 1% of dietary fiber, and a TSS (Total Soluble Sugars) of roughly 4.05 brix. Magnesium and calcium are present in significant amounts. The three main types of carbohydrates in cacao honey are sucrose, fructose, and glucose. The amount of vitamin C is between 8 and 11% (Guirlanda *et al.*, 2021, Anvoh *et al.*, 2009).

Many organic volatile substances were observed in cacao bean shells. It goes into ovens and bread and confectionery like cookies after being roasted. It may substitute vegetable oil in cakes, chocolate muffins, and other baked goods and also has some antioxidant characteristics. Due to its antioxidant properties, it can be added to some foods to avoid lipid oxidation. The presence of polyphenolic content makes the addition of soybean oil more stable, resulting in a low peroxide rate and free fatty acid content. Because it has an alkalizing tendency, cacao beans are utilized in the making of chocolate. Due to their antioxidant capabilities and assistance with active packaging, they were also employed to extend the shelf life of food by integrating them into bio-elastomers (Rojo-Poveda et al., 2020; Soares *et al.*, 2022).

The husk from cacao beans, which is used to make flour and other baking ingredients, is high in iron, calcium, fiber, and vitamins A, E, and C. High levels of antioxidants, electrolytes, and soluble dietary fiber may be found in cacao water. Advanced hydration and health advantages will be available.

Tea made from cacao shells is both delicious and energizing since it has a faint flavor and scent of roasted cacao (Rojo-Poveda et al., 2019; Cantele et al., 2020; Delgado-Ospina et al., 2021; San Siow et al., 2022). Starch degradation somewhat decreased but resistant starch, polyphenol content, and antioxidant activity of the grains rose correspondingly with the addition of cacao husks, demonstrating that cacao husks may be successfully used as a tool for the fortification of nutrients (Jozinović et al., 2019).

The inclusion of phenolic compounds enhances the stability of lipid oxidation in lipid-based and lipid-rich products. According to research on cacao shells' used in cooked beef by Ismail and Yee (2006), adding cacao shells lowers the thiobarbituric acid numbers (TBA numbers), which signifies a slower rate of lipid oxidation. Using ethanol-extracted cacao shell extract and ground beef, the experiment was carried out. After this, the meat was finished in a water bath and kept at 4°C for 14 days. The results showed that natural phenolic extract produced from cacao shells provides superior lipid oxidation stability than synthetic antioxidants when compared to cooked beef treated with butylated hydroxytoluene (BHT), tocopherol, and a control group (Manzano et al., 2017).

11.4.2 Livestock feed

To lessen the conflict between food and feed, using agroindustrial byproducts in animal nutrition is a viable technique. Animals are fed using the husk from cacao pods. After being dried, cut, and pelted, the pod husk can be utilized as feed. It is safe to use as animal feed because it contains concentrated or polymerized flavonoids (anthocyanidins, catechins, etc.), tannins, and minuscule amounts of theobromine (Laconi & Jayanegara, 2015; Oduro-Mensah et al., 2020). Adding alkalis to fibrous feed material increases the feeding value. In this technique, materials such as cacao bean meal, bean shells, and pod husks can be utilized to make animal feed (de Souza Vandenberghe et al., 2022). Cacao shells are a rich source of energy, nutrients, and minerals and have a fiber level that is comparable to that of grass. When it is utilized as roughage, it aids in the intake of 5–6% dry matter (Rojo-Poveda et al., 2019). Crude protein, fiber, fat, calcium, sodium, phosphorous, and magnesium enhance their application as animal feed (Marcel et al., 2011;Makinde et al., 2019). Due to the high fiber content and the components of its cell walls, cacao bone meal is ideally suited for ruminants. When a cacao pod has a high protein and low fiber content, it can also be used as animal feed. After treating with hot water, it is then utilized as a valuable source of animal feed after being detheobrominated by alkali, urea, fungi, and microorganisms.

Due to the presence of theobromine, intake of cacao byproducts is limited in livestock feeding because of its toxicity such as drastic weight reduction in broilers and decrease in daily feed intake and egg production. The constraints are managed by the pretreatment of wastes which will reduce their toxicity to some extent (Balentic *et al.*, 2018). Besides cacao bean, shell together with other agri residues such as *Leucaena* leaf meal, yam peel can be used as cattle feed.

Thus, cacao byproducts can be a viable source of dietary fiber and protein for various organisms such as rabbits, goats, fishes, pigs etc. It was reported that the addition of cacao bean shells to goat feed decreased their consumption of all dry matter. The intervention had no impact on milk production, milk fat, protein, casein contents, or yields (Renna et al., 2022). Theobromine in the cacao pod was removed by using an isolate of *Talaromyces verruculosus* and its safety was tested against SpragueDawley rats. In the treatment without the removal of theobromine, rats exhibited a significant reduction in body weight, thymus weight, and tissue histology whereas in the case of detheobromibated pods, an increase in body weight was observed (Oduro-Mensah et al., 2020).

In a 12-week feeding study, the performance of weaner pigs given increasing quantities of cacao bean shell meal in place of maize (0, 10, 20, 30, and 40%) was examined. Feed intake was unaffected. However, when cacao shell meal made up more than 20% of the diet, live weight growth dropped. When cacao shell meal made up more than 10% of the diet, feed conversion declined (Ogunsipe et al., 2017). The impact of biologically enhanced CPH meal (BCPHM) on the growth performance, hematobiochemical indices, and antioxidant status of broiler chickens was investigated over the course of 49-day feeding studies. At the starting phase, broiler chicks fed meals containing 10% BCPHM gained substantially more weight than controls and diets containing 20% BCPHM (p0.05) in terms of average daily weight gain (ADWG), total weight gain (TWG), and final body weight (FBW). The values of ADWG, TWG, and FBW were comparable to control during the finisher phase. As BCPHM in the diet rose during the finisher phase, feed intake was seen to rise. Broiler chicken carcass characteristics, relative internal organ weight, and hematobiochemical indicators were comparable among dietary treatments (Olugosi *et al.*, 2020). Thus, cacao byproducts after pretreatment processes were encouraged to utilize as animal feed but the concentration must be considered for their toxic effect on the organisms.

11.4.3 In agriculture

If not managed, these wastes might cause environmental issues, but if they are added to growth substrates for oyster mushroom production, that would be advantageous. Due to the cacao pods' high mineral nutrient content, particularly their potassium and nitrogen minerals, the addition of cacao pod waste can raise the substrate's nitrogen concentration. Mudakir et al.(2014) reported that cacao pods contain 61% of the total amount of nutrients. Compost derived from cacao pods contains 1.81% nitrogen by weight. The application of cacaopod compost can boost output by up to 19.48%. The findings indicated that the use of cacao pod waste in the comparative ratio of 75% sawdust and 25% cacao pod waste was optimal for the production and nutritional content of brown oyster mushrooms, and it was advised to be used in the growth of brown oyster mushroom (Mudakir et al., 2014).

Another attempt was made by Fidelis and Rajasekharara Rao in 2017 to utilize pest-infected cacao pods to generate compost after chemical treatment and the effect of compost on the growth of hybrid cacao seedlings. Superphosphate-enriched waste cacao pods had no negative impacts on the growth of cacao seedlings; instead, they increased plant height, dry matter output, and foliar N concentration. Waste cacao pods formed composts of a desirable grade that may be used successfully to fertilize cacao seedlings when combined with triple super phosphate and chicken manure.

According to a study by Rahim *et al.* (2018), compost fermented with *Tremella sp.* had a greater Indole Acetic Acid (IAA) and GA_3 concentration than compost treated with *Pleurotus sp.* A similar outcome was seen in the hemicellulose's capacity for breakdown. *Pleurotus sp.* might, however, create compost with greater quantities of nutrients. The development of cacao seedlings was significantly influenced by compost fermented by rot fungus. However, the development of cacao seedlings was unaffected by the various cacao kinds. *Tremella sp.* and *Pleurotus sp.* composted CPH waste, and this had a substantial impact on the leaf area index (LAI), net assimilation rate (NAR), crop growth rate (CGR), root–shoot ratio, and root dry weight of cacao seedlings.

Organic wastes may be efficiently turned into vermicompost with a recovery of 74.65–87.75% after a three-month composting period using *Eudrilus eugeniae* on cacao pod waste along with arecanut waste. Regardless of the type of organic waste utilized, earthworm biomass increased in just two months. All the vermicompost samples had somewhat greater major nutrient (NPK) and micronutrient (Cu, Zn, Fe, and Mn) levels than standard compost. Regardless of the origin of the organic waste, vermicompost had a lower C:N ratio and pH than regular compost. Vermicompost had a significantly larger microbial population than regular compost (Chowdappa *et al.*, 2013).

Cacao husk was used as a mulching agent and weed suppression was observed. By calculating the correlation between the number of green pixels and the total number of pixels in each experimental plot, weed coverage was calculated using nondestructive image analysis. Within and across mulch types, there were substantial differences in the thickness of the mulch layer needed to achieve an ED50 and ED90 of 50 and 90% weed suppression, respectively. The cacao mulch was excellent in controlling weeds in all but one case (Arentoft *et al.*, 2013).

These waste products are subjected to composting and are applied to soil as bio-compost which increases the nutrient content of the soil, enhances the microbial activity, and increases the water-holding capacity of the soil. In addition to this, cacao husks were employed as mulching agents to suppress the growth of several weeds in agricultural fields, gardens, and urban landscapes. Furthermore, cacao pods are converted into biochar and used for the slow release of nutrients in the soil.

Burning the CPH produces potassium-rich ash. By adding starch, it may be transformed into a fertilizer that is high in potassium. It can be sprayed on fields in the form of pellets. The cacao bean's seed coat must be removed to produce the cacao bean shell. Mulching is one of its main applications. It has 3.5% potassium, 1.5% phosphate, and 3% nitrogen content. When water is added, it may also be utilized as an adhesive agent. It significantly slows the development of weeds. It lessens the amount of water that evaporates. It also guards against slug and snail damage to the plants (Kone *et al.*, 2020).

11.4.4 Biofuels

Regarding overexploitation of fossil fuels and the environmental hazards, there is a need for development of alternate sustainable fuel sources. Nowadays, ethanol is blended with gasoline and is used in operating vehicles. Ethanol can also be obtained from cacao pods by acid hydrolysis and fermentation process since it is rich in lignocelluloses. Moreover, biogas can also be extracted from the wastes during anaerobic digestion (Antwi *et al.*, 2019). The industrial wastes from cacao production include the cacao bean shells, mucilage, and CPHs, all of which include substances with industrial value. The lignocellulose concentration of these byproducts, however, necessitates a pretreatment to properly use them (Mendoza-Meneses *et al.*,2021, Svarc-Gajic *et al.*, 2023). As a result, a variety of biofuels can be created, depending on the conversion process employed to get the maximum biomass yield. However, the most frequent usage recorded is as a direct combustion source, which is utilized to supply the same manufacturing units. Recent research has shown that cacao industrial wastes may be used to produce solid, liquid, and gaseous biofuels (Mendoza-Meneses *et al.*, 2021).

Ionic liquidswere employed in this investigation to pretreat cacao waste. When compared to the alkalis and acids employed in the standard pretreatment medium, it performed better during the treatment procedure. Based on the reduction of biomass, generation of reducing sugar and bioethanol, as well as the use of ionic liquid, H_2SO_4, and NaOH, pretreatment was shown to be beneficial. Biomass loss following ionic liquid pretreatment was found to be extremely low, at only 31%. H_2SO_4 produced 7.9 g/L, while ionic liquid generated a total of 7.885 g/L. At the end of a 24-hour incubation, untreated cacao waste generated 5.116 g/L of ethanol whereas NaOH produced 6.824 g/L (Idi *et al.*, 2012).

Cacao pod ash was employed as a biocatalyst to produce biodiesel from waste chicken fat. At 3 wt% catalyst concentration and two hours of reaction time, the greatest biodiesel production of 75.4% was achieved and gas chromatography confirmed the methyl ester production representing the esterification process. The characteristics of biodiesel, such as density, viscosity, saponification value, iodine value, acid value, and color, showed that CPA might be used as a bio-based catalyst for the transesterification of oil made from used chicken fat (Odetoye and Amusan, 2019).

11.4.5 Adsorbent

Agricultural wastes are transformed into biochar through pyrolysis or any other thermochemical conversion methods and are extensively used in the adsorption of pollutants such as heavy metals, gasses, or industrial dyes (Eletta *et al.*, 2020). Binderless carbon monoliths were extracted from bean husk which are more effective in adsorption due to their large micropore volume and good mechanical properties. Maximum removal of 263.9 mg g^{-1} of methylene blue was achieved by NaOH-treated CPH and adsorption was found to be spontaneous and endothermic (Ling Pua et al., 2013). The removal efficiency was found to be more than 90% in the case of adsorbents derived from the valorization of CPH. In contrast to the findings of Ling Pua et al. (2013), the adsorption was observed to be spontaneous and exothermic. Pesudo-second-order kinetic model was reported to be the best fit for the adsorption of pollutants (Eletta et al., 2020). It is also used as an adsorbent in the removal of metals such as lead and copper (Njoku et al., 2011) and congo red dye (Olakunle et al., 2018) whose absorption is in line with the findings of Ling Pua et al. (2013).

Biochar was made from CPH waste with the intention of using it in artificial wastewater to mimic the actual conditions of textile effluents to lessen disposal and water pollution issues. These findings show that biochar produced using the one-step process (AQ1) was chemically characterized by Boehm titration and the point of zero charge, and utilized to filter a blend of brilliant green (BG), rhodamine B (RB), and methyl orange (MO) as synthetic wastewater (Córdova *et al.*, 2020). In order to withhold harmful crystal violet (CV) dye from an aqueous solution, chemically activated cacao leaves biochar (CLB) was effectively produced out of fallen cacao leaves (CLs) using $ZnCl_2$-activation and pyrolysis at 700°C. With maximal adsorption capacities of 190.70 and 253.3 mg/g, respectively, CLs and CLB successfully compete with adsorbents utilized in the literature to remove CV dye from wastewater. The higher adsorption effectiveness of CLB was shown by the large Brunauer-Emmett-Teller (BET) surface area (957.02 m^2/g) and mean pore diameter (7.21 nm) (Jabar et al., 2022).

Their application does not limit its use in dye removal but also extends toward the removal of antibiotics such as norfloxacin. By calcining discarded cacao shells with urea and sodium bicarbonate ($NaHCO_3$) as nitrogen suppliers and green activators, respectively, a nitrogen-doped biochar (N-A) was created. Biochar has a strong potential for its future use in the real world as an ecologically friendly alternative because it combines a high adsorption capacity with a low inhibitory impact on environmental variables (Zanli et al., 2022). It is also widely employed in the removal of heavy metals from water such as lead, cadmium, mercury, nickel, etc. (Lara et al., 2016).

11.4.6 Dyelbiofilms

Biofilaments can be obtained from cacao shell waste which could be used in 3D printing of objects with potential household and biomedical applications. Biofilaments made from biodegradable polycaprolactone(PCL) and cacao shell waste, a byproduct of the chocolate industry, have been created using a single-screw extruder. The polymer may include up to 50% by weight of powdered cacao shell waste without significantly changing its crystalline structure.

The elastic (Young's) modulus of the biofilaments produced is comparable to that of pure PCL. Results from scanning electron microscopy show that throughout the extrusion process, micronized cacao shell debris is uniformly distributed throughout the polymer, and it can be replicable (Tran *et al.*, 2017).

Besides, biodegradable biofilms were prepared from cacao shells in combination with lactic acid. The produced biofilm mulches also exhibit low water solubility (1–15 wt%) and moisture content (1–3 wt%). Additionally, it was found that the type of vegetable waste used as filler had a substantial impact on both the film's stated qualities and its capacity to decompose. For instance, it was found that films made with 20 wt% starch sucrose (SS)-enhanced polylactide (PLA) composite soil biodegradability rose from 0 to 38 wt% following six months of a soil burial experiment. The created composite's potential as fertilizers is further shown by the fact that they contain varying levels of plant micro- and macronutrients (Merino et al., 2022).

The procedure involved adjusting the final mordant changes (alum, calcium oxide, and ferrous sulfate on silk fabric), the natural dye solution of CPH, and the concentration of maltodextrin as filler powder by 5%, 10%, and 15%. The phytochemistry of the natural dye for cacao pod skin was examined to determine its tannin and flavonoid contents, and the silk batik fabric's dyeing results were examined for their colorfastness, vibrancy, and consistency. The results of the color strength test indicated that the optimum maltodextrin was 5% with a natural dye concentration of 15 g/L, while the results of the colorfastness test indicated a good category with an average value of 4–5.

With a predominance of brown, red, and blackish brown, the effects of alum mordant, calcium oxide, and ferrous sulfate (tunjung) generated colors that led to light and dark hues (Haerudin *et al.*, 2020). Cotton and wool were dyed using extracts of sunflower petals, purple onion peel, and cacao shell since it is currently sought after to utilize natural ingredients sustainably in textiles (Arroyo Figueroa et al., 2022). Thus, cacao shell pigments are used as a natural dye for dyeing fabric and in manufacturing UV-protective cotton fabric (Tu,2016).

11.4.7 Other applications

Waste cacao (*Theobroma cacao L*) fruit skin or waste CPHs was effective in treating alopecia or hair loss. The purpose of the study was to identify the composition and properties of hair care containing cacao pod peel water, which is useful for promoting hair development. The Tanaka technique has been updated for activity testing. The findings indicated that the cacao husk wastewater fraction may be used to create hair care, but there were some variations in viscosity and pH during the freeze-thaw test. Based on hair length data and a notable difference in preparation concentration, the hair cream water fraction gel promotes hair growth activity. At a concentration of 12.5%, the formulation of hair cream performed well. Furthermore, the area of the rabbit's skin where the hair cream preparation was administered showed no evidence of irritation.

The findings of this study suggested that cream made from cacao fruit peels might be utilized to treat alopecia (Mustarichie et al., 2022).

The extracts from second-quality and musty cacao beans had the highest levels of polyphenols and antioxidant properties. The second-quality cacao bean extract had the strongest efficacy in the collagenase and elastase inhibition studies, with 86% inhibition and 36% inhibition, respectively. With 18.15 mg/100 g of sample and 229.8 mg/100 g of sample, respectively, catechin and epicatechin were found in greater concentrations in the extract as compared to other cacao byproducts. These properties govern their entry into the cosmetic industry (Agudelo et al., 2021).

Twelve panelists assessed the gel to see how well cacao pod extract (CPE) in gel form reduced skin wrinkles and improved skin condition using Visioscan. Malic acid, procyanidin B1, rosmarinic acid, procyanidin C1, apigenin, and ellagic acid have all been shown to be present in CPE and may all have contributed to the functional esthetic aspects of CPE. The amount of CPE to be integrated into the gel was determined using the EC50 value of cacao pod extracts so that the designed product could achieve an appropriate concentration of extract while being nontoxic to the skin cells. The findings suggested that CPE is a viable component for wrinkle reduction (Abdul Karim et al., 2016).

Enzyme extraction was also attempted from cacao wastes after pretreatment with citric acid. Pectin recovery accounted for 19.3% of the biomass subjected to pretreatment under ideal circumstances, which included 120 °C, 10 min, and 2% $w.v^{-1}$, with a degree of methyl esterification of 52.2%. The cellulosic fraction's enzymatic conversion efficiency was 58.9%, producing 92.4 kg of glucose for every ton of CPH (Valladares-Diestra et al., 2022).

11.5 Conclusion

Cacao byproducts have a vast potential which could be explored and exploited in various sectors as a means of waste reduction and other beneficial applications. The current chapter highlighted that cacao byproducts have a rich source of nutrients and other bioactive compounds that can be modified and then used for a variety of environmental applications because they are readily available, inexpensive, and adaptable. To effectively utilize the cacao wastes generated, various eco-friendly technologies should be developed in a way that limits the entry of hazardous substances into the environment and results in economically efficient products. Future research should be focused on the complete utilization of cacao wastes resulting in a circular economy that will balance both environmental and economical sustainability.

References

Abdul Karim, A., Azlan, A., Ismail, A., Hashim, P., Abd Gani, S.S., Zainudin, B.H.&Abdullah, N.A. (2016). Efficacy of cacao pod extract as antiwrinkle gel on human skin surface. *Journal of Cosmetic Dermatology*, *15*(3), 283–295.

Agudelo, C., Bravo, K., Ramírez-Atehortúa, A., Torres, D., Carrillo-Hormaza, L. & Osorio, E. (2021). Chemical and skincare property characterization of the main cocoa byproducts: Extraction optimization by RSM approach for development of sustainable ingredients.*Molecules*, *26*(24), 7429.

Agus, B.A.P., Mohamad, N.N.&Hussain, N. (2018). Composition of unfermented, unroasted, roasted cacao beans and cacao shells from Peninsular Malaysia. *Journal of Food Measurement and Characterization*, *12*, 2581–2589.

Ajongbolo, K.(2020). Chemical properties of local black soap produced from cocoa pod ash and palm oil waste.*IJTSRD*, *4*(6), 713–715.

Amir, I. Z., Hanida, H. S., & Syafiq, A. (2013). Development and physical analysis of high fiber bread incorporated with cocoa (Theobroma cacao sp.) pod husk powder. *International Food Research Journal*, *20*(3), 1301.

Antwi, E., Engler, N., Nelles, M.&Schüch, A. (2019). Anaerobic digestion and the effect of hydrothermal pretreatment on the biogas yield of cacao pods residues. *Waste Management*, *88*, 131–140.

Anvoh, K.Y.B., Bi, A.Z.&Gnakri, D. (2009). Production and characterization of juice from mucilage of cacao beans and its transformation into marmalade. *Pakistan Journal of Nutrition*, *8*(2), 129–133.

Arentoft, B.W., Ali, A., Streibig, J.C.&Andreasen, C. (2013). A new method to evaluate the weed-suppressing effect of mulches: A comparison between spruce bark and cacao husk mulches. *Weed Research*, *53*(3), 169–175.

Arroyo Figueroa, G., Dzul Cauich, J.G., Medina Saavedra, T.&Garcia Vieyra, M.I. (2022). Dyeing and colour fastness in cotton and wool, using natural extracts of sunflower petals, onion peel and cacao shell. *Journal of Natural Fibers*, *19*(14), 7470–7479.

Awuah, R.T.&Frimpong, M. (2002). Cocoa-based media for culturing *Phytophthora palmivora* (Butl.) Butl., causal agent of black pod disease of cacao. *Mycopathologia*, *155*, 143–147.

Barbosa-Pereira, L., Guglielmetti, A.&Zeppa, G. (2018). Pulsed electric field assisted extraction of bioactive compounds from cacao bean shell and coffee silverskin. *Food and Bioprocess Technology*, *11*, 818–835.

Bastidas, J.V., Badillo Melo, W.A.&Briones-Bitar, J. (2022). Sustainability of the cocoa industry: Cocoa waste mucilage use to produce fermented beverages. Case study in Los Ríos Province.*International Journal of Sustainable Development & Planning*, *17*(4), 1147–1152.

Bello, O.S., Siang, T.T.&Ahmad, M.A. (2012). Adsorption of Remazol Brilliant Violet-5R reactive dye from aqueous solution by cacao pod husk-based activated carbon: Kinetic, equilibrium and thermodynamic studies. *Asia-Pacific Journal of Chemical Engineering*, *7*(3), 378–388.

Belwal, T., Cravotto, C., Ramola, S., Thakur, M., Chemat, F.&Cravotto, G. (2022). Bioactive compounds from cacao husk: Extraction, analysis and applications in food production chain. *Foods*, *11*(6), 798.

Boateng, M., Okai, D.B., Amoah, K.O., Opoku-Asenso, A., Aye, K.A.&Frimpong, Y.O. (2019). Cacao placenta meal: Yield estimates, nutrient composition and response of albino rats to diets containing graded levels with or without xzymetm an exogenous enzyme-probiotic complex. *Animal Health and Production*, *67*,141–150.

Botella-Martínez, C., Lucas-Gonzalez, R., Ballester-Costa, C., Pérez-Álvarez, J.Á., Fernández-López, J., Delgado-Ospina, J., Chaves-López, C.&Viuda-Martos, M. (2021). Ghanaian cacao (*Theobroma cacao* L.) bean shells coproducts: Effect of particle size on chemical composition, bioactive compound content and antioxidant activity. *Agronomy*, *11*(2), 401.

Campos-Vega, R., Nieto-Figueroa, K. H., & Oomah, B. D. (2018). Cocoa (Theobroma cacao L.) pod husk: Renewable source of bioactive compounds. *Trends in Food Science & Technology*, *81*, 172–184.

Cantele, C., Rojo-Poveda, O., Bertolino, M., Ghirardello, D., Cardenia, V., Barbosa-Pereira, L.&Zeppa, G. (2020). In vitro bioaccessibility and functional properties of phenolic compounds from enriched beverages based on cacao bean shell. *Foods*, *9*(6),715.

Castillo, E., Alvarez, C.&Contreras, Y. (2018). Caracterización fisicoquímica de la cáscara del fruto de un clon de cacao (*Theobroma cacao* L.) cosechados en Caucagua, estado Miranda. Venezuela. *Revistas De Investigación*, *42*(95) 154–175.

Chowdappa, P., Biddappa, C.C.&Sujatha, S. (2013). Efficient recycling of organic wastes in arecanut (*Areca catechu*) and cacao (*Theobroma cacao*) plantation through vermicomposting. The Indian Journal of Agricultural Sciences 69(8).

Čolović, R., Puvača, N., Cheli, F., Avantaggiato, G., Greco, D., Đuragić, O., Kos, J. & Pinotti, L. (2019). Decontamination of mycotoxin-contaminated feedstuffs and compound feed. *Toxins*, *11*(11), 617.

Córdova, B.M., Santa Cruz, J.P., Huamani-Palomino, R.G.&Baena-Moncada, A.M. (2020). Simultaneous adsorption of a ternary mixture of brilliant green, rhodamine B and methyl orange as artificial wastewater onto biochar from cacao pod husk waste. Quantification of dyes using the derivative spectrophotometry method. *New Journal of Chemistry*, *44*(20), 8303–8316.

Cornejal, N., Pollack, E., Kaur, R., Persaud, A., Plagianos, M., Juliani, H.R., Simon, J., Zorde, M., Priano, C., Koroch, A.&Fernández Romero, J.A. (2023). Antimicrobial and antioxidant properties of *Theobroma cacao, Bourreria huanita, Eriobotrya japonica*, and *Elettaria cardamomum*-Traditional plants used in Central America. *Journal of Medicinally Active Plants*, *12*(1),1–17.

Deena, S.R., Vickram, A.S., Manikandan, S., Subbaiya, R., Karmegam, N., Ravindran, B., …&Awasthi, M.K. (2022). Enhanced biogas production from food waste and activated sludge using advanced techniques–A review. *Bioresource Technology*, *355*, 127234.

de Barros, H.E.A., Natarelli, C.V.L., de Abreu, D.J.M., de Oliveira, A.L.M., do Lago, R.C., Dias, L.L.D.C., de Carvalho, E.E.N., Bilal, M., Ruiz, H.A., Franco, M.&Vilas Boas, E.V.D.B. (2021). Application of chemometric tools in the development of food bars based on cacao shell, soy flour and green banana flour. *International Journal of Food Science & Technology*, *56*(10), 5296–5304.

de Barros, H.E.A., Natarelli, C.V.L., de Carvalho Tavares, I.M., de Oliveira, A.L.M., Araújo, A.B.S., Pereira, J., Carvalho, E.E.N., de Barros Vilas Boas, E.V.&Franco, M. (2020). Nutritional clustering of cookies developed with cacao shell, soy, and green banana flours using exploratory methods. *Food and Bioprocess Technology*, *13*,1566–1578.

de Souza Vandenberghe, L.P., Valladares-Diestra, K.K., Bittencourt, G.A., de Mello, A.F.M., Vásquez, Z.S., de Oliveira, P.Z., de Melo Pereira, G.V.&Soccol, C.R. (2022). Added-value biomolecules' production from cacao pod husks: A review. *Bioresource Technology*, *344*, 126252.

Delgado, J.D., Mandujano, J.I., Reátegui, D.&Ordoñez, E.S. (2018). Development of dark chocolate with fermented and non-fermented cacao nibs: Total polyphenols, anthocyanins, antioxidant capacity and sensory evaluation. *Scientia Agropecuaria*, *9*(4), 543–550.

Delgado-Ospina, J., Martuscelli, M., Grande-Tovar, C.D., Lucas-González, R., Molina-Hernandez, J.B., Viuda-Martos, M., Fernández-López, J., Pérez-Álvarez, J.Á.&Chaves-López, C. (2021). Cacao pod husk flour as an ingredient for reformulating frankfurters: Effects on quality properties. *Foods*, *10*(6),1243.

Díaz-González, A., Perez Luna, M. Y., Ramírez Morales, E., Saldaña-Trinidad, S., Rojas Blanco, L., De La Cruz-Arreola, S., …&Robles-Ocampo, J. B. (2022). Assessment of the pretreatments and bioconversion of lignocellulosic biomass recovered from the husk of the cocoa pod. *Energies*, *15*(10), 3544.

Donkoh, A., Atuahene, C.C., Wilson, B.N.&Adomako, D. (1991). Chemical composition of cacao pod husk and its effect on growth and food efficiency in broiler chicks. *Animal Feed Science and Technology*, *35*(1–2), 161–169.

Eletta, O.A., Adeniyi, A.G., Ighalo, J.O., Onifade, D.V.&Ayandele, F.O. (2020). Valorisation of cacao (*Theobroma cacao*) pod husk as precursors for the production of adsorbents for water treatment. *Environmental Technology Reviews*, *9*(1), 20–36.

Fidelis, C.&Rajashekhar Rao, B.K. (2017). Enriched cacao pod composts and their fertilizing effects on hybrid cacao seedlings. *International Journal of Recycling of Organic Waste in Agriculture*, *6*, 99–106.

Figueroa, K.H.N., García, N.V.M.&Vega, R.C. (2020). Cocoa by-products.In*Food Wastes and By-Products: Nutraceutical and Health Potential* (pp. 373–411). John wiley & Sons

Grillo, G., Boffa, L., Binello, A., Mantegna, S., Cravotto, G., Chemat, F., …& Telysheva, G. (2019). Cocoa bean shell waste valorisation; extraction from lab to pilot-scale cavitational reactors. *Food Research International*, *115*, 200–208.

Guirlanda, C.P., da Silva, G.G.&Takahashi, J.A. (2021). Cacao honey: Agro-industrial waste or underutilized cacao by-product?*Future Foods*, *4*, 100061.

Gunasekaran, A.P., Chockalingam, M.P., Padmavathy, S.R.&Santhappan, J.S. (2021). Numerical and experimental investigation on the thermochemical gasification potential of Cocoa pod husk (*Theobroma cacao*) in an open-core gasifier.*Clean Technologies and Environmental Policy*, *23*, 1603–1615.

Haerudin, A., Lestari, D.W., Mandegani, G.B., Satria, Y., Arta, T.K.&Atika, V.(2020). Application of cacao pod husk (*Theobroma cacao*spp.) for natural dyes powder on silk batik cloth. In *IOP Conference Series: Materials Science and Engineering* (Vol. 980, No. 1, p. 012020). IOP Publishing.

Handojo, L., Triharyogi, H., & Indarto, A. (2019). Cocoa bean shell waste as potential raw material for dietary fiber powder. *International Journal of Recycling of Organic Waste in Agriculture*, *8*, 485–491.

Hernández-Mendoza, A. G., Saldaña-Trinidad, S., Martínez-Hernández, S., Pérez-Sariñana, B. Y., & Láinez, M. (2021). Optimization of alkaline pretreatment and enzymatic hydrolysis of cocoa pod husk (Theobroma cacao L.) for ethanol production. *Biomass and Bioenergy*, *154*, 106268.

Hettiarachchi, H., Ryu, S., Caucci, S.&Silva, R. (2018). Municipal solid waste management in Latin America and the Caribbean: Issues and potential solutions from the governance perspective.*Recycling*, *3*(2), 19.

ICCO n.d.– International Cacao Organization Annual Report 2018/2019. Available online: https://www.icco.org/statistics/#production

Idi, A., Salleh, M.M., Ibrahim, Z.&Mohamad, S.E. (2012). Pretreatment of cacao waste for bioethanol production using ionic liquid. *Jurnal Teknologi*, *59*(1) 49–56.

Igbinadolor, R.O. & Onilude, A.A. (2013). Bioprocess systems applied for the production of bio-ethanol from lignocellulosic biomass of cacao pod husk (*Theobroma cacao* L.) and other agricultural residues: A review. *African Journal of Biotechnology*, *12*(35), 5375–5388.

Ikotun, A.A., Faboro, E.O., Kolade, A.S.&Owoseni, A.A. (2018). Studies on enhanced African black soap from *Theobroma cacao* (cacao) and *Elaeis guineensis* (palm kernel oil). *African Journal of Biotechnology*, *17*(24),760–766.

Ismail, A.&Yee, C.L. (2006). Antioxidative effects of extracts of cacao shell, roselle seeds and a combination of both extracts on the susceptibility of cooked beef to lipid oxidation. *Journal of Food Technology*, *4*(1),10–15.

Jabar, J.M., Adebayo, M.A., Owokotomo, I.A., Odusote, Y.A.&Yılmaz, M. (2022). Synthesis of high surface area mesoporous ZnCl2–activated cacao (*Theobroma cacao* L.) leaves biochar derived via pyrolysis for crystal violet dye removal. *Heliyon*, *8*(10), 10873.

Jaimes, Y.L.V., Guerrero, J.S.R.&Castrillo, L.C.L. (2017).Caracterización fisicoquímica, microbiológica y funcional de harina de cáscara de cacao (*Theobroma cacao* L.) variedad CCN-51. *Cuaderno activa*, *9*, 65–75.

Jeong, N.R., Ki, W.J., Kim, M.J.&Nam, M.S. (2022). Anti-oxidant, ant-microbial and anti-inflammatory activity of yogurt with added cacao nibs (*Theobroma cacao* L.). *Korean Journal of Agricultural Science*, *49*(3), 583–593.

Jönsson, L.J.&Martín, C. (2016). Pretreatment of lignocellulose: Formation of inhibitory by-products and strategies for minimizing their effects.*Bioresource Technology*, *199*, 103–112.

Joseph, C., Savoire, R., Harscoat-Schiavo, C., Pintori, D., Monteil, J., Faure, C.&Leal-Calderon, F. (2020). Pickering emulsions stabilized by various plant materials: Cacao, rapeseed press cake and lupin hulls. *LWT*, *130*, 109621.

Jozinović, A., Panak Balentić, J., Ačkar, Đ., Babić, J., Pajin, B., Miličević, B., Guberac, S., Vrdoljak, A.&Šubarić, D. (2019). Cacao husk application in the enrichment of extruded snack products. *Journal of Food Processing and Preservation*, *43*(2), 13866.

Kone, K., Akueson, K.&Norval, G. (2020). On the production of potassium carbonate from cacao pod husks. *Recycling*, *5*(3), 23.

Laconi, E.B.&Jayanegara, A. (2015). Improving nutritional quality of cacao pod (*Theobroma cacao*) through chemical and biological treatments for ruminant feeding: In vitro and in vivo evaluation. *Asian-Australasian Journal of Animal Sciences*, *28*(3), 343.

Lara, J., Tejada, C., Villabona, Á., Arrieta, A.&Granados Conde, C. (2016). Adsorption of lead and cadmium in continuous fixed bed on cacao waste. *Revista ION*, *29*(2), 113–124.

Lessa, O.A., Reis, N.D.S., Leite, S.G.F., Gutarra, M.L.E., Souza, A.O., Gualberto, S.A., de Oliveira, J.R., Aguiar-Oliveira, E.&Franco, M. (2018). Effect of the solid state fermentation of cacao shell on the secondary metabolites, antioxidant activity, and fatty acids. *Food Science and Biotechnology*, *27*, 107–113.

Ling Pua, F., Sajab, M.S., Chia, C.H., Zakaria, S., Rahman, I.A.&Salit, M.S. (2013). Alkaline-treated cacao pod husk as adsorbent for removing methylene blue from aqueous solutions. *Journal of Environmental Chemical Engineering*, *1*(3), 460–465.

Lu, F., Rodriguez-Garcia, J., Van Damme, I., Westwood, N. J., Shaw, L., Robinson, J. S., …&Charalampopoulos, D. (2018). Valorisation strategies for cocoa pod husk and its fractions. *Current Opinion in Green and Sustainable Chemistry*, *14*, 80–88.

Makinde, O.J., Okunade, S.A., Opoola, E., Sikiru, A.B., Ajide, S.O.&Elaigwu, S. (2019). Exploration of cacao (*Theobroma cacao*) by-products as valuable potential resources in livestock feeds and feeding systems. In *Theobroma Cacao-Deploying Science for Sustainability of Global Cacao Economy*(pp. 1–14) Edited by Peter Osobase Aikpokpodion. IntechOpen Publisher.

Mandari, V.&Devarai, S.K. (2022). Biodiesel production using homogeneous, heterogeneous, and enzyme catalysts via transesterification and esterification reactions: A critical review. *BioEnergy Research*, *15*, 935–961.

Manzano, P., Hernández, J., Quijano-Avilés, M., Barragán, A., Chóez-Guaranda, I., Viteri, R.&Valle, O. (2017). Polyphenols extracted from *Theobroma cacao* waste and its utility as antioxidant. *Emirates Journal of Food and Agriculture*, *29*(1), 45.

Marcel,B.K.G., Andre,K.B., Theodore,D.&Seraphin,K.C. (2011). Waste & by-products of cacao in breeding: Research synthesis. *International Journal of Agronomy and Agricultural Research*, *1*, 9–19.

Mariatti, F., Gunjević, V., Boffa, L.&Cravotto, G. (2021). Process intensification technologies for the recovery of valuable compounds from cacao by-products.*Innovative Food Science & Emerging Technologies*,*68*, 102601.

Martínez, R., Torres, P., Meneses, M.A., Figueroa, J.G., Pérez-Álvarez, J.A.&Viuda-Martos, M. (2012). Chemical, technological and in vitro antioxidant properties of mango, guava, pineapple and passion fruit dietary fibre concentrate. *Food Chemistry*, *135*(3), 1520–1526.

Mellinas, A.C., Jiménez, A.&Garrigós, M.C. (2020). Optimization of microwave-assisted extraction of cacao bean shell waste and evaluation of its antioxidant, physicochemical and functional properties. *LWT*, *127*, 109361.

Mendoza-Meneses, C.J., Feregrino-Pérez, A.A.&Gutiérrez-Antonio, C. (2021). Potential use of industrial cacao waste in biofuel production. *Journal of Chemistry*, *2021*, 1–11.

Merino, D., Zych, A.&Athanassiou, A. (2022). Biodegradable and biobased mulch films: Highly stretchable PLA composites with different industrial vegetable waste. *ACS Applied Materials &Interfaces*, *14*(41), 46920–46931.

Meza-Sepúlveda, D.C., Castro, A.M., Zamora, A., Arboleda, J.W., Gallego, A.M.&Camargo-Rodríguez, A.V. (2021). Bio-based value chains potential in the management of cacao pod waste in Colombia, a case study. *Agronomy*, *11*(4), 693.

Mudakir, I., Hastuti, U.S., Rohman, F.&Gofur, A. (2014). The effect of cacao pods waste as a growing media supplement on productivity and nutrient content of brown oyster mushroom (*Pleurotus cystidiosus*). *Journal of Biology, Agriculture and Healthcare*, *4*(26), 134–140.

Murthy, P.S.&Naidu, M.M. (2012). Sustainable management of coffee industry by-products and value addition—A review. *Resources, Conservation and Recycling*, *66*, 45–58.

Mustarichie, R., Hasanah, A.N., Wilar, G., Gozali, D.&Saptarini, N.M. (2022). New hair growth cream formulation with cacao pod peel (*Theobroma cacao* L.). *The Scientific World Journal*, *2022*, 2299725.

Netania, G., Permana, T., Effendy, J.&Santoso, F. (2022). A review on the potential applications of cacao shell in food industry. In *6th International Conference of Food, Agriculture, and Natural Resource (IC-FANRES 2021)* (pp. 126–131). Atlantis Press.

Nikman, K.A. (2018). Preparation, characterization and applications of activated carbon from cacao (*Theobroma cacao*) nibs waste (Doctoral dissertation, Universiti Sains Islam Malaysia).

Njoku, V.O., Ayuk, A.A., Ejike, E.E., Oguzie, E.E., Duru, C.E.&Bello, O.S. (2011). Cacao pod husk as a low cost biosorbent for the removal of Pb (II) and Cu (II) from aqueous solutions. *Australian Journal of Basic and Applied Sciences*, *5*(8), 101–110.

Nsor-Atindana, J., Zhong, F., Mothibe, K.J., Bangoura, M.L.&Lagnika, C. (2012). Quantification of total polyphenolic content and antimicrobial activity of cacao (*Theobroma cacao* L.) bean shells. *Pakistan Journal of Nutrition*, *11*(7), 574.

Nunes, C.S., da Silva, M.L., Camilloto, G.P., Machado, B.A., Hodel, K.V., Koblitz, M.G.B., Carvalho, G.&Uetanabaro, A.P.T. (2020). Potential applicability of cacao pulp (*Theobroma cacao* L) as an adjunct for beer production. *The Scientific World Journal, Volume 2020* (1–14), *Article ID 3192585*.

Oddoye, E. O., Agyente-Badu, C. K., & Gyedu-Akoto, E. (2013). Cocoa and its by-products: Identification and utilization. *Chocolate in Health and Nutrition*, 23–37. https://doi.org/10.1007/978-1-61779-803-0_3

Odetoye, T.E.&Amusan, G.S. (2019). Cacao pod ash as bio-based catalyst for biodiesel production from waste chicken fat. *Agricultural Engineering International: CIGR Journal*, *21*(4), 144–151.

Oduro-Mensah, D., Ocloo, A., Nortey, T., Antwi, S., Okine, L.K. & Adamafio, N.A. (2020). Nutritional value and safety of animal feed supplemented with Talaromyces verruculosus-treated cacao pod husks. *Scientific Reports*, *10*(1), 13163.

Ogunsipe, M.H., Ibidapo, I., Oloruntola, O.D.&Agbede, J.O. (2017). Growth performance of pigs on dietary cacao bean shell meal. *Livestock Research for Rural Development*, *29*(1), 18.

Okiyama, D.C., Navarro, S.L.&Rodrigues, C.E. (2017). Cacao shell and its compounds: Applications in the food industry. *Trends in Food Science & Technology*, *63*, 103–112.

Olakunle, M.O., Inyinbor, A.A., Dada, A.O.&Bello, O.S. (2018). Combating dye pollution using cacao pod husks: A sustainable approach. *International Journal of Sustainable Engineering*, *11*(1), 4–15.

Olugosi, O.A., Agbede, J.O., Onibi, G.E., Adebayo, I.A.&Ayeni, A.O. (2020). Biologically upgraded cacao pod husk: Effect on growth performance, haemato-biochemical indices and antioxidant status of broiler chickens. *Journal of Food, Nutrition and Agriculture*, *3*, 26–32.

Orejuela-Escobar, L.M., Landázuri, A.C.&Goodell, B. (2021). Second generation biorefining in Ecuador: Circular bioeconomy, zero waste technology, environment and sustainable development: The nexus.*Journal of Bioresources and Bioproducts*, *6*(2), 83–107.

Oussou, K.F., Guclu, G., Kelebek, H.&Selli, S. (2023). Valorization of cacao, tea and coffee processing by-products-wastes. *Advances in Food and Nutrition Research*. 107 (91–130).

Panak Balentić, J., Ačkar, Đ., Jokić, S., Jozinović, A., Babić, J., Miličević, B., … &Pavlović, N. (2018). Cocoa shell: A by-product with great potential for wide application. *Molecules*, *23*(6), 1404.

Pinzon-Nuñez, D.A., Adarme-Duran, C.A., Vargas-Fiallo, L.Y., Rodriguez-Lopez, N. &Rios-Reyes, C.A. (2022). Biochar as a waste management strategy for cadmium contaminated cacao pod husk residues. *International Journal of Recycling Organic Waste in Agriculture*, *11*(1),101–115.

Rahim, I., Nasruddin, A., Kuswinanti, T., Asrul, L.&Rasyid, B. (2018). Utilization of cacao pod husk waste composting by *Tremella* sp and *Pleurotus* sp as a medium to growth of cacao seedling. In *IOP Conference Series: Earth and Environmental Science* (Vol. 156, No. 1, p. 012012). IOP Publishing.

Renna, M., Lussiana, C., Colonna, L., Malfatto, V.M., Mimosi, A.&Cornale, P. (2022). Inclusion of cacao bean shell in the diet of dairy goats: Effects on milk production performance and milk fatty acid profile. *Frontiers in Veterinary Science*, 9 Article ID 848452: 137.

Rojo-Poveda, O., Barbosa-Pereira, L., Mateus-Reguengo, L., Bertolino, M., Stévigny, C.&Zeppa, G. (2019). Effects of particle size and extraction methods on cacao bean shell functional beverage. *Nutrients*, *11*(4), 867.

Rojo-Poveda, O., Barbosa-Pereira, L., Zeppa, G.&Stévigny, C. (2020). Cacao bean shell—a by-product with nutritional properties and biofunctional potential. *Nutrients*, *12*(4), 1123.

Rousseaux, C.G. (2023). Herbal remedies. In *Haschek and Rousseaux's Handbook of Toxicologic Pathology*, Volume 3 (pp. 183–303). Academic Press Edited by Wanda M. Haschek, Colin G. Rousseaux, … Brad Bolon.

San Siow, C., Chan, E.W.C., Wong, C.W.&Ng, C.W. (2022). Antioxidant and sensory evaluation of cacao (*Theobroma cacao* L.) tea formulated with cacao bean hull of different origins. *Future Foods*, *5*,100108.

Shahbaz, M., Al-Ansari, T., Aslam, M., Khan, Z., Inayat, A., Athar, M., Naqvi, S.R., Ahmed, M.A.&McKay, G. (2020). A state of the art review on biomass processing and conversion technologies to produce hydrogen and its recovery via membrane separation. *International Journal of Hydrogen Energy*, *45*(30),15166–15195.

Shahbendaz, M. (2023). Cocoa bean production worldwide 2020/2021-2022/2023, by country. Retrieved from https://www.statista.com/statistics/263855/cacao-bean-production-worldwide-by-region/

Soares, T.F.&Oliveira, M.B.P. (2022). Cocoa by-products: Characterization of bioactive compounds and beneficial health effects. *Molecules*, *27*(5), 1625.

Souza, F.N.S., Vieira, S.R., Campidelli, M.L.L., Rocha, R.A.R., Rodrigues, L.M.A., Santos, P.H., Carneiro, J.D.D.S., de Carvalho Tavares, I.M.&de Oliveira, C.P. (2022). Impact of using cacao bean shell powder as a substitute for wheat flour on some of chocolate cake properties. *Food Chemistry*, *381*, 132215.

Švarc-Gajić, J., Brezo-Borjan, T., Dzedik, V., Rodrigues, F., Morais, S.&Delerue-Matos, C. (2023). ESG approach in the valorization of cacao (*Theobroma cacao*) by-products by subcritical water: Application in the cosmetic industry. *Sustainable Chemistry and Pharmacy*, *31*, 100908.
Syamsiro, M., Saptoadi, H.&Tambunan, B.H. (2011). Experimental investigation on combustion of bio-pellets from Indonesian cacao pod husk. *Asian Journal of Applied Sciences*, *4*(7), 712–719.
Tran, M. (2017). Effect of high pressure jet processing on the physicochemical and foaming properties of whole milk and its impact on cocoa stability in chocolate milk.
Tsai, C.H., Tsai, W.T., Liu, S.C.&Lin, Y.Q. (2018). Thermochemical characterization of biochar from cacao pod husk prepared at low pyrolysis temperature. *Biomass Conversion and Biorefinery*, *8*, 237–243.
Tu, C. (2016). Study about stability of cacao husk pigment and its dyeing properties on cotton. In *Key Engineering Materials* (Vol. 671, pp. 133–138). Trans Tech Publications Ltd.Edited by Jianzhong Shao and Jiping Wang
Tursi, A. (2019). A review on biomass: Importance, chemistry, classification, and conversion. *Biofuel Research Journal*, *6*(2), 962.
Valladares-Diestra, K.K., de Souza Vandenberghe, L.P., Torres, L.A.Z., Zandoná Filho, A., Woiciechowski, A.L.&Soccol, C.R. (2022). Citric acid assisted hydrothermal pretreatment for the extraction of pectin and xylooligosaccharides production from cacao pod husks. *Bioresource Technology*, *343*, 126074.
Vásquez, Z.S., de Carvalho Neto, D.P., Pereira, G.V., Vandenberghe, L.P., de Oliveira, P.Z., Tiburcio, P.B., Rogez, H.L., Neto, A.G.&Soccol, C.R. (2019). Biotechnological approaches for cacao waste management: A review. *Waste Management*, *90*, 72–83.
Zanli, B.L.G.L., Tang, W.&Chen, J. (2022). N-doped and activated porous biochar derived from cacao shell for removing norfloxacin from aqueous solution: Performance assessment and mechanism insight. *Environmental Research*, *214*, 113951.

12 Multiple factors influencing *Theobroma cacao* and their impact on the chocolate market worldwide

Jyoti Sarwan, Shreya Sahani, Mithila V Nair, Shruti Bhargav, Shivam Kumar, Robin Singh, Komal Mittal, Pankaj Jaglan, Nazim Uddin, and Jagadeesh Chandra Bose K

12.1 Introduction

12.1.1 Background of Cacao Production

Cacao and chocolate have rich histories, particularly in Central America, where they have been produced and consumed for centuries. The knowledge of cacao reached the rest of the world when Spanish explorers arrived in Central America in the fifteenth and sixteenth centuries. Initially, cacao manufacturing remained localized to its original production areas, but with the rise in demand for chocolate in Europe and North America, production spread to other regions, including Latin America, Asia, and Africa. West Africa, in particular, emerged as a dominant force in global cacao production by the end of the 1960s (Squiccirini et al., 2016).

The demand for cacao surged in Europe and North America, especially during the Industrial Revolution. Scientific innovations during this period transformed cacao from a drink consumed by the elite into a mass-produced source of energy for the general population. The temperance movement, which advocated for alternatives to alcoholic beverages, further boosted the demand for chocolate toward the end of the nineteenth and early twentieth centuries (Squiccirini et al., 2016).

12.1.2 Popularity of Chocolate Worldwide

Switzerland and Belgium have established themselves as leaders in chocolate manufacturing, each with its own rich tradition. The Swiss chocolate industry, for instance, traces its roots back to a culture of small craftsmen. Despite facing stiff competition from large multinational corporations, small-scale Swiss chocolate makers have maintained their niche in the market, with consumers valuing artisanal production and being willing to pay premium prices for quality. Moreover, the Swiss chocolate industry has responded to calls for

DOI: 10.4324/9781003381761-16

socially responsible consumption by implementing transparency and corporate social responsibility (CSR) initiatives (Squiccirini et al., 2016).Similarly, Belgian chocolates have gained worldwide recognition, with pralines, desserts, and other delicacies originating from Belgium. While Belgium became a major chocolate exporter after 1960, the industry has since seen significant changes, with many Belgian chocolate companies now owned by international holdings. This globalization, coupled with increasing demand and counterfeiting issues, has prompted efforts to protect and regulate the 'Belgian chocolate' label (Squiccirini et al., 2016) (Figure 12.1).

12.2 Consumption

12.2.1 Source of Caffeine

Europe was one of the continents where caffeine consumption originated many hundreds of years ago, and now coffee and other caffeine-containing beverages have become ingrained in our daily routines and hold an eminent place in our cultures. The same can be said for most nations around the world. The worldwide consumption of caffeine by people can be relatively summarized in a table based mainly on formal enumeration (Battig et al., 1999).

Figure 12.1 This image is a collection of six photographs that demonstrate the cacao production process, from the raw pods to the chocolates.

The photographs show how cacao is harvested, processed, and manufactured (Blenkinsop et al., 2013).

12.2.2 Caffeine Absorption, Distribution, and Pharmacokinetics

Caffeine absorption from the gastrointestinal tract is very rapid and reaches 99% in humans within 45 minutes after intake. Caffeine can also be absorbed completely by animals. Pharmacokinetics are verbally comparable after caffeine administration in humans and animals, resulting in congruent, superimposable plasma curves. Absorption may not be complete when caffeine is ingested as coffee, and unexpectedly large doses (around LD-50) can cause hazardous effects. A few hundred micromoles per liter in the body can lead to acute caffeine poisoning and may require hospitalization. Due to its hydrophobic properties, caffeine can easily pass through any biological membrane. Additionally, there is no blood–brain barrier or placental barrier for caffeine (Nehlig et al., 1999).

12.2.3 Tolerance to the Effects of Caffeine

Humans exhibit tolerance to caffeine, although not uniformly across all its effects.

A. **Cardiovascular Effects**
Studies have shown that high intake of caffeine, especially from coffee, can lead to tachycardia and an increased risk of elevated blood pressure. The caffeine content in coffee, tea, chocolate, and soft drinks contributes to cardiovascular impact, which is mitigated by reducing caffeine-induced increases in plasma adrenaline, noradrenaline, and renin levels. However, tolerance to the effects of caffeine on the cardiovascular system may diminish after a relatively short period of caffeine abstinence.

B. **Effect on Sleep**
Sleep, being a vital physiological function, is highly sensitive to the effects of caffeine in humans. Consuming caffeine before bedtime can negatively affect sleep patterns. Analysis of various studies comparing sleep difficulties among heavy and light caffeine consumers yields inconclusive results. Furthermore, the relationship between coffee consumption and sleep disturbances remains unclear.

C. **Effects of Coffee on Mood**
Mood, as an undefined psychic phenomenon, plays a crucial role in psychological and behavioral functions. Caffeine consumption is believed to have positive effects on mood, leading consumers to feel more energetic, enthusiastic, inventive, devoted, and motivated in their work, as well as increasing their desire for social interaction (Holmen et al., 1999).

The Impact of Caffeine on Special Biochemical Targets: caffeine can significantly block adenosine outcomes on A2A (most effective) and A1 receptors even at low concentrations, such as those found after

> consuming a single cup of espresso. Inhibition of phosphodiesterase requires concentrations 20 times higher to inhibit cyclic nucleotide breakdown, while 40 times higher concentrations are needed to block GABAA receptors. Mobilizing intracellular calcium depots requires concentrations 100 times higher.
>
> The prehistoric civilizations of the Aztecs and Mayas in the Mesoamerican region were the first to cultivate *Theobroma cacao* between 250 and 900 AD. In the Mesoamerican culture, cacao had various uses ranging from therapeutic to currency. In New Spain and Europe, more than a hundred therapeutic applications have been documented (Dillinger et al., 2000).
>
> Cacao seeds are acquired in dried, fermented, and unfermented forms. Cacao has always been a remarkable crop for many countries and has been widely used to treat numerous diseases and disorders.
>
> (Amin et al., 2004; Keen et al., 2001)

Cacao has been widely recognized for its therapeutic properties in treating various diseases and disorders. Historically, it has been utilized to address cardiovascular health issues, angina, and certain cancer treatments, such as stomach cancer and hemorrhoid tumors. Additionally, it has been employed in treating liver diseases, including infirmities and hot distempers. Dark chocolates, in particular, are rich in phenolic acids and flavonoids, making them valuable for health benefits (Waterhouse et al., 1996).

Figure 12.2 illustrates the diverse therapeutic uses of cacao, showcasing its potential in managing various health conditions, such as liver diseases, cancer, angina, and cardiovascular health. Flavanols, the primary flavonoids found in cacao, include flavanol monomers like epicatechin and catechin, as well as dimer to decamer procyanidin oligomers, which are renowned for their antioxidant properties (Kerimi et al., 2015; Williamson et al., 2016).

The unique flavor of chocolate is predominantly derived from cacao. The volatile aroma fraction, believed to contain the essence of cacao-specific flavor, undergoes complex post-harvesting processes. Key intrinsic components contributing to the flavor and aroma of cacao beans include epicatechin, catechin, procyanidin, alkaloids, theobromine, and caffeine. Other polyphenols present in cacao, such as luteolin, naringenin, isoquercitrin, quercitrin, and apigenin, along with trace amounts of caffeine and theobromine, further enrich its flavor profile (Lamuela-Raventos et al., 2005; Martin et al., 2016).

Polyphenol oxidases play a crucial role in cacao flavor development by catalyzing the oxidation or polymerization of polyphenols, which interact with carbohydrates and proteins, influencing the flavor, color, and texture of cacao beans (Jalil et al., 2008). Cacao polyphenols are primarily composed of 4% anthocyanins, 37% flavan-3-ols, and 58% proanthocyanidins, with catechins representing a significant portion (29–38%) of polyphenols present in

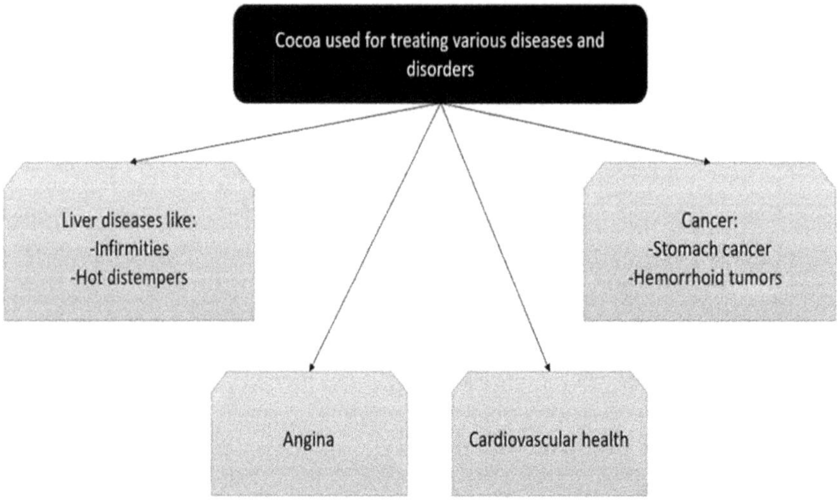

Figure 12.2 A diagram that illustrates the uses of cacao for treating various diseases and disorders. It shows connections between cacao and specific health issues, including liver diseases, cancer, angina, and cardiovascular health (Amin et al., 2004; Keen et al., 2001).

cacao and chocolate (Aprotosoaie et al., 2016; Arunkumar et al., 2019; Barisic et al., 2019).

Catechins, natural phenols, and antioxidants found in cacao beans and other plants play a crucial role in flavor development. Fermentation and drying processes are essential steps in enhancing the flavor precursors of chocolates. During fermentation, various metabolites, such as organic acids and alcohols, are released, leading to the breakdown of cacao bean pulp by microorganisms like lactic acid bacteria, yeasts, and acetic acid bacteria (Lopez et al., 2022).

Roasting is a crucial process that significantly impacts the quantity of polyphenols and flavor in cacao beans. Intense roasting not only reduces polyphenol levels but also hinders the potential health benefits of cacao products and impairs the flavor of chocolate. This process also affects the growth domain and genotype of cacao, highlighting its importance in determining the quality of chocolate. Therefore, the composition and variety of aroma compounds, crucial for chocolate quality, depend not only on the origin and genotype of cacao but also on the processing methods used for refining cacao beans. Consequently, both roasting and fermentation play pivotal roles in determining the flavor of chocolates or cacao beans (Lin et al., 2021).

Research indicates a reduction in the absolute content of polyphenols during the roasting process, accompanied by changes in the ratio of catechin and epicatechin. Elevated levels of (-)-catechin and decreased levels of (-)-epicatechin are observed due to epimerization under increased temperatures during roasting. Furthermore, roasting leads to an increase in procyanidin B1 and

(+)-catechin levels, while decreasing procyanidin B2 and (-)-epicatechin levels (Aprotosoaie et al., 2016; Ioannone et al., 2014; Stanley et al., 2018).

The refining and transformation phases of cacao beans, especially after fermentation, harvesting, roasting, and drying, involve complex changes in chemical profiles. Postharvest alteration processes and industrial transformation into the endmost cacao products significantly influence the chemical profile of cacao. Fermentation initiates the accumulation of flavor precursors through chemical reactions involving endogenous compounds like polyphenols, proteins, and carbohydrates. These reactions, including dehydration, condensation, cyclization, and fragmentation, transform various flavor precursors such as short-chain peptides, free amino acids, and reducing sugars. Subsequent processing steps, notably roasting, further modify the chemical composition, influenced by factors such as high temperatures and pH levels. The flavor of cacao beans is a crucial quality aspect, with various factors contributing to its composition, including postharvest transformation, bean genotype, fermentation, industrial operations, drying, age of the cacao tree, and soil type (Kongor et al., 2016).

12.3 Factors Influencing Quality Variations in *Theobroma cacao* and Chocolate Manufacturing

12.3.1 Influence of Seed Coat

The seed coat of *Theobroma cacao* plays a crucial role in mediating the exchange of substances and materials between the seed embryo, the maternal plant, and the external environment. While other seed characteristics may be influenced by various factors, the seed coat and the pulp of the fruit are uniquely sourced from the maternal plant. Consequently, the flavor quality of *Theobroma cacao* can be closely linked to the transport properties of the seed coat, which extends beyond mere physical protection of the embryo (M. Andersson et al., 2006).

12.3.2 Impact of Fermentation

Fermentation constitutes a pivotal stage in the production of *Theobroma cacao* (see Fig. 12.3), serving as a cornerstone in the commercial processing of cacao beans.

Another critical phase in cacao production involves the formation of cacao mass, also referred to as cacao liquor. Figure 12.3 depicts fermentation as a fundamental stage in the commercial processing of *Theobroma cacao*. Cacao mass represents the initial fluid stage of cacao bean processing, which undergoes grinding across two to three levels to achieve the desired refinement. It serves as a primary ingredient in the production of various chocolate products.

Research conducted in laboratories in Ghana and Malaysia has demonstrated that the quality of produced chocolate is enhanced when cacao liquor is allowed to ferment for periods of 48 to 72 hours (Lima et al., 2011).

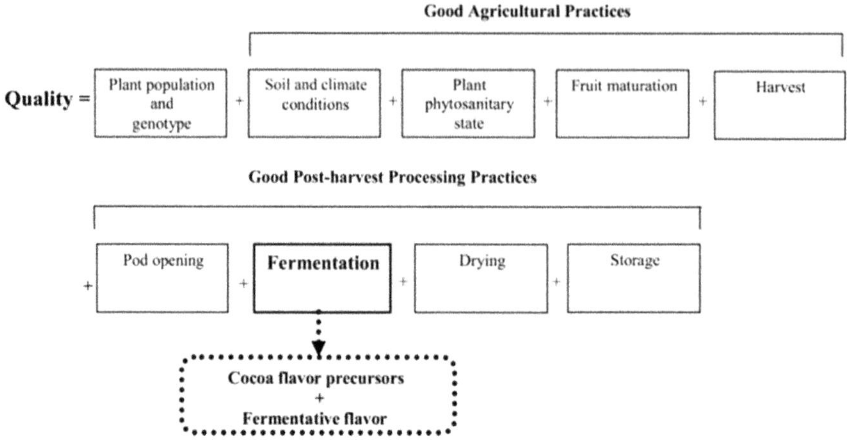

Figure 12.3 Fermentation is one of the core stages in the commercial production of *Theobroma cacao* (Lima et al., 2011).

12.3.3 pH Variations

On one hand, acidification within cacao seeds during fermentation can potentially enhance flavor due to the optimal pH range for proteolysis, typically between pH 3.5 and pH 4.5. However, excessive acidity in cacao beans is known to unfavorably impact flavor, resulting in a weakened chocolate taste (Biehl et al., 1985). Experimental findings revealed significant differences in pH levels among dry cacao beans during fermentation, with longer fermentation times leading to greater pH fluctuations and subsequent decreases in pH values (see Table 12.1). These fluctuations are attributed to the release of organic acids, such as lactic and acetic acids. Nevertheless, it is crucial to strike a balance, as overly acidic beans can diminish flavor. pH levels above 5.0 tend to yield an overall improved flavor profile for cacao beans (Dewandari et al., 2021).

12.3.4 Variation in Fat Content

With prolonged fermentation of cacao beans, there is a concurrent increase in fat content. This can be attributed to the production of fatty acids from various compounds such as carbohydrates, amino acid chains, and phenols by microbial activity.

Table 12.1 Changes in the pH during fermentation (Dewandari et al., 2021)

Fermentation Type	*Fermentation time at 0^0 Celsius*			
	0 hr	*72 hr*	*120s hr*	*168 hr*
Spontaneous	5.37 +/− 0.13	4.43 +/− 0.13	4.97 +/− 0.04	6.54 +/− 0.21
Controlled	5.69 +/− 0.32	4.61 +/− 0.32	5.52 +/− 0.34	6.62 +/− 0.35

12.3.5 Fluctuations in Moisture Content

The moisture content in cacao beans directly correlates with the duration of fermentation. As fatty acids are generated from pulp degradation, pores develop in the cacao beans, facilitating an increase in moisture content. Cacao butter, characterized by its yellowish hue, is the fat content released post-fermentation of cacao. It serves as a vital ingredient in commercial chocolate, significantly influencing its physical properties (Lima et al., 2011). With a melting point close to body temperature (32°C to 35°C), cacao butter readily melts in the mouth, imparting a cooling sensation that intensifies the release of flavor (Tables 12.2 and 12.3).

12.3.6 Bioactive Compounds in Theobroma cacao

Among the essential bioactive compounds found in *Theobroma cacao* are polyphenols, which are categorized into at least ten classes within the plant kingdom (Figure 12.4).

12.3.7 Factors Influencing Theobroma cacao *Cultivation and Their Impact Worldwide*

12.3.7.1 Chocolate Market

Theobroma cacao, an indigenous evergreen plant, yields beans utilized in chocolate production (see Figure 12.4 for major polyphenol classes found in *Theobroma cacao*). The global popularity of chocolate necessitates its cultivation across various regions worldwide. However, several essential conditions crucial for successful cacao harvesting must be considered:

Table 12.2 Changes in the Fat Content during Fermentation (Dewandari et al., 2021)

Type of Fermentation	*Fermentation Time at 0^0 Celsius*		
	72 hr	*120 hr*	*168 hr*
Spontaneous	26.77 +/− 0.52[c]	27.04 +/− 1.12[c]	31.72 +/− 0.56[b]
Controlled	31.46 +/− 0.61[b]	32.41 +/− 0.75[b]	37.20 +/− 1.39[a]

Table 12.3 Changes in the moisture content during fermentation (Dewandari et al., 2021)

Type of Fermentation	*Fermentation Time at 0^0 Celsius*		
	72hr	*120 hr*	*168 hr*
Spontaneous	4.27 +/− 0.22[b]	4.48 +/− 0.30[b]	4.58 +/− 0.51[b]
Controlled	2.97 +/− 0.07[c]	4.35 +/− 0.13[b]	5.29 +/− 0.31[a]

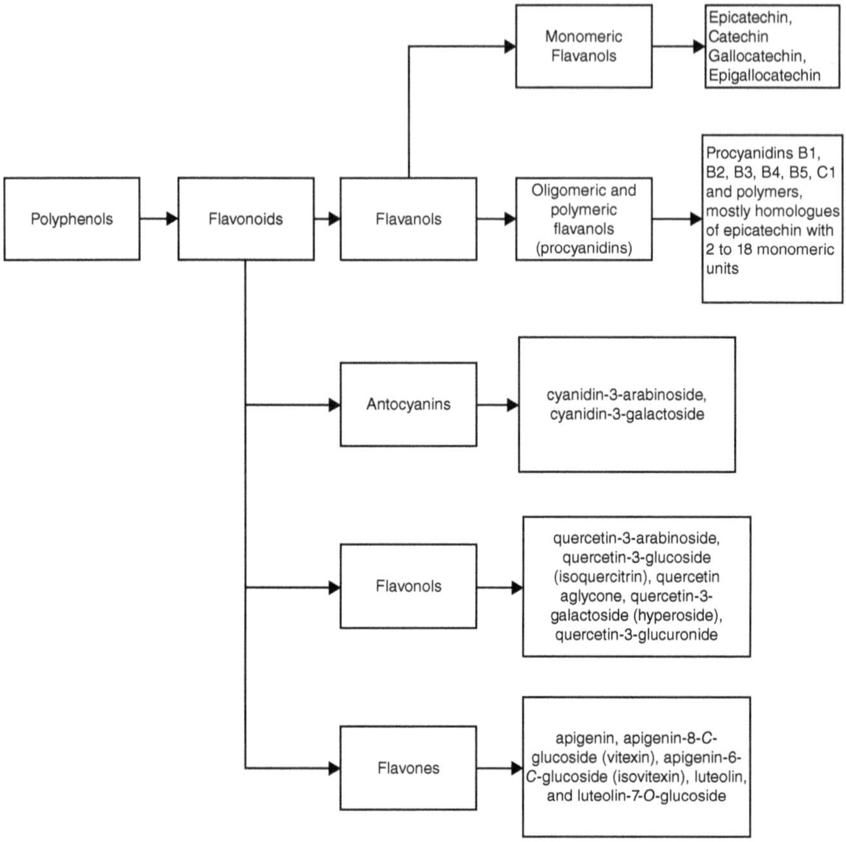

Figure 12.4 Major polyphenol classes found in *Theobroma cacao* (Vásconez et al., 2020).

- Suitable Climatic Conditions: cacao thrives in tropical regions such as Ghana and the Ivory Coast due to their favorable environmental conditions. Equatorial climates with hot temperatures, high humidity, and ample rainfall ranging from 200 to 300 cm annually, along with an average temperature of 27°C, are optimal for high yields. While cacao is cultivated worldwide, its cultivation is preferred within 10° North and South of the equator due to the ideal growth conditions.
- Harvesting Time: cocoa plantations typically commence either at the onset or conclusion of the monsoon season, with specific months varying by region based on local seasonal patterns.
- Appropriate Soil Parameters: cacao can be grown in various soil types, but clay and sandy loam soils with coarse particles and high humus content are preferred. The plant can tolerate both acidic and alkaline soils but thrives best in a pH range of 6.5 to 7.0. Adequate soil ionic balance, with 35% cation exchange capacity, is essential, along with sufficient levels of iron,

potassium, nitrogen, and phosphorus. Proper water-holding capacity and drainage systems are also crucial to prevent waterlogging.

- Adequate Shade: although cacao plants require warmth, direct sunlight can be detrimental. Optimal growth occurs when plants receive only 50% of available light. Additional taller plants are often planted to provide shade, particularly for young cacao plants.
- Protection from Strong Winds: storms and strong winds can damage cacao plants, causing mechanical stress and detachment of unripened pods. Protecting plants from strong winds and hurricanes is essential for healthy development.
- Labor Requirements: cacao cultivation and processing are labor-intensive tasks. The evergreen nature of the plant results in year-round fruit production, with bean harvesting typically occurring during dry months. Harvesting involves chopping down pods, splitting them, fermenting, drying, and finally bagging them, requiring skilled manual labor.
- Transportation: given that cacao is primarily cultivated in humid equatorial regions, beans must be transported overseas for global markets. Adequate transportation networks are necessary to ensure the safe transit of goods, protecting them from heavy rainfall and other logistical challenges.

Chocolate's global popularity has led to its widespread expansion in the chocolate business. It has become an integral part of people's lives worldwide, regardless of age. Table 12.4 illustrates the production of cacao in various countries. Chocolate exerts a significant influence on people's health, mood, and nutritional intake (Szogyi, 1997). While cacao was originally cultivated in Central America, it is now grown and consumed in many other parts of the world. Despite numerous countries producing cacao beans, some nations like Cote d'Ivoire stand out for their significant contribution, accounting for 30% of global chocolate production, which is

Table 12.4 Production of cacao in different countries (Weathers et al., 1983)

Rank	*Country*	*Production of Cacao (metric tons)*
1	Ivory Coast	20,34,000
2	Ghana	8,83,789
3	Indonesia	6,59,828
4	Nigeria	3,38,278
5	Cameroon	2,95,160
6	Brazil	2,35,899
7	Ecuador	2,05,990
8	Peru	1,21,850

reflected in the country's GDP. Renowned brands such as Nestle and Cadbury source their cacao from these nations.

Over time, the global chocolate business has experienced exponential growth, driven by its distinctive taste that captivates consumers worldwide. This industry holds a strong position in the global market, evident from the billions of dollars in sales generated by these popular ventures. The multibillion-dollar global chocolate market has been extensively reviewed in several papers (Bloss, 2010; International Cocoa Organization, 2020; Technavio, 2020), which highlight the increasing demand and consumption of chocolate over the years. However, the chocolate market faced a downturn during the COVID-19 pandemic due to disruptions in exports, labor shortages, and soaring commodity prices (Cadby et al., 2020). Lockdown measures significantly impacted the business, particularly in the second half of the year since 2014 (Terazono et al., 2020), resulting in a sharp decline in chocolate prices.

In addition to plummeting prices, the demand for cacao also decreased during the pandemic. Producers faced challenges as transportation limitations hindered market demand despite ample production. Border closures and export restrictions further exacerbated the crisis, leaving cultivators in a precarious situation. However, as lockdown restrictions eased, several countries witnessed a rebound in the chocolate market, leading to significant economic growth, as reported by AIPC.

12.4 Conclusion

Chocolate stands as the world's most popular dessert, consumed not just occasionally but as an essential treat in daily life. It has become ubiquitous in supermarkets globally, alongside bakery and confectionery products, owing to its extended shelf life. While sweet chocolates are favored by children, dark chocolate, known as cacao, is highly popular among adults due to its positive health impacts and lower calorie content. This chapter explores various factors related to *Theobroma cacao* cultivation and its significance in the chocolate market, shedding light on how cacao transforms into chocolate and is sold worldwide. Factors such as pH, temperature, seed quality, and processing techniques play crucial roles in cacao processing, while the impact of the COVID-19 pandemic on the chocolate market underscores the industry's resilience and adaptability to external challenges.

Ethical Permission

Not Applicable

Declaration of Conflict

The authors don't have any conflicts of declaration.

Acknowledgment

The authors would like to pay gratitude to Chandigarh University and Dr. Jagadeesh Chandra Bose K* for sharing knowledge with us.

References

Amin, I., Faizul, H. A., & Azli, R. (2004). Effect of cacao powder extract on plasma glucose levels in hyperglycaemic rats. *Nutrition & Food Science*, *34*, 116–121.

Aprotosoaie, A. C., Luca, S. V., & Miron, A. (2016a). Flavor chemistry of cacao and cacao products—An overview. *Comprehensive Reviews in Food Science and Food Safety*, *15*(1), 73–91.

Aprotosoaie, A. C., Miron, A., Trifan, A., Luca, V. S., & Costache, I. I. (2016b). The cardiovascular effects of cacao polyphenols—An overview. *Diseases*, *4*(4), 39.

Barišić, V., Flanjak, I., Križić, I., Jozinović, A., Šubarić, D., Babić, J., … & Ačkar, Đ. (2020). Impact of high-voltage electric discharge treatment on cacao shell phenolic components and methylxanthines. *Journal of Food Process Engineering*, *43*(1), e13057.

Bloss, R. (2010). Palletizing candy orders and never squeezing the chocolates. *Assembly Automation*, *30*, 32–35.

Cadby, J., & Araki, T. (2021). Towards ethical chocolate: Multicriterial identifiers, pricing structures, and the role of the specialty cacao industry in sustainable development. *SN Business & Economics*, *1*(3), 1–36.

Counet, C., Callemien, D., & Collin, S. (2006). Chocolate and cacao: New sources of trans-resveratrol and trans-piceid. *Food Chemistry*, *98*(4), 649–657.

Dave, B. S., Amin, A. F., & Patel, M. M. (2004). Gastroretentive drug delivery system of ranitidine hydrochloride: Formulation and in vitro evaluation. *Aaps PharmSciTech*, *5*(2), 77–82.

Decker, H., Dillinger, R., & Tuczek, F. (2000). How does tyrosinase work? Recent insights from model chemistry and structural biology. *Angewandte Chemie International Edition*, *39*(9), 1591–1595.

Dewandari, K. T., Rahmawati, R., & Munarso, S. J. (2021). The effect of techniques and fermentation time on cacao beans quality (*Theobroma cacao* L.). In *IOP Conference Series: Earth and Environmental Science* (Vol. 653, No. 1, p. 012046). IOP Publishing.

Dillinger, T. L., Barriga, P., Escárcega, S., Jimenez, M., Lowe, D. S., & Grivetti, L. E. (2000). Food of the gods: Cure for humanity? A cultural history of the medicinal and ritual use of chocolate. *The Journal of Nutrition*, *130*(8), 2057S–2072S.

Fraga, C. G., Actis-Goretta, L., Ottaviani, J. I., Carrasquedo, F., Lotito, S. B., Lazarus, S., … & Keen, C. L. (2005). Regular consumption of a flavanol-rich chocolate can improve oxidant stress in young soccer players. *Clinical and Developmental Immunology*, *12*(1), 11–17.

Fredholm, B. B., & Hedqvist, P. (1980). Modulation of neurotransmission by purine nucleotides and nucleosides. *Biochemical Pharmacology*, *29*(12), 1635–1643.

Fredholm, B. B., Bättig, K., Holmén, J., Nehlig, A., & Zvartau, E. E. (1999). Actions of caffeine in the brain with special reference to factors that contribute to its widespread use. *Pharmacological Reviews*, *51*(1), 83–133.

Goya, L., Martín, M. Á., Sarriá, B., Ramos, S., Mateos, R., & Bravo, L. (2016). Effect of cacao and its flavonoids on biomarkers of inflammation: Studies of cell culture, animals and humans. *Nutrients*, *8*(4), 212.

Halib, N., Amin, M. C. I. M., & Ahmad, I. (2012). Physicochemical properties and characterization of nata de coco from local food industries as a source of cellulose. *Sains Malaysiana*, *41*(2), 205–211.

Hurst, W. J., Stanley, B., Glinski, J. A., Davey, M., Payne, M. J., & Stuart, D. A. (2009). Characterization of primary standards for use in the HPLC analysis of the procyanidin content of cacao and chocolate containing products. *Molecules*, *14*(10), 4136–4146.

Ioannone, F., Di Mattia, C. D., De Gregorio, M. I. R. I. A. M., Sergi, M., Serafini, M. A. U. R. O., & Sacchetti, G. (2015). Flavanols, proanthocyanidins and antioxidant activity changes during cacao (*Theobroma cacao* L.) roasting as affected by temperature and time of processing. *Food Chemistry*, *174*, 256–262.

Jalil, A. M. M., & Ismail, A. (2008). Polyphenols in cacao and cacao products: Is there a link between antioxidant properties and health? *Molecules*, *13*(9), 2190–2219.

Jalil, A. M. M., Ismail, A., Pei, C. P., Hamid, M., & Kamaruddin, S. H. S. (2008). Effects of cacao extract on glucometabolism, oxidative stress, and antioxidant enzymes in obese-diabetic (Ob-db) rats. *Journal of Agricultural and Food Chemistry*, *56*(17), 7877–7884.

Kenny, T. P., Keen, C. L., Jones, P., Kung, H. J., Schmitz, H. H., & Gershwin, M. E. (2004). Cacao procyanidins inhibit proliferation and angiogenic signals in human dermal microvascular endothelial cells following stimulation by low-level H2O2. *Experimental Biology and Medicine*, *229*(8), 765–771.

Kerimi, A., & Williamson, G. (2015). The cardiovascular benefits of dark chocolate. *Vascular Pharmacology*, *71*, 11–15.

Kondo, K., Hirano, R., Matsumoto, A., Igarashr, O., & Ltahura, H. (1996). Inhibition of LDL oxidation by cacao. *The Lancet*, *348*(9040), 1514.

Kongor, J. E., Hinneh, M., Van de Walle, D., Afoakwa, E. O., Boeckx, P., & Dewettinck, K. (2016). Factors influencing quality variation in cacao (*Theobroma cacao*) bean flavour profile—A review. *Food Research International*, *82*, 44–52.

Lamuela-Raventós, R. M., Romero-Pérez, A. I., Andrés-Lacueva, C., & Tornero, A. (2005). Health effects of cacao flavonoids. *Food Science and Technology International*, *11*(3), 159–176.

Lima, L. J., Almeida, M. H., Nout, M. R., & Zwietering, M. H. (2011). *Theobroma cacao* L., "the food of the gods": Quality determinants of commercial cacao beans, with particular reference to the impact of fermentation. *Critical Reviews in Food Science and Nutrition*, *51*(8), 731–761.

Lin, Y. C., Choong, Y. M., & Chu, H. L. (2021). Effects of processing time and temperature on flavanol and procyanidin, proanthocyanidin and antioxidant activity of cacao bean in Taiwan. *Journal of Food and Nutrition Research*, *9*(1), 10–17.

Lorig, K. R., Sobel, D. S., Stewart, A. L., Brown Jr, B. W., Bandura, A., Ritter, P., ... & Holman, H. R. (1999). Evidence suggesting that a chronic disease self-management program can improve health status while reducing hospitalization: A randomized trial. *Medical Care*, *37*, 5–14.

Martín, M. A., & Ramos, S. (2016). Cacao polyphenols in oxidative stress: Potential health implications. *Journal of Functional Foods*, *27*, 570–588.

Martin, M. Á., Goya, L., & Ramos, S. (2016). Antidiabetic actions of cacao flavanols. *Molecular Nutrition & Food Research*, *60*(8), 1756–1769.

Muhammad, D. R. A., Fibri, D. L. N., Saputro, A. D., & Sunarharum, W. B. (2022). Investigating the elicited emotion of single-origin chocolate towards sustainable chocolate production in Indonesia. *Open Agriculture*, *7*(1), 382–391.

Myers, A. n.d.Brazil's cacao production bounces back after Covid problems. (2009) https://www.confectionerynews.com/Article/2022/02/28/brazil-s-cacao-production-bounces-back-after-covid-problems

Othman, A., Ismail, A., Ghani, N. A., & Adenan, I. (2007). Antioxidant capacity and phenolic content of cacao beans. *Food Chemistry*, *100*(4), 1523–1530.

Othman, A., Jalil, A. M. M., Weng, K. K., Ismail, A., Ghani, N. A., & Adenan, I. (2010). Epicatechin content and antioxidant capacity of cacao beans from four different countries. *African Journal of Biotechnology*, *9*(7), 1052–1059.

Persson, I. A. L., Persson, K., Hägg, S., & Andersson, R. G. (2011). Effects of cacao extract and dark chocolate on angiotensin-converting enzyme and nitric oxide in human endothelial cells and healthy volunteers–A nutrigenomics perspective. *Journal of Cardiovascular Pharmacology*, *57*(1), 44–50.

Quelal-Vásconez, M. A., Lerma-García, M. J., Pérez-Esteve, É., Talens, P., & Barat, J. M. (2020). Roadmap of cacao quality and authenticity control in the industry: A review of conventional and alternative methods. *Comprehensive Reviews in Food Science and Food Safety*, *19*(2), 448–478.

Sarmadi, B., Aminuddin, F., Hamid, M., Saari, N., Abdul-Hamid, A., & Ismail, A. (2012). Hypoglycemic effects of cacao (*Theobroma cacao* L.) autolysates. *Food Chemistry*, *134*(2), 905–911.

Sarwan, J., Uddin, N., Sahani, S., Mittal, K., Maurya, D. D. K. P., & Jagadeesh Chandra Bose, K. (2024b). Isolation and characterization of novel Bacillus with dual enzymes, biodegradation potential: a special reference to cellulolytic and proteolytic activities. *Biomass Conversion and Biorefinery*, 1–17.

Bose K. J. C., & Sarwan, J. (2023). Multi-enzymatic degradation potential against wastes by the novel isolate of Bacillus. *Biomass Conversion and Biorefinery*, 1–14.

Sarwan, J., Bose, J. C., Kumar, S., Bhargav, S. S., Dixit, S. K., Sharma, M., ... & Uddin, N. (2024a). Biodegradation of Cellulosic Wastes and Deinking of Colored Paper with Isolated Novel Cellulolytic Bacteria. *Nature Environment and Pollution Technology*, *23*(2), 761–773.

Soares, T. F., & Oliveira, M. B. P. (2022). Cacao by-products: Characterization of bioactive compounds and beneficial health effects. *Molecules*, *27*(5), 1625.

Squicciarini, M. P., & Swinnen, J. (Eds.). (2016). *The Economics of Chocolate*. Oxford University Press.

Szogyi, A. (Ed.). (1997). *Chocolate: Food of the Gods (No. 14)*. Greenwood Publishing Group.

Terazono, E., & Munshi, N. (2020). Choco waves: How coronavirus shook the cacao market. *Economic Times*.

Thompson, S. S., Miller, K. B., Lopez, A. S., & Camu, N. (2012). Cocoa and coffee. *Food microbiology: fundamentals and frontiers*, 881–899. Book Editor(s):Michael P. Doyle, Robert L. Buchanan- Wiley Online.

Toma, S. G., & Săseanu, A. S. (2020). Chocolate, a global business. *New Trends in Sustainable Business and Consumption, 1233.*

Voigt, J., & Biehl, B. (1995). Precursors of the cacao-specific aroma components are derived from the vicilin-class (7S) globulin of the cacao seeds by proteolytic processing. *Botanica Acta*, *108*(4), 283–289.

Waterhouse, A. L., Shirley, J. R., & Donovan, J. L. (1996). Antioxidants in chocolate. *The Lancet*, *348*(9030), 834.

Williamson, G. (2017). The role of polyphenols in modern nutrition. *Nutrition Bulletin*, *42*(3), 226–235.

Zhu, Q. Y., Schramm, D. D., Gross, H. B., Holt, R. R., Kim, S. H., Yamaguchi, T., ... & Keen, C. L. (2005). Influence of cacao flavanols and procyanidins on free radical-induced human erythrocyte hemolysis. *Clinical and Developmental Immunology*, *12*(1), 27–34.

13 Post-harvest quality of cacao (*Theobroma cacao* L.) for the chocolate industry

Frank Intriago Flor, Kerly Alvarado Vásquez, Jaime Vera Chang, Luis Vásquez Cortez, Solanyi Tigselema Zambrano, and Cristhian Verduga López

13.1 Introduction

The domestication, cultivation, and consumption of *Theobroma cacao* L., were carried out by the Toltecs, Aztecs, and Mayas about 2,000 years ago; however, recent research indicates that at least one variety of cacao has its origin in the Upper Amazon, 5,000 years ago (Cruz and Pereira, 2009).

Cacao is a tree of economic interest, of tropical origin, and its domestication has allowed obtaining different purposes for agroindustry, food, and chemical industries, and in history this fruit has been considered medicinal and even had a more expensive value than gold by numerous cultures throughout the Ancestral times (Tornés, 2013). Although it is still found in the wild, cacao has been domesticated since time immemorial and has been cultivated in Central and South America and is now cultivated worldwide. Its name *Theobroma cacao* comes from the Greek words 'theo' and 'broma' meaning 'food of the Gods', while the epithet cacao is a term that comes from Nahuatl and is considered Latinized (Attolini, 2011).

This type of crop is divided into three genetic groups: criollo, forastero, and trinitario. The first group integrates a diversity of species that are currently uncommon due to the decrease in the areas cultivated with this variety; on the other hand, the group of the forastero variety is characterized by high yields and greater resistance to pests and diseases; and finally, the trinitario variety is characterized by its high resistance to pests and diseases (Maldonado and Sánchez, 2021).

According to Anecacao, within the Ecuadorian territory, fine aroma cacao represents only 5% of world cacao production, making it a producer par excellence of fine and aroma Cacao Arriba (63% of world production) from the Nacional variety whose flavor has been recognized for centuries in the international market. This type of bean is used in all refined chocolates. However, what many do not know is that fine chocolate is distinguished by its purity, specifically, the flavor and fragrance that the cacao has (Córdova et al., 2022).

DOI: 10.4324/9781003381761-17

The fine aroma cacao represents the presentation card of high-quality cacao producers. Known for its floral and fruity aromas depending on the place of origin, it is one of the cacao varieties that is most appreciated by renowned manufacturers because of its organoleptic and sensory properties being unique coming from the Ecuadorian regions (Mata, 2018). The morphological development of cacao begins in a green cob that changes to yellow when it matures, hence the name 'pepa de oro' (Larrea, 2008); in past times, the known 'cacao boom' made this the main source of wealth and livelihood of many families throughout Ecuador. As it highlights the flavors of chocolate, it is highly desired by producers, chefs, and consumers to prepare high-end chocolates (Abad et al., 2020).

From the beans or grains of this species (also called almonds), fermented and dried or unfermented cacao paste or liquor, cacao butter, cacao powder, and traditional or gourmet chocolates are prepared in addition to being able to create new nontraditional products such as dark chocolate based on dehydrated carambola fruits and made with raw material from cacao crosses (Dueñas et al., 2018).

The recognition of fine and aroma cacao is in constant demand and is recognized worldwide, mainly in the United States and Europe, witnessing a significant increase in exports in the Asian market (Abad et al., 2020). By 2023, chocolate had become a luxury food, which favored the country significantly and was another opportunity for growth (Asanza et al., 2019).

According to the report of the Ministry of Agriculture ([MAG], 2021a), during this period at the territorial level cacao production reached a total of 302,094 tons (t) with a yield of 0.56 t per hectare (ha) harvested, mostly centered in the provinces of Los Ríos (83. 677 t; 27.70%), Guayas (72,335 t; 23.94%), Manabí (44,825 t; 14.84%), Esmeraldas (36,671 t; 12.14%), Santo Domingo de los Tsachilas (14,512 t; 4.80%) and the remaining production represented by 16.58% from the provinces of El Oro, Zamora Chinchipe, Sucumbíos, Orellana, Cotopaxi, and Pichincha.

In this sense, Lema (2019), mentions that during 2019 cacao production contributed 5.9% to the agricultural Gross Value Added (GVA). The GVA was positive because it generated an increase in exports, of which 270,940 tons were of raw and roasted cacao, while the imports were only 60 tons of raw and roasted cacao, which represented USD 657,256 (value of the goods placed on board a maritime transport).

In this scenario, cacao exports for 2021 reached a total of 764,926.1 t, proportionally increasing the USD to 1,848,797.6 thousand. These exports were distributed in countries that integrate the ANSA+3 (38.10%), NAFTA (31.00%), the EU (28.20%), and other countries with lower export representation with 2.30% (MAG, 2021b).

Cacao almonds are a rich source of polyphenols corresponding to about 15% of the dry weight and in total alkaloids of 4%, which determine the flavor and aroma characteristics of cacao. These abovementioned polyphenols give a perception of bitterness and astringency and enhance the green and fruity aroma of almonds. Alkaloids are known for their bitterness and there is a relationship between drinkability and their quantity but there is a vector in the quality and its molecular group conformation. There is a possibility of changing the flavor or aroma, in addition to its sensory organoleptic characteristics, both being genetic and environmental consequences and not due to contamination. This would affect the antioxidant quality and phenolic content of cacao (Vasquez et al., 2016).

The excellent quality of almonds is reflected in the wide production and export for its demand to international markets being this fundamental for the livelihood of small and medium farmers who are dedicated to the cultivation of this fruit (Villamar et al., 2016). However, there is an emphasis on the time of exporting to European countries because they are regulated by the EU by Regulation No. 2021/1323 which mentions the maximum levels of cadmium (Cd). The levels of Cd should not be greater than 0.80 mg/kg of chocolate. For this reason, cacao exports in the EU correspond to $256.9 million dollars, being the most important export item and the first exportable product to foreign countries either as cacao beans. This has a direct influence on the different families dedicated to fieldwork and cacao production, who in the eagerness to speed up production processes have inappropriately used chemical products that over time have accumulated on the cultivated soils (Cardona et al., 2016).

This chapter emphasizes the effects of heavy metals on cacao production and the alternatives established to mitigate the effect of heavy metal accumulation.

13.1.1 Nacional cacao

This variety of cacao is unmatched and unique in the entire planet. It is characterized as having an excellent quality being unique in the market as well as being preferred by the industry for its short fermentation. It is used as a raw material in making one of the best chocolates in the world due to its aroma and eccentric flavor that are internationally recognized (Fine cacao flavor). One of the drawbacks of the production of Nacional cacao is they are very sensitive to pests, which affects its production (Mendoza et al., 2021).

13.1.2 Criollo cacao

It is a cylindrical shaped cob, composed of five grooves that have a yellow or reddish hue. The kernels are of a cylindrical white coloration. They are sensitive to diseases and pests (Arvelo et al., 2017).

13.1.3 Forastero cacao

It has a greenish hue with a smooth fruit coating. In some warty cobs, there are two terminals in a circular shape in addition to a bottleneck at the base. Its almonds are of purple color, triangular, small, and flattened. A primary advantage of this type is its stronger resistance to pests or diseases, being very different from the Nacional and Criollo types (Arvelo et al., 2017).

13.1.4 Trinitario cacao

Trinitario cacao occupies a high percentage of world production. The CCN-51 clone has high production and disease tolerance characteristics but does not have the aroma of the Nacional clone. CCN-51 cacao plants reach a height between 6 to 8 meters, can be planted at a distance of 3 x 3 meters, and have a weighted amount of aerial biomass and root system (Nivela, 2020).

13.1.5 CCN-51 cacao

This variety originated through cross-cloning multiple types of cacao. In the 1890s, a crisis paralyzed Suriname with a serious disease that affected cacao production and caused huge losses. This so-called witches' broom, a fungal biological agent, affected some countries of South America such as Peru, Trinidad, the Colombian Republic, Guyana, Tobago, Brazil, Panama, and Peru multiple times during the twentieth century. Therefore, Homero Castro, an Ecuadorian independent research scientist and agronomist, developed Castro Naranjal 51 and had his name marked in the history of the cacao field. This variety was of preference and gained popularity for being more resistant than the Nacional cacao and began to dominate the scene of fine cacao in all regions of Ecuador (Anzuelos et al., 2019).

13.1.6 Cacao clones

There are several clones of cacao since they are experimental investigations of the Finca La Represa of the State Technical University of Quevedo for improving their organoleptic and sensory characteristics.

13.1.7 Harvest and cob selection

The harvest is the most important stage because this stage will have much emphasis on the fermentative part. The first instance is the hue of the cob because harvesting will be performed depending on the shades of cacao. According to the variety, the color can vary from green to red, yellow, and orange. Ancestrally, maturity is also understood by the sound obtained from a blow to the cob.

Harvesting is the process that allows the pruning step to the collection of ripe fruits in good condition separated by varieties. To have better results at the time of harvesting, it is advised to do it with tweezers in order to avoid damage to the cut and avoid injury to the tree.

At the time of extraction of the cacao beans, it is recommended to make the cuts longitudinally to the cob to avoid hurting the almonds because injured almonds will not ensure good fermentation and health of the almond.

The almonds that have not reached harvest maturity are not suitable because they would not have adequate and/or optimal brix degrees for the fermentation of cacao. This will affect the final result in terms of sensory quality.

It is important to perform constant monitoring to avoid losing organoleptic and sensory properties that highlight the aroma, flavor, and color of the cacao bean as poor handling would affect the quality and result in an unpleasant final product for the consumer (Sanchez et al., 2008). It is advisable to harvest ripe fruits for the reason that they provide almonds efficiently for excellent fermentation in such a way as to manage an excellent fermentative stage. At the time of harvesting, immature cobs will yield low-quality beans with the presence of molds, slate, and violet tone besides being wrinkled, which would affect the industrialization process and result in undesirable flavors and organoleptic qualities (Quiñonez et al., 2018).

At the time of selecting the pods, they should be classified according to the variety they belong to and the kernels should also be classified according to the type of cacao. At the stage of cacao processing, the genetic material, whether National, Trinitario, or Forastero, of CCN-51 should not be mixed and kept separate at the time of marketing because each one has its characteristics and aroma quality (Ahmed et al., 2019).

Cacao pods are harvested at different stages of maturity (from grade zero to grade 5). A grade 0 cacao fruit has two distinctive colors of the degree of maturity which are dark violet colors and greenish-yellowish grooves. The slime has a very solid silky characteristic, and the kernels are not developed. Grades 1 and 2 have the presence of a less violet color, have a yellowish hue on the outside of the cob, and the cacao slime is less silky and begins to be juicy. At the same time, for grade 3, the groove on the back of the cacao begins to have a distinctive shade of a mature cob, and for grade 4, the cacao fruit is almost ready to be harvested with the early end of the physiological development and genotype in a complete way; the cacao slime is about to finish its development and the internal characteristic of the bean is rougher and less poorly done. For the final stage corresponding to grade 5, the cacao fruit would have completed its physiological development, and corresponding to the genotype, the external and internal qualities are finalized and are recommended for harvesting (Figure 13.1.) (Mora et al., 2021).

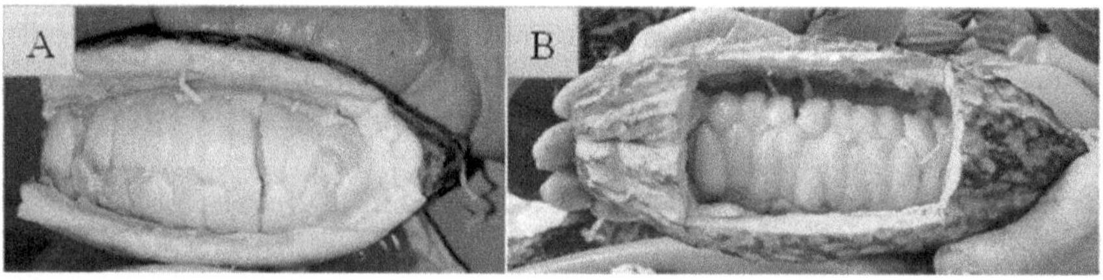

Figure 13.1 A) Cacao pods, physiological maturity stage 3 and grade 3. B) Maturity stage grade 5.

It is advisable to classify the green and brownish cobs in order to change their physiological state of maturity to 100% so that they can be fermented. If this is not done, the cobs of this origin are known as slaty almonds, while the overripe cobs generate sprouted and moldy almonds. So, this practice is not recommended and the addition of whole cobs is avoided due to the effect that they lose the product quality (Figure 13.2) (Quiñonez et al., 2018).

13.1.8 Extraction of cacao beans

After having collected the cacao pods, we proceed to remove the seed in a non-craft and artisanal way. The procedure is to slide through the cavities of the centric vein of the fruit with the help of the fingers or with an instrument. To facilitate the work and in order to prevent taking into account the mixture with the beans, it is advisable to select the cobs that are not sick and/or have the presence of any parasites because they will lower the quality of cacao beans. If the cob is too ripe, we will get germinated seeds and the presence of molds (Figure 13.3) (Garcia et al., 2021).

13.1.9 Post-harvest stage

Also called cacao processing, this procedure is to dry the almonds from the ripe, fermented, and dried fruit in an appropriate manner with the goal of achieving almonds of excellent quality and suitable for sale so that it is acceptable and can also be sold at a good price in the regional market and the best abroad. Therefore, these activities are as important as harvesting, fermentation, and drying (Vera et al., 2008).

The sourdough or fresh cacao mass the main step in the processing of cacao is the development of flavors and aromas in such a way that the characteristics of the almonds begin to swell and have a brownish color, besides being striking

Figure 13.2 A) Harvest of healthy cobs of Nacional variety. B) Collection of mature cobs.

Figure 13.3 Extraction of cocoa beans separated by variety.

for its appearance. As previously mentioned, optimal fermentation proceeds to originate cacao that later is manufactured into very appetizing chocolate for the palate and smell of the consumer. This process has its vital importance that corresponds to the separation of cacao slime and fixing flavors and aromas so that the embryo of the almond dies, and therefore, we can obtain the chocolate that the consumer desires so much (Figure 13.4) (Machuca et al., 2019).

13.1.10 Fermentation time

Time is important and depends on the type of cacao. The cacao beans should be between three to four days but this depends on the amount of mucilage; it would be around 72 to 96 hours from the deposit of the beans in the containers. It is advisable not to mix cacao beans of various kinds or from other days of harvesting because this would affect fermentation as the process would not be complete and we would have a poor-quality product. It is advisable to collect well-organized cobs which is feasible in small volumes and to ferment the entire harvest on the same day (Rivera et al., 2012).

Figure 13.4 Fermentative stage A). Day one of fermentation B). The fourth day of fermentation.

Fermentation is carried out in three stages:

1 The stage begins with the cacao full of mucilage that previously was removed from the cacao pods. This mucilage has a lot of soluble solids.
2 After the aforementioned stage, the soluble cacao solids in the mucilage are converted into alcohol by beneficial microorganisms and fungi. This process is called the alcoholic stage.
3 The acetic or acid stage is where the alcohol mentioned in the previous stage is chemically converted into vinegar and acetic acid.

Fermentation can be carried out in various types of containers, the main thing is the exit of the mucilage, preserving the temperature in a controlled manner and avoiding the existence of strange odors or flavors of cacao. Several types of fermentation are widely used such as: in fibers, jute sacks to piles on the ground, or wooden crates built of wood. The cacao world farmers ancestrally possess fermentative management of cacao beans (Freire and Díaz, 2016).

13.1.11 Methods for the fermentation of cacao beans

In some places, the ancestral fermentation method is still practiced, which consists of putting the grain in fermenters. One of them is to put them in a plastic tub covered in such a way that air does not enter and air filtration does not occur, avoiding contamination by pathogenic microorganisms. Another method to ferment is the heap or rumos of grains. This is one of the oldest methods that exists and is based on piling the beans on the ground or on plastic to facilitate draining and covering with banana leaves or jute or plastic bags in order to have the internal heat. The result of the fermentation is optimal. These rums must be exposed to sunlight and must be removed one to two times a day. These fermentation techniques are still used in order to reduce costs and obtain productive gains (Teneda, 2016).

13.1.11.1 Rohan box or vat fermentation

It is considered as one of the most common methods that have been experimented with in order to obtain quality results. Depending on climatic factors and the variety of cacao that is subjected to fermentation, the amount of beans can vary between 100 to 200 kg of cacao beans (Suarez et al., 1994).

13.1.11.2 Bulk fermentation

The heap fermentation method is an ancestral method. For this treatment, we use banana leaves which are used as stretchers to facilitate the drainage of the slime or mucilage. This treatment generates excellent responses and without having to generate major expenses, the base for this fermentation is made with chopped cane or wooden planks. It should be taken into account that this pile should not be exposed to rainwater or water channels, and as far as possible it should be kept in dry places without exposure to the soil since all these factors will affect the final product, even interrupting the fermentation process. The optimum temperature required is between 40° to 45°C (Suárez et al., 1994).

13.1.11.3 Fermentation by vat or baskets

The main characteristics of this fermentation treatment are baskets that are normally woven by the Africans. The material is taken from cane or sometimes you can find baskets made from palm trees. These baskets can contain up to 40 kg of cacao beans at the time of fermentation and must be covered with banana or plantain leaves for proper fermentation. This process takes up to five days before being removed for the homogenization of the beans (Suarez et al., 1994).

13.1.11.4 Fermentation in wooden boxes

Around the world, this treatment is used. These boxes can contain up to 300 kilograms. This method requires certain techniques: one of them is to place the boxes suspended in order to drain the mucilage. Another technique is to place the boxes at ground level. Finally, there is a procedure that involves suspending the boxes as steps. When fermentation is carried out, the appropriate method must be taken into account so that the results are as expected and optimal homogenization and standardization of the cacao beans are obtained (Suárez et al., 1994).

13.1.11.5 Fermentation in trays

This method is the most satisfactory since it corrects the errors of the other fermentation methods that do not achieve the correct homogeneity of the cacao beans. In this type of fermentation, only the center of the fermentation tray is filled so that the fermentation reaches the optimum temperature (Suárez et al., 1994).

13.1.11.6 Rotary or barrel fermentation

This fermentation method consists of fermenting in a rotating drum made from white wood, which allows fermenting up to 200 to 250 kg of cacao. Depending on the size of the drum, its system qualities are simple and coupled to an external remission system facilitating the overall homogenization of cacao beans. The advantage is that a large amount of cacao can be removed once every 24 hours due to its drainage potential (Ureña and Sandoval, 2021).

13.1.11.7 Drying process

The purpose of drying is to gradually reduce the humidity of the cacao almonds to reach an ideal temperature between 6% and 7%. If the values are less than 6%, the quality is low and the almond is very brittle. If the humidity is higher than 8%, the almonds develop molds and acidity, which would directly damage the quality of aroma and flavor and be a detriment to the industry (Rivera et al., 2012).

When performing drying, some producers recommend performing it in two stages: the pre-drying and the main drying. The first stage is done with the point of removing the almonds using evaporation. The acidity that the cacao beans have when leaving the fermenter, later in the drying stage, would be ready to be removed. As a result of fermentation, the acidity of cacao beans is reduced and its flavor and aroma would be optimal for the production of good chocolates (Tinoco and Ospina, 2013).

Already ready and dry fermented almonds should be processed as raw materials suitable and of excellent quality for the production of cacao paste, cacao butter, and cacao; the first process should be done by roasting the almonds to change the color tone and change their flavor and then proceed to the elimination of the testa. So, this process is carried out after roasting and shelling (Alvarez et al., 2010).

13.1.11.8 Different drying methods

✔ **Sun-drying**: Cement roofs, cane or wood slats, canopy.
✔ **Artificial drying**: Metal dryers (rotary) fuel, gas, and firewood (Pérez, 2017).

13.1.11.9 Ecuadorian cacao raw material and its efficiency in the food industry

The word quality is important for Ecuadorian cacao production, and the management of quality the recognized and satisfactory journey is known throughout the world. Hernandez et al. (2015) mention in entirety that it is necessary to maximize and have greater improvement so that the raw material is optimal and results in quality industrialized products and most significantly minimizes losses that are detrimental both for the country and the industry as an agricultural economy is one of the most important incomes of Ecuador.

The National Institute of Agricultural Research of Ecuador suggests that the result of high-quality cacao depends on several vectors such as the genetic material, providing improvements for the production of resistance to diseases suffered by cacao, care, and good post-harvest management because it is one of the best in Latin America that highlights its organoleptic properties (Medina et al., 2009). INIAP mentions that the use of agrochemicals and phosphate fertilizers should be avoided in every possible way because it has negative effects and contracts with itself through the increase of heavy metal 'Cd' in raw materials and the final product (Rubio and Carrillo, 2018).

Thus, quality is closely linked to the agroindustrial value chain because of its traceability in cacao processing.

The word quality in Ecuadorian cacao is for the production and management of quality throughout the recognized and satisfactory journey known throughout the world, the meaning of quality. Hernandez et al. (2015) mention in entirety that it is necessary to maximize and have greater improvement so that the raw material is optimal and has quality industrialized products in the most significant way to minimize losses that are detrimental both for the country, the industry, and the agricultural economy, being one of the most important incomes of Ecuador.

The National Institute of Agricultural Research of Ecuador suggests that the result of high-quality cacao depends on several vectors such as the genetic material that is grown, providing improvements in the production of resistance to diseases suffered by cacao, the care and good post-harvest management is a priority because it is one of the best in Latin America which highlights its organoleptic properties (Medina et al., 2009), INIAP mentions that the use of agrochemicals and phosphate fertilizers should be avoided in every possible way because it has its negative effect and contracts with itself the increase of heavy metals "Cd" in raw materials and the final product (Rubio and Carrillo, 2018).

Thus, quality is closely linked to the agroindustrial value chain because of its traceability in cacao processing: in order to be classified as a quality product, it must comply with certain INEN standards currently in force in Ecuador:

INEN 175 (Cut test): It mentions the following: 500 almonds correspond to a batch that should be cut in half. The practice used in cutting test trials to know, observe, and analyze the quality of the raw material is by means of a visual examination and then 100 cacao beans that are cut in half should be taken randomly by the quartile sampling method in order to know directly the fermentation quality and corroborate it with for any failure in the fermentation process by injuries or insects or molds and moths. Inside the beans that are cut into two parts, are beans that give rise to strange flavors and poor quality (INEN, 1986).

INEN 176 (cacao bean quality conditions): It decrees the requirements of excellent quality being fulfilled by the cacao beans to be sold and the different criteria applicable for its classification. We have the cut test that is always used to observe if the degree of fermentation of a statistical sample of the cacao bean is excellent. Therefore, this diagnosis allows us to trust if the measures or parameters of the degree of fermentation are optimal. Jointly, we would rule out the presence of foreign bodies and therefore contamination of both the physicochemical existence of the proliferation of microorganisms that cause mold or any defect that influences its properties and the presence of any insect as mentioned to determine the fermentation of almonds that is essential to enhance the presence of sensory properties and organoleptic characteristics of both its flavor and aroma that will reflect in the final product, if there is nefarious fermentation that will alter its property and notes that will make it astringent (INEN, 2018).

The standard mentions must be met regarding the percentage of fermentation, and fermented cacao beans must be in the range of at least 70 to 85%, the moisture content in cacao beans should not exceed 7%, beans with defects should not be more than 1% of almonds cut or cracked. To be of optimal quality, beans must be free of foreign agents or external odors that affect the organoleptic properties (Alvarez et al., 2010).

The cut test of cacao almonds is a visual examination with the naked eye that serves as a fundamental tool to distinguish and corroborate whether there has been an optimal fermentation as well as commercial parameters or witnessing the quality. The beans should have optimum size, weight, percentage of moisture, and be free of external agents that contaminate the beans. The beans should be cut by making a longitudinal cut to classify them according to the degree of fermentation (Mundaca, 2016).

The quality regulations for cacao beans must strictly comply with the physical and chemical parameters according to INEN standards, which are presented as the main requirements for fine cacao. Table 13.1 shows the classification of the beans processed and the quality required according to current standards.

Table 13.1 Quality regulations for cacao beans according to INEN standards

Requirements	*Cacao Fine*		*Cacao CCN51*			*Test methods*
	A.S.S.S	*A.S.S*	*A.S.E*	*C.S.S*	*C.S.C*	
Humidity, máx %*	7	7	7	7	7	NTE INEN-ISO 2291
Weight of 100 grams, g (>)	130	120–130	100–120	125	110–125	a
Fermented grains min, %	75	65	53	68	55	NTE INEN-ISO 1114
Violet grains, máx %	15	21	25	18	26	NTE INEN-ISO 1114
Slate grains, máx %	9	12	18	12	15	NTE INEN-ISO 1114
Moldy grains, máx %	1	2	4	2	4	NTE INEN-ISO 1114
TOTALS (analysis over 100 grains) min	100	100	100	100	100	
Defective grains máx %	0	1	3	1	3	b
cacao-related material, máx %	1	1	1	1	1	b
Foreign material máx %	0	0	0	0	0	b

* The percentage symbol in the notation 0.01 means the mass fraction.
[a] The mass must be measured by a weighing instrument.
[b] A 500-gram sample of cacao almonds.

13.1.11.10 Breakdown of cacao classification by acronyms

- A.S.S.S Top Superior Summer Select
- A.S.S Top Superior Select
- A.S.E Top Superior Epoca
- C.S.S Superior Select cacao
- C.S.C Superior cacao current

Observations: The existence of granza is allowed for A.S.E cacao and C.S.C cacao at a maximum of 1.5% (INEN, 2018).

INEN Standard 177: Establishes that a lot of cacao should be properly sampled by quartile, where it is divided into four parts. We must choose two of them and repeat the same principle only. The two we chose should be the opposite in order to reduce the size of the sampling and see their respective quality (INEN, 1995).

NTE INEN 623: This standard mentions the requirements and parameters as well as the steps for the production of cacao liquor and chocolates (INEN, 1988).

13.1.11.11 Properties and characteristics of cacao

The almonds of *Theobroma Cacao* L. have embryos with two cotyledons, and the dry weight of the almonds ranges between 86 to 90%. The pulp with soluble solids called 'mucilage' has a dry weight of 10 to 14%. This value can change depending on the origin of the cacao type. As mentioned previously, the pulp contains soluble solids composed of fructose, sucrose, and glucose, in addition to citric acid and pectins used as food for pathogens involved in the fermentation process and its high content of polyphenols ranging between 15 to 20%. Fresh seeds without fat have an extrinsic bitter flavor. The fruit has a total of approximately 300 volatile components including esters, manocarbonyls, hydrocarbons, and pyroles among others. The important compounds that give rise to the flavor are the aliphatic esters, carbonyls, and alpha imino esters. Good aroma is established by polyphenols, diketopiperazines, pyrazines, and theobromine, 18% of protein, of which eight are digestible, amines and alkaloids, trigonelline, nicotinic acid, and free amino acids. Fats correspond to cacao butter, amines and alkaloids include theobromine of 0.5–2.27%, and caffeine similar to coffee with 0.25–1.43%. Next, cacao butter is composed of triglycerides that are monounsaturated fatty acids consisting of 37% oleic acid, 34.4% stearic acid, 26% palmitic acid, and 73% of glycerides (Pazmiño, 2005). Table 13.2 shows the chemical components of the cacao bean.

13.1.11.12 Chocolates

In the development and manufacture of chocolate, the final product must have proper farm management production with good agricultural practices. Additional post-harvest process should be very optimal and appropriate as this section should not be mixed with other varieties of cacao or fresh almonds

Table 13.2 Cacao characteristics

Components	*%p/p*
Caffeine	0.2
Sugars	1
Theobromine	1.2
Mineral salts	2.6
Water	5
Tannic acids and color	6
Cellulose	9
Organic acids and aroma	9.5
Protein	11.5
Cacao butter	54

source taken from (Alvarado et al., 2022).

from the harvest of other days. We will proceed to fermentation with the main point to obtain chocolate of number one quality. With good manufacturing practices and sun drying, the selection of cacao beans according to their size and weight, good logistics of packaging and packing, and finally storage in jute bags for conservation until its industrialization into chocolate, this process continues in the food industry, with the roasting of almonds and removal of shelling being the removal of cotyledon testa (Valenzuela, 2017).

In the manufacture of chocolate bars, it is necessary to carry out the mixing and refining of raw material, to such a point that it will be able to obtain the minimization of particles to an approximate and appropriate size of 18 microns which is ideal for the consumer and his palate (Vásquez and Almeida, 2019).

13.1.11.13 Cacao liquor

En la fabricación de barras de chocolate es necesario realizar el mezclado y refinado de su materia prima, a tal punto de poder obtener la minimización de partículas de un tamaño aproximado y apropiado de 18 micras el cual es ideal para el consumidor y su paladar (Vásquez y Almeida, 2019).

13.1.11.14 Oil of Theobroma Cacao L

Also known as cacao butter, it is a fat of natural origin that is edible for human beings. Its extraction is from the manufacture of chocolate. This butter has a slight aroma and flavor of chocolate, and it is used commercially and its yield is around 30% for by-products (Codiní et al., 2004).

13.1.11.15 Sugar in the manufacture of chocolate

The sugar of natural origin extracted from sugar beet and sugarcane provides many benefits essential for the manufacture of chocolate and for the body, such as energy (Chire et al., 2019).

13.1.11.16 Lecitina

The use of natural phospholipid foods is very diverse in bakery products. Chocolate in addition to instant foods in the chocolate industry is known for the addition of lecithin that allows modifying the rheological characteristics of chocolate in addition to being used for homogenization of the texture. It is found that the application of 0.1 to 0.3% reduces the viscosity, which is 10 times its weight as the same cacao butter (Valenzuela, 2017).

13.1.11.17 Manufacturing suitable for quality chocolate

13.1.11.17.1 ROASTING PROCESS

After a thorough cleaning of fresh cacao, it is passed through sieves in order to lose impurities that are not typical of cacao, which further proceeds to roasting in order to develop the important and unique aromatic characteristics and appetizing flavor. This roasting process is carried out at a temperature of 120°C for 45 minutes, and this temperature is used in the chocolate industry (Valenzuela, 2017).

13.1.11.18 Separation of the testa from the cotyledon

Its purpose is to expose the cotyledon that is the source of the production of chocolate. Once shelled, the shell remains as a residue and has other purposes. It is used as a by-product such as cacao husk tea infusions and tobacco production (Álvarez et al., 2010).

13.1.11.19 Milling

In this phase, cacao is already presented in particles no longer than grains with a size of millimeters of diameter. This processing has several forms, but all have the same criteria as the solid cacao, sugar, and any other solid milk, but not having a good grinding damages the final quality of the product. The particle size should not exceed 40 microns (Erazo et al., 2021).

13.1.11.20 Refined

Refining is a mixing process using a series of rollers to obtain a smooth and suitable paste in order to achieve a suitable chocolate texture (Erazo et al., 2021).

13.1.11.21 Spotting

In the presence of undesirable chemical components that give rise to acidic flavors as well as astringency on the palate, it is necessary to conch. This means eliminating unpleasant flavors so that the pleasant flavors of chocolate can develop. Thus, refining the cacao paste and losing the acidic flavors can contribute to the quality and smoothness of the final product, which takes approximately 24 to 72 hours (Erazo et al., 2021).

13.1.11.22 Temperate

This step is called tempering which has the function of cooling the mixture to a maximum of around 4°C so that crystallization of cacao butter is stable. These must pass through a system of heating, cooling, and reheating as they provide for discoloration in addition to the blooming of fat in the product according to the prevention of certain crystalline forms of cacao butter in development. This chocolate will have a shiny characteristic and a homogeneous texture (Erazo et al., 2021).

13.1.11.23 Molding

One of the simplest methods of shaping chocolate is to use solid shaped tablets; there are several types of molds that are used according to the needs of each person, such as bonbon molds, large, medium, and small figures, or a variety of other shapes (Erazo et al., 2021).

13.1.11.24 Cooling

The molded chocolate is placed in a cooling chamber at a temperature of 14 to 20°C, so that when it is cold it is packaged and distributed (Erazo et al., 2021).

13.1.11.25 Packaged chocolate

Packaging of chocolate is basically simple. Chocolate needs protection against handling, dirt, stains, insect damage, and moisture. Hence, packaging has been used. It is surrounded by a foil or aluminum foil or tin foil predecessor, but its use is not recommended. The use of a strip of paper 'sleeve' should be considered for the necessary information of the product (Erazo et al., 2021).

13.1.11.26 Types of chocolates

This chocolate is made from cacao greater than 35% and is decorated as a normal chocolate in the form of paste, powder, and butter. Sugar is added as well as an emulsifier. In case a flavor is desired, flavorings are used. The 50% cacao is the lowest quality, while 70 to 99% is of excellent cacao quality and purity. If we have a higher percentage amount of this raw material, the result will be maximum quality, a more bitter and intense chocolate (Pérez and Legarda, 2003).

13.1.11.27 White chocolate

This has cacao butter with a minimum percentage of 20%, powdered milk, sugar, an emulsifier, and sometimes an aroma. It is white in color because it does not have cacao powder or paste (Duque and Caballeros, 2020).

13.1.11.28 Milk chocolate

Between white and dark chocolate, milk chocolate is an intermediate formula. It has cacao with a minimum of 25% sugar, whole milk powder, whey, emulsions, and sometimes many flavorings. These chocolate bars are used for desserts, they can be black, white, or milk. In addition to white or milk, to the traditional classic chocolate bars are added hazelnuts, praline, orange citrus zest usually used, coconut, chocolate chip, mint or even chilies, pepper, and others (Alvis and Arrazola, 2011).

In this sense, Ochoa (2019) in his research work entitled the sensory quality of four experimental crosses of cacao by adding levels of dried fruit paste carambola (Averrhoa carambola) and coconut (Cocos nucifera) to obtain dark chocolate found through organoleptic tests that the preparation of chocolate bars by adding dried fruit (DIRCY- C114) expressed the best flavor profile than the above and sweetness It should be noted that no significant effects were found between the values of the physicochemical properties. Economically, it was found that there were no statistical differences between the treatments with respect to cost, due to the fact that in the elaboration of the bars equal levels were used for the control and cultural labors (harvesting, fermentation, and drying) and also the hours of work and cost did not vary between treatments.

13.1.11.29 Study of the presence of heavy metals in cacao beans

In the last ten years, a massive problem has been observed with respect to Cd contamination in agricultural soils, associated with energy production activities, industrial activity, production and use of pesticides, treatment, and deposit/disposal of waste, sludge from wastewater used in agriculture (among which sewage sludge and compost made from municipal solid waste or industrial waste are important sources of metals in agricultural soils), vehicle fleet, etc. (Fonseca, 2021).

Metals enter plants from the soil through the roots. In general, most heavy metals are poorly soluble and available for plant uptake, and although they are found in high concentrations, only a small fraction is phytoavailable (Albán, 2023).

Plants exposed to high levels of Cd show a decrease in photosynthesis and water and nutrient absorption, which results in chlorosis, reduced growth rates, and gradual death. In Ecuador, soils have increased Cd content, which can cause several crops to reach critical values, including cacao, due to anthropogenic activities (Intriago et al., 2019).

Cacao exports to European markets are threatened by indications of heavy metal contamination in exportable cacao, due to the higher levels found compared to the values established by the European community. Therefore, heavy metal contamination in food has become a concern in many countries (Rosales et al., 2021).

In Ecuador, heavy metal contamination in cacao has been demonstrated; however, the impact of this contamination on chocolate powder for national consumption has not been studied (Sánchez et al., 2021).

Gaibor (2021) studied the minimization of Cd concentrations by exogenously inducing efficient native microorganisms (MEN) of the genus *Rhizobium japonicum* in the fermenting mass of experimental cacao hybrids (*Theobroma cacao* L.). Its decrease of Cd, improvement in the post-harvest, and enhancement of organoleptic properties describes that in relation to the physicochemical parameters of the % of testa and S.I., it was favorable to induce 3% of *Rhizobium japonicum* for good results and to achieve compliance with the parameters of acceptability that govern the INEN:176 standards that accepts a minimum of 65% in CCN-51. The induction of 5% *Rhizobium japonicum* in the fermentative mass of cacao presented a decrease of Cd of 0.01 mg/kg and 0.02 mg/kg, being within the parameters required by the EU. In this case, it is notorious to appreciate increases in the production costs of fermented cacao beans, which is due to the additional cost involved in the use of microorganisms.

These results show that it is possible to use bacteria that reduce the presence of heavy metals in cacao beans, which is favorable for their commercialization in various international markets and could increase their economic value and cover a large part of the production costs generated during post-harvest treatment.

References

Abad, A., Acuña, C., & Naranjo, E. (2020). Cacao in the Ecuadorian Coast: a study of its cultural and economic dimension. Estudios de La Gestión. *International Journal of Management*, 7, 59–83. https://doi.org/10.32719/25506641.2020.7.3

Ahmed, S., González, A., & Flores, A. (2019). Acceleration of cacao fermentation by the action of bacteria (*Acetobacter aceti*) and yeasts (*Saccharomyces cerevisiae*). *Espirales Multidisciplinary Journal of Research*, 3(28), 1–20. https://doi.org/10.31876/er.v3i28.572

Albán, O. (2023). Determination of heavy metals in crop soils and beans of cacao plantations (*Theobroma cacao* L.) belonging to the central coastal zone of Ecuador [Pregraduate thesis, Universidad Técnica de Ambato]. https://repositorio.uta.edu.ec/bitstream/123456789/37932/1/CBT%20021.pdf

Aldas, J., Neira, J., Revilla, K., & Sánchez, S. (2020). Fermentation methods of domestic cacao (*Theobroma cacao*) and their influence on physicochemical characteristics, cadmium content and sensory profiles. *Alternatives UCSG*, 21(3), 42–48. https://doi.org/10.23878/alternativas.v21i3.339

Alvarado, K., Vera, J., Tuarez, D., & Intriago, F. (2022). Fermentation of cacao (*Theobroma cacao* L.) with addition of yeast (*Saccharomyces cerevisiae*) and enzyme (PPO's) in the decrease of heavy metals. *Centrosur*, 2022. https://centrosuragraria.com/index.php/revista/article/view/191

Álvarez, C., Tovar, L., García, H., Morillo, F., Sánchez, P., Girón, C., & Farias, A. (2010). Evaluation of the commercial quality of cacao beans (*Theobroma cacao* L.) using two types of fermenters. *Dialnet*, 10(1), 76–87. https://dialnet.unirioja.es/servlet/articulo?codigo=3909942

Alvis, A., & Arrazola, G. (2011). Determination of textural properties of chocolate bars using instrumental techniques. *Informacion Tecnologica*, 22(3), 11–18. https://doi.org/10.4067/S0718-07642011000300003

Anzuelos, V., Borjas, R., Alvarado, L., Castro, V., & Julca, A. (2019). Cultural, biological and chemical control of *Moniliophthora roreri* and *Phytophthora* spp on Theobroma cacao "CCN-51." *Scientia Agropecuaria*, 10(4), 511–520. https://doi.org/10.17268/sci.agropecu.2019.04.08

Asanza, M., Alvarado, R., Peñafiel, G., & Fernández, F. (2019). Economic growth of Ecuadorian cacao. *Observatorio de La Economia Latinoamericana*, 10–20. https://www.eumed.net/rev/oel/2019/12/crecimiento-cacao-ecuador.pdf

Attolini, A. (2011). Cuentas, dares y tomares dl cacao: delicia, convite, rito mesoamericano. Anthrological aspects. *Revista Digital Universitaria*, 12(4), 3–22. http://www.revista.unam.mx/vol.12/num4/art38/art38.pdf

Cardona, M., Rodríguez, E., & Cadena, E. (2016). Diagnosis of cacao processing practices in the department of Arauca. *Revista Lasallista de Investigación*, 13(1), 94–104. https://doi.org/10.22507/rli.v13n1a8

Codiní, M., Díaz, F., Ghirardi, M., & Villavicencio, I. (2004). Obtaining and utilization of cacao butter. *UCEL Magazine*, 7(0329-3475), 143–148. https://www.redalyc.org/pdf/877/87701213.pdf

Córdova, E., Cornejo, A., Cruz, A., & Alejandro, G. (2022). Systemic appropriation of the consumption of Ecuadorian products at the national level: Cacao Arriba case. *Revista Ibérica de Sistemas e Tecnologias de Informação*, (E53), 400–412. https://sga.unemi.edu.ec/media/evidenciasiv/2023/02/24/solicitudpublicacion_2023224154213.pdf

Cruz, E., & Pereira, I. (2009). Histories, knowledge and flavors around cacao (*Theobroma cacao* L.) in the subregion of Barlovento, Miranda State. *Sapiens*, 10(2), 97–120. https://www.redalyc.org/pdf/410/41021266005.pdf

Chire, G., Ureña, M., & García, S. (2019). Optimization of dark chocolate formulation from cacao bean blend and cacao content applying response surface method. *Enfoque UTE*, 10(3), 42–54.

Dueñas, A., Sacón, E., Párraga, E., Zambrano, M., & Moreira, E. (2018). Protein fortification of cacao liquor using spirulina (Spirulina platensis) in chocolate processing. *Pro Sciences: Journal of Production, Science and Research*, 2(11), 1–15. https://doi.org/10.29018/issn.2588-1000vol2iss11.2018pp22-27

Duque, C., & Caballeros, L. (2020). Effect of temperature in the refining and mixing stages on the polyphenol content of a white chocolate. *Revista Colombiana De Tecnologias De Avanzada (Rcta)*, 2(36), 44–51. https://doi.org/10.24054/16927257.v36.n36.2020.19

Erazo, C., Bravo, K., Tuárez, D., Fernández, Á., Torres, Y., & Vera, J. (2021). Effect of fermentation of cacao (*Theobroma cacao* L.), national and trinitario varieties, in unconventional wood boxes on the physical and sensory quality of cacao liquor. *Revista de Investigación Talentos*, 8(2), 42–55. https://doi.org/10.33789/talentos.8.2.153

Fonseca, J. A. (2021). Heavy metals (Cd++, Pb++) in cacao beans (*Theobroma cacao*) and phytoextraction from alley crops, Upper Huallaga Valley-Peru. [Undergraduate thesis, Hermilio Valdizán National University]. https://repositorio.unheval.edu.pe/bitstream/handle/20.500.13080/6885/PGA00120F74.pdf?sequence=1&isAllowed=y

Freire, D., & Díaz, D. (2016). Fermentation and drying method for the benefit of obtaining white chocolate from Ecuadorian Criollo cacao (*Theobroma cacao* L.). 14(S2), 323–329.

Gaibor, A. E. (2021). Exogenous induction of *Rhizobium japonicum*, in fermentative mass of experimental cacao (*Theobroma cacao* L.) hybrids and its possible effect on cadmium reduction. [Undergraduate thesis, Universidad Técnica de Babahoyo]. https://repositorio.uteq.edu.ec/handle/43000/6135

García, P., Nuñez, J., & Bahamón, A. (2021). Physicochemical and sensory characteristics of fermented almonds of national cacao (*Theobroma cacao* L.) with addition of probiotics in the amazonic research center, Cimaz Macagual (Caquetá, Colombia). *Engineering and Competitiveness*, 23(2), 1–13. https://doi.org/10.25100/iyc.23i2.10885

Gómez, R., Villanueva, S., & Henríquez, M. (2019). Global trend in the processing of cacao-derived products. *Revista INGENIERÍA UC*, 26(2), 213–222. https://www.redalyc.org/journal/707/70760276010/html/
Hernández, A., Paz, M., & García, L. (2015). Taguchi's methodology in statistical quality control. *Revista De La Escuela De Perfeccionamiento En Investigacion Operativa*, 23(37), 65–83. https://revistas.unc.edu.ar/index.php/epio/article/view/11986
INEN (1986). Cacao in grams, cut tests NTE INEN 175. Instituto Ecuatoriano de Normalización, 1–3. https://www.normalizacion.gob.ec/buzon/normas/175.pdf
INEN (1988). Norma Técnica Ecuatoriana Voluntaria de pasta (masa, licor) de cacao (mass, liquor) INEN 623, 2–7. https://www.normalizacion.gob.ec/buzon/normas/623.pdf
INEN (1995). Norma Técnica Ecuatoriana NTE INEN 177 Cacao en grano Muestreo. Instituto Ecuatoriano de Normalización, 2, 1–10. https://www.normalizacion.gob.ec/buzon/normas/177.pdf
INEN (2018). Cacao beans. Requisitos NTE INEN 176. Instituto Ecuatoriano de Normalización, 5, 1–8. https://www.normalizacion.gob.ec/buzon/normas/nte_inen_176-5.pdf
Intriago, F., Talledo, V., Cuenca, G., Macías, R., Álvarez, J., & Menjívar, J. (2019). Evaluation of heavy metal content in cacao beans (*Theobroma cacao* L) during the beneficiation process. *Pro Sciences: Journal of Production, Science and Research*, 3(26), 17–23.
Larrea, M. (2008). The national cacao crop: A generous forest. "Field manual for the implementation of biodiversity friendly practices in National Cacao crops". In A. Benítez (Ed.), *CORPEI* (1st ed.). https://biblio.flacsoandes.edu.ec/libros/digital/43804.pdf
Lema, V. (2019). Report on cacao target yields (dry kernel) 2019. Ministry of Agriculture and Livestock. http://sipa.agricultura.gob.ec/descargas/estudios/rendimientos/cacao/rendimiento_cacao_2019.pdf
Machuca, J., Suárez, E., Motte, E., & Mialhe, E. (2019). Molecular characterization of microorganisms during cacao-bean fermentation process (*Theobroma cacao*). *Revista Peruana de Biologia*, 26(4), 535–542. https://doi.org/10.15381/rpb.v26i4.17220
Maldonado, P., & Sánchez, A. (2021). Competitiveness and sustainability in cacao exports. A comparative analysis of the cases of Ecuador and Côte d'Ivoire. https://n9.cl/w8w7s
Mata, D. (2018). Commercialization process of Aroma fine cacao in Los Rios province, Ecuador. *IDICT*, 20(4), 385–400. https://www.redalyc.org/journal/6378/637869147003/html/
Medina, C., Ortiz, V., & Coronel, A. (2009). Cacao post-harvest operations. Food and Agriculture Organization of the United Nations, 1–78. https://www.fao.org/3/au995s/au995s.pdf
Mendoza, E., Boza, J., & Manjarrez, N. (2021). Socioeconomic impact of cacao production and commercialization of small cacao producers in the canton of Quevedo. *Revista Científica Ecociencia*, 255–272. https://doi.org/10.21855/ecociencia.80.603
Ministry of Agriculture [MAG] (2021a). Comercio exterior del sector agropecuario y agroindustrial (Cacao). http://sipa.agricultura.gob.ec/index.php/comext-productos
Ministry of Agriculture [MAG] (2021b). Ficha del cultivo de cacao. http://sipa.agricultura.gob.ec/index.php/cacao
Mora, C., Quevedo, J., Zhiminaicela, J., Herrera, S., Morocho, A., & León, J. (2021). Influence of cacao pod maturity: Nutritional and sensory quality of cultivar CCN-51. *Revista Bases de La Cienci*, 6(2), 27–40. https://doi.org/10.33936/rev_bas_de_la_ciencia.v6i2.2706
Mundaca, G. (2016). Analysis of cacao bean quality by hyperspectral imaging using machine vision techniques [Universidad de Piura]. In Repositorios Universidad de

Piura. https://pirhua.udep.edu.pe/bitstream/handle/11042/2756/MAS_IME_AUT_013.pdf?sequence=1&isAllowed=y

Nivela, D. J. (2020). Allometric relationships to estimate aerial biomass in cacao (*Theobroma cacao* L.) cultivars of Trinidadian origin (CCN-51) and national type in Los Ríos Province [Undergraduate thesis, Universidad Técnica de Quevedo]. https://repositorio.uteq.edu.ec/bitstream/43000/5365/1/T-UTEQ-0264.pdf

Pazmiño, K. (2005). Study of the behavior of fructose, glucose and sucrose in cacao beans of national production during fermentation [Escuela Superior Politécnica de Chimborazo]. https://repositorio.iniap.gob.ec/bitstream/41000/747/1/iniapsctP348e.pdf

Pérez, B., & Legarda, T. (2003). Ochratoxin a in different types of chocolate and cacao powder preparations marketed in Spain and other foreign countries. *Alimentaria*, 5(347), 143–153. https://doi.org/10.1007/BF00408017

Pérez, R. (2017). Cacao processing. 68.

Quiñonez, M., Espinoza, E., Yovera, F., Cuchilla, Y., & Castro, D. (2018). Identification, georeferencing and morphological characterization of superior trees of *Theobroma cacao* L.1753 cultivar cacao blanco from Piura, Peru. *The Biologist (Lima)*, 16(1), 1–151. https://doi.org/10.24039/rtb2018161225

Rivera Fernández, R. D., Barrera Álvarez, A. E., Guzmán Cedeño, Á. M., Medina Quinteros, H. N., Casanova Ferrín, L. M., Peña Galeas, M. M., & Nivela Morante, P. E. (2012). Effect of fermentation type and time on the physical and chemical quality of cacao (*Theobroma cacao* L.) Nacional type. *Science and Technology*, 5(1), 7–12. https://doi.org/10.18779/cyt.v5i1.77. https://doi.org/10.18779/cyt.v5i1.77

Rosales, J., Centeno, L., Cajacuri, J., Luis, J., & Chávez, C. (2021). Identification of cadmium and lead in cacao crops located in the Satipo-Junin area. *Tecnia*, 31(2), 83–89. http://www.scielo.org.pe/pdf/tecnia/v31n2/2309-0413-tecnia-31-02-83.pdf

Rubio, G., & Carrillo, M. (2018). Evaluation of tropical soil analysis methods to determine cadmium availability [Universidad Tecnica Equinoxial]. In Universidad UTE. https://repositorio.iniap.gob.ec/bitstream/41000/5243/1/INIAPBEETPTR896.pdf

Sanchez, A., Castellanos, O., & Domínguez, K. (2008). Cacao postharvest improvement from roadmapping. *Engineering and Research Journal*, 28(3), 150–158. http://www.scielo.org.co/pdf/iei/v28n3/v28n3a22.pdf

Sánchez, C., Zambrano, D., Arevalo, O., Macas, V., Pernía, B., & Jiménez, D. (2021). Baseline study of heavy metal concentrations in chocolate powder commercialized in Guayaquil, Ecuador. *INVESTIGATIO*, 16, 62–77. https://revistas.uees.edu.ec/index.php/IRR/article/view/574/520

Suárez, C., Moreira, M., & Vera Barahona, J. (1994). *Manual del cultivo de cacao* (Q. E. E. I. E. E. E. T. Pichilingue, Ed.), 2nd.

Teneda, W. (2016). Improvement of the fermentation process of cacao (*Theobroma cacao* L.) variety Nacional and variety CCN51. In Universidad Internacional de Andalucía (1st ed.). https://core.ac.uk/download/pdf/223061502.pdf

Tinoco, H., & Ospina, D. (2013). Analysis of the cacao dehydration process for the decrease of the drying time. *EIA Journal*, 7(13), 53. https://doi.org/10.24050/reia.v7i13.232

Tornés, J. (2013). Cacao, a contribution of Mexico to the world. *Revista Ciencia*, 66(3), 1–8. https://www.revistaciencia.amc.edu.mx/images/revista/66_3/PDF/Cacao.pdf

European Union (2021). Evaluation of Ecuador's exports to the European Union. 14–17. https://www.ekosnegocios.com/articulo/las-exportaciones-de-ecuador-a-la-union-europea-en-la-pandemia-crecieron-un-1-6

Ureña, M., & Sandoval, A. (2021). Technological innovation in Andean cacao. Fontagro, 1–83. file:///C:/Users/uZer/Downloads/16109_-_Product_6.pdf

Valenzuela, A. (2017). Chocolate, a healthy pleasure. *Revista Chilena de Nutrición*, 34(3), 180–190. https://doi.org/10.4067/S0717-75182007000300001

Vásquez, A., Ovando, I., Lourdes, A., Betancur, A., & Salvador, M. (2016). Cacao alkaloids and polyphenols, mechanism regulating their biosynthesis and their implications in flavor and aroma. *Archivos Latinoamericanos de Nutrición*, 66(3), 1–25. https://www.alanrevista.org/ediciones/2016/3/art-10/

Vásquez, M., & Almeida, V. (2019). Evaluation of the physicochemical and sensory characteristics of a chocolate bar, formulated with non-caloric sweetener [Univerdad de las Americas]. In Repositorio udla. https://doi.org/.1037//0033 2909.I26.1.78

Vera, J., Vallejo, C., Párraga, D., Morales, W., Macías, J., & Ramos, R. (2008). Physical-chemical and sensory attributes of almonds from fifteen clones of Nacional cacao (*Theobroma cacao* L.) in Ecuador. *Science and Technology Magazine*, 7(2), 21–34. https://doi.org/10.18779/cyt.v7i2.139

Villamar, F., Calderón, J., & Mayorga, E. (2016). Strategies for the cultivation, marketing and export of fine aroma cacao in Ecuador. *Ciencia Unemi*, 9(18), 45–55. https://www.redalyc.org/pdf/5826/582663825007.pdf

14 Potential use of by-products of cacao industrialization

Freddy Zambrano Gavilánez, Viviana Suárez Navarrete, Marina García, Roberto Bravo Zamora, Soraya Peñarrieta Bravo, Rolando León Aguilar, and Patricia Zambrano Gavilanes

14.1 Introduction

Throughout history, it has been observed how cacao has been one of the most economically and socially important crops in different cultures across the Americas, from Inca times when it was used as currency for exchange with other indigenous peoples and internal consumption.

Theobroma cacao L. is native to South America but was domesticated in Central America. Its beans are the raw material for making chocolate (Castillejos et al., 2022). Whether fermented or dried, these beans have driven economic growth for various countries that produce and trade cacao, both within their territories and abroad (Lachenaud & Motamayor, 2017).

In general, the warm, humid climates similar to cacao's place of origin are characteristic of the main producing countries; however, the largest producing continent is Africa (Achaw and Danso-Boateng, 2021). That is why, according to reports by Voora et al. (2019), Ecuador, Mexico, Colombia, Peru, the Dominican Republic, Venezuela, Brazil, Ivory Coast, Ghana, Nigeria, Cameroon, Uganda, Indonesia, Togo, and Papua New Guinea have been identified as the top 15 cacao-producing countries. In a recent publication, Jarrín et al. (2023) state that Ecuador is now the fifth largest cacao producer worldwide, generating hundreds of tons of waste from this fruit annually.

The cocoa industry accounts for 30% of Ghana's total export earnings. The main cocoa products are classified as processed, semi-finished, and raw products. These include cacao beans, cacao paste, cacao butter, cacao husks, and cacao powder (Bangmarigu et al., 2018).

Cacao beans are considered sources of economy, as they provide raw materials for global food industries valued at over $47 billion in world exports. However, after cacao processing, around 80% of the fruit is discarded as residual biomass, including cacao pod husks, cacao bean shells, and cacao sweats (mucilage). These by-products have high added value with applications in cosmetics, food, and pharmaceutical industries (Vásquez et al., 2019).

Industry waste, which includes cacao pod shell, mucilage, and cacao bean shell, is generated in large volumes annually. These by-products contain

DOI: 10.4324/9781003381761-18

lignocellulose and other valuable compounds that can be used for the production of biofuels and other value-added products. Recent studies have explored the use of these wastes for the production of biochar, bioethanol, biogas, biohydrogen, bio-oil, and liquid smoke. Thermochemical, biological, and physicochemical conversion technologies can be used depending on the type of biofuel desired (Mendoza Meneses et al., 2021)

Due to the aftermath of the COVID-19 pandemic, cacao production faces a major challenge, since prices have been negatively affected. This is due to international price fluctuations in commodities that have interfered with the efficiency of the circular economy. However, by 2026, the world market for the production of confectionery products for the Asian continent is expected to increase, while also raising demand for high-quality cacao products and organic chocolates (Guirlanda et al., 2021).

14.2 Parts of the cacao fruit and their characteristics

According to Indiarto et al. (2021), processing cacao generates by-products including pod husks, cacao beans, cacao bean shells, and mucilage (Figure 14.1). The same authors explain that despite cacao production generating various by-products, only 10% is commercialized while the remaining 90% is discarded.

Pod husks: They are the outer part of the fruit, with a rough, thick texture and are in different colors depending on the cacao variety. Their main function is to protect against mechanical damage, pests, and environmental conditions. Considered an organic material with high nutritional contribution (Vásquez et al., 2019), but still considered waste (Komlaga et al., 2019). Indiarto et al. (2021) explain that they consist of three layers: endocarp, mesocarp, and epicarp, which contain carbohydrates, proteins, and minerals such as nitrogen (N), potassium (K), calcium (Ca), magnesium (Mg), phosphorus (P), sodium (Na), and iron (Fe). Pod husks are used as organic fertilizers for soil. The sustainable transformation of organic waste through processes such as hydrolysis, extraction, pyrolysis, anaerobic digestion, and fermentation enables the production of ingredients, biopesticides, biofuels, organic fertilizers, and other high-impact products for various industries. This transformation represents a strategic solution from both environmental and socioeconomic perspectives.

Beans: The beans constitute 19% of the fresh fruit weight, while the remaining 81% corresponds to the fresh husk and seed pulp juice, which are discarded (Komlaga et al., 2019). According to Achaw and Danso-Boateng (2021), cacao beans consist of crude fiber, proteins, starch, ash, fat (cacao butter and shell fat), theobromine, and caffeine, with composition (%) varying by storage conditions, drying level, bean type, and analysis method. Chocolate, the key ingredient in various foods, beverages, candy bars, cookies, and cereals, is made from cacao beans (Achaw and Danso-Boateng, 2021). Cacao butter, used to formulate cosmetics,

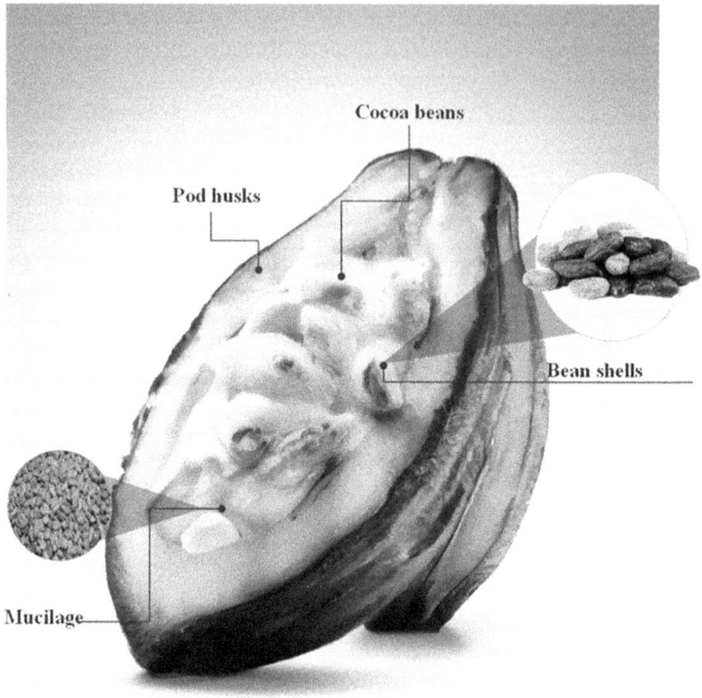

Figure 14.1 Parts of the cacao fruit.

creams, soap, and lotions, can be produced from discarded cacao beans (Komlaga et al., 2019).

Bean shells: They represent 10 to 17% of the total bean weight and 10% to 20% of the total amount of volatile organic compounds, whose aromatic composition is key in cacao and chocolate (Rojo et al., 2020). It is also a source of nutrients, dietary fiber, proteins, minerals, vitamins, methylxanthines, and a wide spectrum of polyphenols similar to cacao, which have antidiabetic capacity. Due to the fiber fractions, they inhibit different glucose degradation enzymes and enhance sugar retention (Rojo et al., 2020; Gutiérrez, 2020).

Gutiérrez (2020) agrees that the seed coat has various uses including it being used as:

- additive in bakery and meat industry
- chemical adsorbent and inoculant
- cooking oil stabilizer
- animal feed

Mucilage: It is formed by the cacao pulp, white in color, composed of soft parenchyma cells containing cell sap, giving it a viscous texture on the outside. It can produce a whitish liquid known as cacao pulp juice or sweat (Indiarto et al., 2021).

According to Komlaga et al. (2019), this by-product is used to make fresh cacao juice by squeezing the freshly harvested ripe beans and extracting the juice, which is then pasteurized, retaining all its natural flavors. Guirlanda et al. (2021) state that cacao honey, with a translucent coloration, is obtained from the mucilage during cacao fermentation, with chemical and sensory characteristics similar to the pulp.

Komlaga et al. (2019) and Guirlanda et al. (2021) agree various by-products can be obtained from cacao mucilage:

- processing into soft drinks, wine, vinegar, alcohol, and pectin.
- honey can be used as a natural substitute for refined sugar in food preparations.

14.3 Use in the pharmaceutical industry

According to what was stated by Sulley (2022), cacao belongs to an important link in the pharmaceutical industry. It is currently considered a viable alternative to replace codeine in the manufacture of cough syrups. It is important to understand that natural cacao contains chemical substances similar to antidepressants, so excessive consumption can cause memory loss, drowsiness, breathing difficulties, fatigue, and confusion, and could even trigger problems in babies of lactating or pregnant women (Sulley, 2022).

Pectin is a complex polysaccharide known for its gelling properties and has been found in large quantities in the cacao pod husk, which is why the food, pharmaceutical, and cosmetic industries use it as a source of pectin. For this, several investigations have been conducted to analyze different extraction conditions and their chemical, physicochemical, and antimicrobial properties (Vásquez et al., 2019). Jarrín et al. (2023) confirm that the pectin generated from cacao pod husk residue is used in the food and pharmaceutical industries.

Similarly, Cádiz et al. (2020) and Mendoza-Meneses et al. (2021) describe that cacao bean shell has a greater richness of phenolic acids for development in industries such as food, pharmaceutical, and cosmetic.

14.4 Uses in human and animal food

By-products derived from cacao processing, such as mucilage, have the potential for use in human food, offering opportunities to generate valuable products. Mucilage, the component that wraps the seeds, is rich in polysaccharides, sugars (mainly sucrose, glucose, and fructose), vitamins, and other compounds (Vera et al., 2014; Arcila & Mendoza, 2016). This material has pleasant organoleptic characteristics, such as taste and smell, which makes it suitable for use in the production of handcrafted and industrial products intended for human consumption (Quimbita et al., 2013; Largo & Yugcha, 2016).

It contains 12% to 16% sugar, 2% pectin, and 1.80% citric acid (Quimbita et al., 2013). Its components make the mucilage an appetizing part that can be used as a component for the production of products such as jams, juices, marmalades, and wines. It can also be used in the production of alcohol and acetic acid (Largo & Yugcha, 2016).

The handcrafted production of drinks and foods such as juices, liqueurs, cocktails, ice cream, marmalades, and pastries from mucilage are very appealing products for consumers. So much so that as enterprises in these products move forward, the diversity of options that can be obtained from the mucilage expands even more (Acevedo et al., 2013).

Some cacao producers are using mucilage in a handcrafted way, in order to obtain other economic income because this raw material does not generate any cost. Alaniz et al. (2013) highlight that natural jelly is one of those products that can be worked with. This is a semisolid food made up of 45% by weight of the ingredients from the pod juice and 55% by weight of sugar, being a concentrated mixture with 66% soluble solids.

The process of obtaining jelly from mucilage juice involves mixing the filtered mucilage with sugar, followed by cooking and packaging (Carvajal Guzmán and Vega García, 2017). Physicochemical analyses of the jelly show that it meets quality requirements for human consumption, including pH within ranges allowed by health standards (FAO, 2009).

The sensory results of the jelly, evaluated using a hedonic scale by a panel of tasters, reflected varied preferences in a study by Cedeño-Bravo et al. (2022). The samples from different locations presented differences in color, smell, texture, and taste. The jelly originating from El Milagro was better evaluated in terms of taste, while other locations showed diverse preferences in different attributes. These findings indicate that cacao mucilage represents an economically viable alternative for producers in Manabí when processed as jelly. In addition, the physicochemical and sensory results support the viability of this by-product within the food supply. This opens up opportunities to improve the sustainability of farmers and their insertion into the food market.

The shell, husk, seed meal, and pod of cacao are rich in nutrients and bioactive compounds. The shell contains mainly insoluble fiber (39.25–66.33%) (Rojo-Poveda et al., 2020), as well as carbohydrates such as NFC (17.5–47%), proteins (2.1–9.1%), and lipids (0.6–4.7%) (Campos-Vega et al., 2018). It also has phenolic and antioxidant compounds (Lu et al., 2018; Okiyama et al., 2017) as well as minerals (Rossi et al., 2021).

Shell and husk provide energy, protein, and minerals such as P and Mg, necessary in the diet of ruminants (Rojo-Poveda et al., 2020; Makinde et al., 2019). However, its limitation is the presence of theobromine, a toxic compound in high doses, so detheobromination treatments are recommended (Renna et al., 2022). In birds, treated husk can replace up to 20% of corn in layer diets without adverse effects (Olubamiwa et al., 2006). Seed meal has a high protein and fiber content, useful in ruminants (Meffeja et al., 2006). Its disadvantage is its high theobromine level (20–33 g/kg), which is reduced with heat treatment (Makinde et al., 2019). Pod has abundant fiber and lignin, limiting its digestibility (Laconi et al., 2015). So, it is advisable to mix it with more digestible ingredients.

Excess cacao by-products can cause vomiting, diarrhea, tremors, and other symptoms due to theobromine and tannins (Makinde et al., 2019). At high doses, it affects the growth and reproduction of animals. Dogs are especially sensitive, with gastrointestinal, cardiovascular, and neurological disorders (Soto-Ramírez et al., 2018).

14.5 Uses for biofuel

Historically, the cacao industry has focused on the commercial use of cacao beans for the production of chocolate and other products, wasting about 90% of the fruit, including the husk and mucilage (Gutiérrez et al., 2020). These residues are rich in carbohydrates, fiber, lipids, and fermentable compounds, with potential for biogas generation through anaerobic digestion (Sánchez, 2013; Moncada et al., 2016).

Cacao husk also contains high levels of potassium oxide (average 59.25%), useful as a fertilizer or in thermochemical processes such as pyrolysis and gasification for the production of biofuels. Through these processes, it is possible to efficiently convert the high carbon content of the cacao husk into liquid, solid, and gaseous fuels (Vivanco et al., 2017). Considering its composition and in view of the urgent need to develop sustainable technologies due to the growing CO_2 pollution, it is proposed to evaluate the biogas generation capacity from mucilage, husk, and a mixture of both (Bah et al., 2016). The resulting biosolid would have application as biofertilizer, reducing the use of synthetic sources (Bernal-Martinez et al., 2020; Gavilanes-Terán et al. 2016).

In a study carried out by Vélez et al. (2021), biogas was produced with husk and mucilage from CCN-51 and nacional cacao using cattle manure as inoculum for 70 days. T6 National Husk (20% ST) was the most outstanding in accumulated biogas production (1297.66 mL) and methane gas percentage (76%).

The waste generated during industrial cacao processing, such as husks, pulp, and mucilage, represents an attractive source for bioethanol production. These by-products are composed mainly of carbohydrates such as starch, cellulose, and hemicellulose, susceptible to hydrolysis and fermentation to obtain ethanol. Through pretreatments, enzymatic hydrolysis, and fermentation with efficient microorganisms, it is feasible to convert the sugars present in these residues into ethanol. Given the high volume of by-products generated by the cacao agro-industry and their rich content of fermentable carbohydrates, their use for the production of biofuels such as bioethanol is a promising alternative both technically and economically (Mendoza-Meneses et al., 2021).

The implementation of these strategies allows adding value and reducing the environmental impact of the cacao production chain, in line with the circular economy and the paradigm of the integral and sustainable use of residual agro-industrial biomass.

14.6 Uses in agriculture

As already mentioned, the cacao agro-industry generates significant waste, such as husk, shell, and mucilage. Recent studies reveal valuable properties in these by-products for use in agriculture (Adejumo et al., 2020).

Husk, rich in phenolic compounds and alkaloids, exhibits allelopathic properties, inhibiting weed growth. Aqueous extracts of cacao husk work as natural herbicides, effective in cabbage, tomato, and cucumber crops, being less toxic than chemical herbicides (Bello-Bello et al., 2019; Oyinlola et al., 2020).

Husk stands out as a biofertilizer due to its macro and micronutrient contents. Its application improves soil properties, increases moisture retention, and gradually releases nutrients, improving crop growth, yield, and quality (Adejumo et al., 2020; Agele et al., 2021). From May to September 2015, Chaniago & Ramlan (2016) carried out a study in the Luwuk Subdistrict, Banggai District, and in the Soil Chemistry and Fertility Laboratory of the Faculty of Agriculture at Hasanuddin University, Makassar. The objective was to evaluate the feasibility of using compost generated from cacao husks mixed with cattle waste. They found that the proportion of 50 kg of cacao husks and 20 kg of cattle waste exhibited the highest content of macronutrients with N = 0.25%, P = 3.91%, and K = 5.23%, highlighting its potential as an effective option for the production of organic fertilizers.

Recent research explores the composting of cacao husks, observing improvements in soil fertility, especially when enriched with *Tithonia diversifolia* biomass. The resulting composts are rich in N, P, and K, with promising microbial diversity. More studies are needed to fully understand the role of each microorganism and perform agronomic field testing (Adu-Dapaah et al., 1994; Maboune et al., 2017).

The use of cacao industrialization waste not only helps reduce accumulation and pollution problems but also offers valuable opportunities to improve sustainability and agricultural productivity.

14.7 Conclusions

The agro-industrial processing of cacao generates millions of tons of by-products such as husks, shells, and mucilage annually, representing around 90% of the fruit. Traditionally discarded as waste, these organic residues entail pollution problems, nutrient waste, and loss of economic opportunities.

However, the bioactive composition of these by-products containing carbohydrates, proteins, fiber, antioxidants, minerals, and other valuable compounds opens up broad possibilities for their integral use and conversion into more useful goods. Their sustainable transformation through processes such as hydrolysis, extraction, pyrolysis, anaerobic digestion, and fermentation allows obtaining ingredients, biopesticides, biofuels, organic fertilizers, and other high-impact products in various industries. While representing an environmentally and socioeconomically strategic solution.

Advancing this vision of a circular economy in the global cacao agroindustry will require further research to adopt appropriate technologies and scalable processes specific to the characteristics of waste in each producing country as well as incentive policies for investment and adoption of optimal practices for the integral management of these by-products with great potential for recovery.

References

Acevedo, I., García, O., Contreras, J., & Acevedo, I. (2013). Elaboración y evaluación de las características de sensoriales de un yogurt de leche caprina con jalea semifluida de piña. *Revista UDO Agrícola*, 9, 442–448.

Achaw, O., & Danso-Boateng, E. (2021). Cocoa processing and chocolate manufacture. In Osei-Wusu Achaw and Eric Danso-Boateng eds. *Chemical and Process Industries: With Examples of Industries in Ghana* (pp. 267–292). Springer International Publishing.

Adejumo, O. V., Ladapo, O. M., & Famoso, A. O. (2020). Heavy metal concentration in cacao bean and by-products of cacao production in Ekiti and Edo States, Southwest Nigeria. *Bulletin of Environment, Pharmacology and Life Sciences*, 9(2), 102–109. https://doi.org/10.9734/bepils/2019/v9i230144

Adu-Dapaah, H. K., Cobbina, J., & Asare, E. O. (1994). Effect of cacao pod ash on the growth of maize. *The Journal of Agricultural Science*, 132, 31–32.

Agele, S., Ayandokun, O. S., Ogunyemi, T., Obanla, S., & Adesina, B. (2021). Utilization of pod husk as organic manure in agricultural production: A review on nutrient composition and benefits to soil and crops. *Ife Journal of Agriculture*, 33(2), 247–258. https://doi.org/10.4314/ijag.v33i2.4

Alaniz, E., Arvisú, S., & González, K. (2013). Producción de postres y vinagre a partir del exudado de cacao de la cooperativa de servicios múltiples “Ríos de agua viva, 21 de junio” Rancho Grande, Matagalpa.

Arcila, N., & Mendoza, Y. (2016). Elaboración de una bebida instantánea a base de semillas de amaranto (*Amaranthus cruentus*) y su uso potencial en la alimentación humana. *Revista de la Facultad de Agronomía de la Universidad del Zulia (LUZ)*, 11(4), 114–124.

Bah, C. S., Bekhit, A. E. D. A., Carne, A., & McConnell, M. A. (2016). Opportunities and challenges for developing pineapple waste biorefineries in Ghana–A review. *Waste and Biomass Valorization*, 7(3), 527–542.

Bangmarigu, E., & Qineti, A. (2018). *Cocoa production and export in Ghana. European Association of Agricultural Economists (EAAE) > 162nd Seminar*, April 26-27, 2018, Budapest, Hungary https://doi.org/10.22004/ag.econ.272051

Bello-Bello, J. J., Jimenez, R. L., Gonzales, A. A., & Fuentes, D. G. (2019). Allelopathic potential of cocoa (*Theobroma cacao* L.) shell as a model for growth of Guinea Grass (*Panicum maximum* Jacq.) seedlings. *Journal of Botanical Research*. 3604–3608 https://doi.org/10.25081/jbr.2019.v2.5448

Bernal-Martinez, A., González-López, G. I., & Cuevas-Rodriguez, G. (2020). Codigestión anaerobia como alternativa para el tratamiento de aguas residuales lácteas y la generación de biogás y biosólidos. En *Mujeres en la Ciencia Ciencias ambientales, uso de recursos* (p. 115). https://doi.org/10.35429/h

Cádiz, M., Fernández, Á., Leyva, F., Guerrero, N., Villegas, M., Pimentel, S., Ramos, F., & Segura, A. (2020). LC-MS and spectrophotometric approaches for evaluation of bioactive compounds from Peru cacao by-products for commercial applications. *Molecules*, 25(14), 3177. https://doi.org/10.3390/molecules25143177

Campos-Vega, R., Nieto-Figueroa, K., & Oomah, B. (2018). Cocoa (*Theobroma cacao* L.) pod husk: Renewable source of bioactive compounds. *Trends in Food Science & Technology*, 81, 172–184. https://doi.org/10.1016/j.tifs.2018.09.022

Carvajal Guzmán, J. C., & Vega García, W. R. (2017). Evaluación de las propiedades nutricionales de la Harina del Mucílago de Cacao con la adición de la placenta y su utilización en panadería (Bachelor’s thesis, Universidad Estatal de Bolívar. Facultad de Ciencias Agropecuarias, Recursos Naturales y del Ambiente. Escuela Ingeniería Agroindustrial).

Castillejos, L., Acosta, A., Gallardo, T., Osorio, G., & Jiménez, C. (2022). Uses of FT-MIR spectroscopy and multivariate analysis in quality control of coffee, cocoa, and commercially important spices. *Foods*, 11(4), 579. https://doi.org/10.3390/foods11040579

Cedeño-Bravo, J. J., Vélez Rivas, J. A., Santana-Parrales, F. A., & Bravo Zamora, R. C. (2022). Mucilago de cacao (*Theobroma cacao* L.), como alternativa económica para productores cacaoteros de Manabí, Ecuador. *Revista Científica MQRinvestigar*, 6(01), 241–256.

Chaniago, R., & Ramlan, W. (2016). Utilizing cocoa rind as organic fertilizer to support sustainable agriculture. *Journal of Tropical Soils*, 21(1), 33–39. https://doi.org/10.5400/jts.2016.21.1.33

Gavilanes-Terán, I., Jara-Samaniego, J., Idrovo-Novillo, J., Bustamante, M. A., Moral, R., & Paredes, C. (2016). Substrate mixtures composed of agro-industrial wastes for the simultaneous production of lipase and biosurfactants by *Candida ingens* isolated from the Amazon rainforest. *Bioresource Technology*, 219, 53–60.

Guirlanda, C., da Silva, G., & Takahashi, J. (2021). Cocoa honey: Agro-industrial waste or underutilized cacao by-product? *Future Foods*, 4, 100061. https://doi.org/10.1016/j.fufo.2021.100061

Gutiérrez, L. F., Riaño, N. M., & Colorado, H. A. (2020). Aprovechamiento integral de subproductos de cacao para la producción de biogás y la obtención de compuestos antioxidantes. *Revista UIS Ingenierías*, 19(1), 91–103.

Indiarto, R., Raihani, Z., Dewi, M., & Zsahra, A. (2021). A review of innovation in cocoa bean processing by-products. *International Journal of Emerging Trends in Engineering Research*, 9(8), 1162–1169. https://doi.org/10.30534/ijeter/2021/22982021

Jarrín, J., Núñez, J., Espín, R., Manosalvas, L., Rodríguez, H., & Pais, J. (2023). Pectin extraction from residues of the cocoa fruit (*Theobroma cacao* L.) by different organic acids: A comparative study. *Foods*, 12(3), 590. https://doi.org/10.3390/foods12030590

Komlaga, G., Oduro, I., & Essel, E. (2019). Utilization of selected tropical crops (cocoa, kola nuts, sorghum, millet, and shea butter). In *Byproducts from Agriculture and Fisheries: Adding Value for Food, Feed, Pharma, and Fuels* (pp. 563–580), John Wiley & Sons. https://doi.org/10.1002/9781119383956.ch24

Lachenaud, P., & Motamayor, J. (2017). The Criollo cacao tree (*Theobroma cacao* L.): A review. *Genetic Resources and Crop Evolution*, 64(8), 1807–1820. https://doi.org/10.1007/s10722-017-0563-8

Laconi, E. B., & Jayanegara, A. (2015). Improving nutritional quality of cocoa pod (*Theobroma cacao*) through chemical and biological treatments for ruminant feeding: In vitro and in vivo evaluation. *Asian-Australasian Journal of Animal Sciences*, 28(3), 343–350. https://doi.org/10.5713/ajas.13.0798

Largo, S., & Yugcha, J. (2016). Elaboración de néctar de cacao a partir del mucílago. https://www.dspace.espol.edu.ec/handle/123456789/32357

Lu, F., Rodriguez-Garcia, J., Van Damme, I., Westwood, N., Shaw, L., Robinson, J. S., Warren, G., Chatzifragkou, A., McQueen Mason, S., Gomez, L., Faas, L., Balcombe, K., Srinivasan, C., Picchioni, F., Hadley, P., & Charalampopoulos, D. (2018). Valorisation strategies for cacao pod husk and its fractions. *Current Opinion in Green and Sustainable Chemistry*, 14, 80–88. https://doi.org/10.1016/j.cogsc.2018.07.007

Maboune, T. S. A., Tchinmegni, F. I., Mounjouenpou, P., Tchana, E. K., & Temgoua, E. (2017). Valorisation of cacao pod's shells production in the area of south-central Cameroon by composting with addition of *Tithonia diversifolia*. *International Journal of Research in Pharmacy and Biosciences*, 4(4), 32–42.

Meffeja, F., Njifutié, N., Manjeli, Y., Dongmo, T., Tchakounté, J., & Fombad, R. B. (2006). Digestibilité et influence des rations contenant des niveaux croissants de coques de cacao sur les performances des porcs en croissance finition. *Cellulose*, 17, 8.

Mendoza-Meneses, C. J., Feregrino-Pérez, A. A., & Gutiérrez-Antonio, C. (2021). Potential use of industrial cocoa waste in biofuel production. *Journal of Chemistry*, 2021, 1–11.

Moncada, J., Aristizábal, V., & Cardona, C. A. (2016). Design strategies for sustainable biorefineries. *Biochemical Engineering Journal*, 116, 122–134.

Okiyama, D. C., Navarro, S. L., & Rodrigues, C. E. (2017). Cocoa shell and its compounds: Applications in the food industry. *Trends in Food Science and Technology*, 63, 103–112. https://doi.org/10.1016/j.tifs.2017.03.007

Makinde, O. J., Okunade, S. A., Opoola, E., Sikiru, A. B., Ajide, S. O., & Elaigwu, S. (2019). Exploration of cocoa (Theobroma cacao) by-products as valuable potential resources in livestock feeds and feeding systems. *Theobroma Cacao-deploying science for sustainability of global cocoa economy*, 1–14. https://doi.org/10.5772/intechopen.87871

Olubamiwa, O., Ikyo, S. M., Adebowale, B. A., Omojola, A. B., & Hamzat, R. A. (2006). Efecto del tiempo de ebullición sobre la utilización de la cáscara de cacao en alimentos para gallinas ponedoras. *Revista Internacional de Ciencias Avícolas*, 5, 1137–1139.

Oyinlola, E. Y., Ojo, A. O., Adelani-Akande, T. A., & Ogunyemi, S. O. (2020). Efficacy of cacao pod husk extract on weed suppression and crop productivity. *Journal of Environmental Management*, 255, 109861. https://doi.org/10.1016/j.jenvman.2019.109861

Quimbita, F., Rodríguez, P., & Vera, E. (2013). Uso del exudado y placenta del cacao para la obtención de subproductos. *Revista Tecnológica ESPOL-RTE*, 26(1), 8–15.

Renna, M., Lussiana, C., Colonna, L., Malfatto, V. M., Mimosi, A., & Cornale, P. (2022). Inclusion of cocoa bean shell in the diet of dairy goats: Effects on milk production performance and milk fatty acid profile. *Frontiers in Veterinary Science*, 9, 848452. https://doi.org/10.3389/fvets.2022.848452

Rojo, O., Barbosa, L., Zeppa, G., & Stévigny, C. (2020). Cocoa bean shell—a by-product with nutritional properties and biofunctional potential. *Nutrients*, 12(4), 1123.

Rojo-Poveda, O., Barbosa-Pereira, L., Zeppa, G., & Stévigny, C. (2020). Cocoa bean shell—A by-product with nutritional properties and biofunctional potential. *Nutrients*, 12(4), 1123. https://doi.org/10.3390/nu12041123

Rossi, I., Zaidhiya, R. R., Mega, P., & Zsahra, R. (2021). Una revisión de la innovación en los subproductos del procesamiento del grano de cacao. *International Journal of Emerging Trends in Engineering Research*, 9(8), 1162–1169.

Sánchez, M. (2013). Biometanización de residuos del cultivo de cacao (*Theobroma cacao* L) como una opción para la generación de energía limpia. *Ambiente y Desarrollo*, 17(32), 31–45.

Soto-Ramírez, L., Garduño-Ríos, M., Millán-Carbajal, K., Moreno-López, A., & Valladares-Carranza, B. (2018). Intoxicación por teobromina en perros. Una revisión. *Revista Electrónica de Veterinaria*, 19(3), 1–5 . http://www.veterinaria.org/revistas/redvet/n030318.htm

Sulley, Y. (2022). Cocoa and its health benefits.Ghscientific. https://ghscientific.com/cocoa-and-its-health-benefits/

Vásquez, Z., de Carvalho Neto, D., Pereira, G., Vandenberghe, L., de Oliveira, P., Tiburcio, P., Rogéz, H., Goéz, A., & Soccol, C. (2019). Biotechnological approaches for cacao waste management: A review. *Waste Management*, 90, 72–83. https://doi.org/10.1016/j.wasman.2019.04.030

Vélez, J. C. M., Zambrano-Gavilanes, F., & Ponce-Saltos, W. (2021). Metanización de la biomasa residual de dos variedades de cacao y caracterización nutricional del sustrato biodigerido. *Biotempo*, 18(2), 167–176.

Vera, J., Vallejo, C., Párraga, D., Macías, J., Ramos, R., & Morales, W. (2014). Propiedades físico químicas y sensoriales de las almendras de 15 clones de Cacao Nacional. *Revista de Ciencia y Tecnología*, XX, 21–27.

Vivanco, V., Matute, G., & Campo, A. (2017). Panela bagasse hydrolysis using CO2 as catalyst. *Chemical Engineering Transactions*, 57, 887–892.

Voora, V., Bermúdez, S., & Larrea, C. (2019). Global market report: Cacao. International Institute for Sustainable Development. https://www.jstor.org/stable/pdf/resrep22025.pdf

Index

Note that page references in **bold** represent tables, while those in *italics* represent figures.

For Product Safety Concerns and Information please contact our EU
representative GPSR@taylorandfrancis.com
Taylor & Francis Verlag GmbH, Kaufingerstraße 24, 80331 München, Germany

www.ingramcontent.com/pod-product-compliance
Ingram Content Group UK Ltd.
Pitfield, Milton Keynes, MK11 3LW, UK
UKHW020153200926
475702UK00016B/2695

9781032464596